T0073164

A FRIENDLY APPROACH TO
COMPLEX ANALYSIS

Second Edition

A FRIENDLY APPROACH TO
COMPLEX ANALYSIS

Second Edition

Sara Maad Sasane
Lund University, Sweden

Amol Sasane
London School of Economics, UK

World Scientific

NEW JERSEY · LONDON · SINGAPORE · BEIJING · SHANGHAI · HONG KONG · TAIPEI · CHENNAI

Published by

World Scientific Publishing Co. Pte. Ltd.

5 Toh Tuck Link, Singapore 596224

USA office: 27 Warren Street, Suite 401-402, Hackensack, NJ 07601

UK office: 57 Shelton Street, Covent Garden, London WC2H 9HE

British Library Cataloguing-in-Publication Data
A catalogue record for this book is available from the British Library.

Cover image: The chameleons are conformal images of one another.

A FRIENDLY APPROACH TO COMPLEX ANALYSIS
Second Edition

ISBN 978-981-127-280-6 (hardcover)
ISBN 978-981-127-410-7 (paperback)
ISBN 978-981-127-281-3 (ebook for institutions)
ISBN 978-981-127-282-0 (ebook for individuals)

For any available supplementary material, please visit
https://www.worldscientific.com/worldscibooks/10.1142/13321#t=suppl

Desk Editor: Tan Rok Ting

Contents

Overview

We give an overview of what complex analysis is about and why it is important. For the discussion in the overview, we assume some familiarity with complex numbers, but in Chapter 1, we will start from scratch again. So the reader should not worry about being lost in this overview!

What is complex analysis?

In *real* analysis, calculus is studied rigorously in the setting of real numbers. Thus concepts such as the convergence of real sequences, continuity of real-valued functions, differentiation and integration are studied. Based on this, one might guess that *complex* analysis is all about studying similar concepts in the setting of complex numbers. This is partly true, but it turns out that up to the point of studying differentiation, there are no new features in complex analysis as compared to the real analysis counterparts. But the subject of complex analysis departs radically from real analysis when one studies differentiation. Thus, complex analysis is not merely about doing analysis in the setting of complex numbers, but rather, much more specialised:

| Complex analysis is the study of 'complex differentiable' functions. |

In real analysis, a function $f : \mathbb{R} \to \mathbb{R}$ is *differentiable at* $x_0 \in \mathbb{R}$ if there exists a real number L such that

$$\lim_{x \to x_0} \frac{f(x) - f(x_0)}{x - x_0} = L,$$

that is, for every $\epsilon > 0$, there is a $\delta > 0$ such that whenever $0 < |x - x_0| < \delta$,

$$\left| \frac{f(x) - f(x_0)}{x - x_0} - L \right| < \epsilon.$$

In other words, given any distance ϵ, we can make the difference quotient $\frac{f(x) - f(x_0)}{x - x_0}$ lie within a distance of ϵ from the real number L for all x sufficiently close to, but distinct from, x_0.

In the same way, $f : \mathbb{C} \to \mathbb{C}$ is *complex differentiable at $z_0 \in \mathbb{C}$* if there exists a complex number L such that

$$\lim_{z \to z_0} \frac{f(z) - f(z_0)}{z - z_0} = L,$$

that is, for every $\epsilon > 0$, there is a $\delta > 0$ such that whenever $0 < |z - z_0| < \delta$,

$$\left| \frac{f(z) - f(z_0)}{z - z_0} - L \right| < \epsilon.$$

The only change from the previous definition is that now the distances are measured with the *complex* absolute value, and instead of multiplication/division of real numbers, we now have multiplication/division of complex numbers — so this is a straightforward looking generalisation.

But we will see that this innocent looking generalisation is actually quite deep, and the class of complex differentiable functions looks radically different from real differentiable functions. Here is an instance of this.

Example 0.1. Let $f : \mathbb{R} \to \mathbb{R}$ be given by $f(x) = \begin{cases} x^2 & \text{if } x \geqslant 0, \\ -x^2 & \text{if } x < 0. \end{cases}$

Then f is differentiable everywhere, and $f'(x) = \begin{cases} 2x & \text{if } x \geqslant 0, \\ -2x & \text{if } x < 0. \end{cases} = 2|x|$.

Indeed, the above expressions for $f'(x)$ are immediate when $x \neq 0$, and $f'(0) = 0$ can be seen as follows. For $x \neq 0$,

$$\left| \frac{f(x) - f(0)}{x - 0} - 0 \right| = \left| \frac{f(x)}{x} \right| = \frac{|x|^2}{|x|} = |x| = |x - 0|,$$

and so given $\epsilon > 0$, taking $\delta = \epsilon$ (> 0), we have that whenever $0 < |x - 0| < \delta$,

$$\left| \frac{f(x) - f(0)}{x - 0} - 0 \right| = |x - 0| < \delta = \epsilon.$$

However, it can be shown that f' is not differentiable at 0; see Exercise 0.1. This is visually obvious since the graph of f' has a 'corner' at $x = 0$.

Summarising, we gave an example[1] of an $f : \mathbb{R} \to \mathbb{R}$, which is differentiable everywhere in $\mathbb{R}$, but whose derivative f' is not differentiable on $\mathbb{R}$.   ◇

---

[1]Here f' only failed to be differentiable at one point (namely, 0), but it can get much worse. There exist differentiable $f : \mathbb{R} \to \mathbb{R}$ for which f' is differentiable nowhere! For example we can start with a continuous function $g : \mathbb{R} \to \mathbb{R}$ which is nowhere differentiable, and define f by $f(x) = \int_0^x g(\xi)d\xi$, so that $f' = g$ everywhere. Examples of such g can be found for example in [**7**, Part I,Chap.3, §8].

Exercise 0.1. Let $g = f' : \mathbb{R} \to \mathbb{R}$ be given by $g(x) = 2|x|$ for all $x \in \mathbb{R}$. Prove that g is not differentiable at 0.

In contrast, we will later learn that if $f : \mathbb{C} \to \mathbb{C}$ is a complex differentiable function in $\mathbb{C}$, then it is infinitely many times complex differentiable! In particular, its complex derivative f' is also complex differentiable in $\mathbb{C}$. Clearly this is an unexpected result if all we are used to is real analysis. We will later learn that the reason this miracle takes place in complex analysis is that complex differentiability imposes some 'rigidity' on the function which enables this phenomenon to occur. We will also see that this rigidity is a consequence of the special geometric meaning of multiplication of complex numbers[2].

Why study complex analysis?

Although it might seem that complex analysis is just an exotic generalisation of real analysis, this is not so. Complex analysis is fundamental in all of mathematics. In fact real analysis is actually inseparable with complex analysis, as we shall see, and complex analysis plays an important role in the applied sciences as well. Here is a list of a few reasons to study complex analysis:

Partial Differential Equations (PDE). Associated with a function $f : \mathbb{C} \to \mathbb{C}$, there are two real-valued functions $u, v : \mathbb{R}^2 \to \mathbb{R}$, namely the real and imaginary parts of f: for $(x, y) \in \mathbb{R}^2$, $u(x, y) := \mathrm{Re}(f(x, y))$ and $v(x, y) := \mathrm{Im}(f(x, y))$.

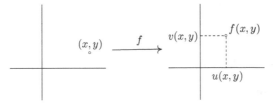

It turns out that if f is complex differentiable, then its real and imaginary parts u, v satisfy an important PDE, called the *Laplace equation*:

$$\Delta u := \frac{\partial^2 u}{\partial x^2} + \frac{\partial^2 u}{\partial y^2} = 0.$$

Similarly $\Delta v = 0$ in $\mathbb{R}^2$ as well. The Laplace equation arises in many applied problems, e.g., in electrostatics, steady-state heat conduction, incompressible fluid flow, Brownian motion, etc.

---

[2]Given a function $f : \mathbb{C} \to \mathbb{C}$, and a point $z_0 \in \mathbb{C}$, we could just think of real differentiability of $f : \mathbb{R}^2 \to \mathbb{R}^2$ at $z_0 = (x_0, y_0) \in \mathbb{R}^2$; see e.g. [1, §8.18]. But this is *not* the same as complex differentiability at z_0. Instead, we use the *multiplicative structure* of $\mathbb{C}$ to define the complex derivative, as limit of difference quotients, just as in the case of functions from $\mathbb{R}$ to $\mathbb{R}$. This turns out to be a much stronger notion: We shall see (p.45) that only when the matrix of the (real analysis) derivative at (x_0, y_0) is of a special type (namely corresponding to a 'rotation' followed by a 'scaling/dilation') will f be complex differentiable at z_0!

Real analysis. Using complex analysis, we can calculate some awkward integrals in real analysis, for example $\int_{-\infty}^{\infty} \frac{1}{1+x^4} dx$. The problem is set in the reals, but one can solve it using complex analysis.

Moreover, sometimes complex analysis helps to clarify some matters in real analysis. Here is an example. Let $f(x) := \frac{1}{1-x^2}$, $x \in \mathbb{R} \backslash \{-1, 1\}$. Then f has a 'singularity' at $x = \pm 1$, i.e. it is not defined there. It is, however defined in the interval $(-1, 1)$. The geometric series $1 + x^2 + x^4 + x^6 + \dots$ converges for $|x^2| < 1$, or equivalently for $|x| < 1$, and $1 + x^2 + x^4 + x^6 + \dots = \frac{1}{1-x^2} = f(x)$ for all $x \in (-1, 1)$. From the formula for f, it is not a surprise that the power series expansion of f is valid only for $x \in (-1, 1)$, since f itself has singularities at $x = 1$ and at $x = -1$. Let $g(x) := \frac{1}{1+x^2}$, $x \in \mathbb{R}$. The geometric series $1 - x^2 + x^4 - x^6 + - \dots$ converges for $|-x^2| < 1$, or equivalently for $|x| < 1$, and $1 - x^2 + x^4 - x^6 + - \dots = \frac{1}{1+x^2} = g(x)$ for all $x \in (-1, 1)$. So the power series expansion of g is again valid only for $x \in (-1, 1)$, despite there being no obvious reason from the formula for g for the series to break down at $x = -1$ and $x = 1$. The mystery will be resolved later: Consider the *complex* extensions of the functions, $f(z) = \frac{1}{1-z^2}$ and $g(z) = \frac{1}{1+z^2}$ (whose restriction to $\mathbb{R}$ are the given functions f and g, respectively). Then g has singularities at $z = \pm i$, and we will see that the power series expansion is valid in the biggest disk centred at $z = 0$ which does not contain any singularity of g.

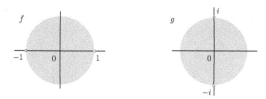

Applications. Many tools used for solving problems in applications, such as the Fourier/Laplace/z-transform, rely on complex function theory. These tools in turn are useful for example to solve differential equations which arise from applications. Thus complex analysis plays an important role e.g. in physics and engineering.

Analytic number theory. Perhaps surprisingly, many questions about the natural numbers can be answered using complex analytic tools. For example, consider the Prime Number Theorem, which gives an asymptotic estimate on $\pi(n)$ for large n, where $\pi(n) :=$ number of primes $\leqslant n$:

$$\lim_{n \to \infty} \frac{\pi(n)}{\frac{n}{\log n}} = 1.$$

A proof of the Prime Number Theorem can be given using complex analytic computations with a certain complex differentiable function called the *Riemann zeta function* ζ. Associated with ζ is also a famous unsolved problem in analytic number theory, namely the *Riemann Hypothesis*, saying that all the 'nontrivial' zeros of the Riemann zeta function lie on the line $\operatorname{Re} s = \frac{1}{2}$ in the complex plane. We will meet the Riemann zeta function in Exercise 4.6.

What will we learn in complex analysis. The central object of study in this course is: Holomorphic functions in a domain, that is, complex differentiable functions $f : D \to \mathbb{C}$, where D is a 'domain' (the precise meaning of what we mean by a domain will be given in Subsection 1.3.4). The bulk of the course is in Chapters 2, 3, 4, where the following three 'lanterns' shed light on holomorphic functions in a domain:

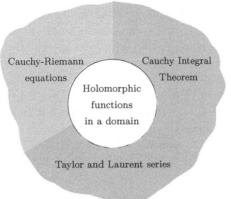

The core content of the book can be summarised in the following Main Theorem[3]:

Main Theorem. *Let D be an open path connected set and let $f : D \to \mathbb{C}$. Then the following are equivalent:*

(1) *f is complex differentiable in D.*

(2) *f is infinitely many times complex differentiable in D.*

(3) *$u := \operatorname{Re} f$, $v := \operatorname{Im} f$ are real differentiable in D, and moreover, we have the Cauchy-Riemann equations $\frac{\partial u}{\partial x} = \frac{\partial v}{\partial y}$, and $\frac{\partial u}{\partial y} = -\frac{\partial v}{\partial x}$ in D.*

(4) *For each simply connected domain $S \subset D$, there exists an $F : S \to \mathbb{C}$, complex differentiable in S, such that $F' = f$ in S.*

(5) *f is continuous on D, and for all piecewise smooth closed paths γ in each simply connected domain $S \subset D$, $\int_\gamma f(z)dz = 0$.*

(6) *If $z_0 \in D$ and $r > 0$ is such that $D(z_0, r) := \{z \in \mathbb{C} : |z - z_0| < r\} \subset D$, then $f(z) = \sum\limits_{n=0}^{\infty} \frac{f^{(n)}(z_0)}{n!}(z - z_0)^n$ for all $z \in D(z_0, r)$.*

Complex analysis is not complex analysis! It is not very complicated, and there is not much messy analysis. Once core properties of complex differentiable functions are established, it is 'softer' than real analysis, and there are fewer cumbersome ϵ-δ estimates. The Main Theorem above tells us that the subject is radically different from real analysis. Indeed, a real-valued differentiable function on an open interval (a, b) need not have a continuous derivative. In contrast, a complex

---

[3]Do not worry about the unfamiliar terms/notation here. That is what we will learn, besides the proof!

differentiable function on an open subset of $\mathbb{C}$ is infinitely many times complex differentiable! This happens because the geometric meaning of complex multiplication implies that complex differentiable functions behave in a controlled manner locally infinitesimally. This controlled behaviour makes these functions rigid and we will see this in Section 2.3. Nevertheless there are enough of them to make the subject nontrivial and interesting!

Prerequisites. We assume familiarity with multivariable calculus and linear algebra, e.g. at the level of [1, Chapters 1,2,8]. Even if the student is unfamiliar with proofs, we hope that at least the definitions and statements of results are familiar. For example, for a function $f : \mathbb{R}^n \to \mathbb{R}^m$, the reader knows the meaning of continuity at a point $\mathbf{x} \in \mathbb{R}^n$, and of real differentiability and the real analysis derivative. Also, we assume knowledge of the fact that if a function has continuous partial derivatives in an open subset $U \subset \mathbb{R}^n$, then it is real differentiable in U, and the fact if a function has continuous partial derivatives up to order 2, then the mixed partial derivative with respect to x and y can be taken in any order. At the end of the book, an appendix is included, where we have summarised some of the relevant results from real analysis.

Title, first edition versus the second edition. The title of the book is meant to indicate that we aim to cover the bare bones of the subject with minimal prerequisites. It does not mean that it is merely a book with 'drill' kind of material. In fact it contains many challenging exercises. As the solutions to all the exercises are included in the book, we hope that the book will be useful for self-study.

The first edition was a much revised version of lecture notes written when the second author gave a cour for third year students of the BSc programme in Mathematics and/with Economics. In the second edition we have tried to correct all the mistakes we could find, and also added many exercises, some of which cover auxiliary topics that were omitted in the first edition. We have added an appendix to Chapter 3 containing a detailed proof of the Cauchy Integral Theorem, and a chapter called 'Some real analysis background', recalling the needed real analysis.

Acknowledgements. This book relies heavily on some of the sources mentioned in the bibliography. At some instances detailed references are given in a chapter called 'Notes', but no claim to originality is made in case there is a missing reference. Thanks are due to Raul Hindov, Raymond Mortini, Adam Ostaszewski, Rudolf Rupp and Gabriel Winberg for useful comments. Finally, it is a pleasure to thank the World Scientific team, especially executive editor Rochelle Kronzek for her support, enthusiasm and prompt help.

Sara Maad Sasane and Amol Sasane

Chapter 1

Complex numbers and their geometry

In this chapter, we set the stage for doing complex analysis. First, we introduce the set of complex numbers, and their arithmetic, making $\mathbb{C}$ into a field, 'extending' the usual field of real numbers.

Points in $\mathbb{C}$ can be depicted in the plane $\mathbb{R}^2$, and we will see that the arithmetic in $\mathbb{C}$ has geometric meaning in the plane. This correspondence between $\mathbb{C}$ and points in the plane also allows one to endow $\mathbb{C}$ with the usual Euclidean topology of the plane.

We will study a fundamental function in complex analysis, namely the exponential function (and some elementary functions related to the exponential function, namely trigonometric functions and the logarithm). Thus besides polynomial and rational functions, these will serve as important examples of complex differentiable functions in certain domains.

1.1. The field of complex numbers

By definition, a *complex number* is an ordered pair of real numbers. For example, $(1,0)$, $(0,1)$, $(0,0)$, $(-\frac{3}{4}, \sqrt{2})$ are all complex numbers. The set $\mathbb{R} \times \mathbb{R} = \mathbb{R}^2$ of all complex numbers is denoted by $\mathbb{C}$. Thus $\mathbb{C} = \{z = (x,y) : x \in \mathbb{R} \text{ and } y \in \mathbb{R}\}$. For $z = (x,y) \in \mathbb{C}$, where $x, y \in \mathbb{R}$, the real number x is called the *real part of* z, and y is called the *imaginary part of* z. The operations of *addition* '+' and *multiplication* '·' on $\mathbb{C}$ are defined by:

$$(x_1, y_1) + (x_2, y_2) = (x_1 + x_2, y_1 + y_2),$$
$$(x_1, y_1) \cdot (x_2, y_2) = (x_1 x_2 - y_1 y_2, x_1 y_2 + x_2 y_1),$$

for complex numbers (x_1, y_1), (x_2, y_2). With these operations, $\mathbb{C}$ is a field, that is,

(F1) $(\mathbb{C}, +)$ is an 'Abelian group',

(F2) $(\mathbb{C}\backslash\{0\}, \cdot)$ is an Abelian group, and

(F3) the distributive law holds: for $a, b, c \in \mathbb{C}$, $(a+b) \cdot c = a \cdot c + b \cdot c$.

(F1) means that $+ : \mathbb{C} \times \mathbb{C} \to \mathbb{C}$ is associative and commutative, that there exists an 'identity element' $(0,0)$, such that $(x,y) + (0,0) = (x,y) = (0,0) + (x,y)$ for all $(x,y) \in \mathbb{C}$, and that each element $(x,y) \in \mathbb{C}$ has a corresponding 'additive inverse' $(-x,-y)$: $(x,y) + (-x,-y) = (0,0) = (-x,-y) + (x,y)$.

Similarly, in (F2), the multiplicative identity is $(1,0)$, and the multiplicative inverse of a complex number $(x,y) \in \mathbb{C}\backslash\{(0,0)\}$ is given by $(\frac{x}{x^2+y^2}, \frac{-y}{x^2+y^2})$.

Exercise 1.1. Check that $(\frac{x}{x^2+y^2}, \frac{-y}{x^2+y^2})$ is indeed the inverse of $(x,y) \in \mathbb{C}\backslash\{(0,0)\}$.

Proposition 1.1. $(\mathbb{C}, +, \cdot)$ *is a field.*

$\mathbb{R}$ is 'contained' in $\mathbb{C}$. We can embed $\mathbb{R}$ inside $\mathbb{C}$, and view $\mathbb{R}$ as a 'subfield' of $\mathbb{C}$, that is, the map $x \mapsto (x,0)$ sending the real number x to the complex number $(x,0)$ is an injective field homomorphism. This means that the operations of addition and multiplication are preserved by this map, and distinct real numbers are sent to distinct complex numbers.

$\mathbb{R}$		$\mathbb{C}$
x	$\mapsto$	$(x,0)$
$x_1 + x_2$	$\mapsto$	$(x_1 + x_2, 0) = (x_1, 0) + (x_2, 0)$
$x_1 \cdot x_2$	$\mapsto$	$(x_1 \cdot x_2, 0) = (x_1, 0) \cdot (x_2, 0)$
1	$\mapsto$	$(1,0)$
0	$\mapsto$	$(0,0)$

So we can view real numbers as if they are complex numbers, via this identification. For example, the real number $\sqrt{2}$ can be viewed as the complex number $(\sqrt{2}, 0)$. If this makes one uneasy, one should note that we have been doing such identifications right from elementary school, where for instance, we identified integers with rational numbers, for example, $\mathbb{Z} \ni 3 = \frac{3}{1} \in \mathbb{Q}$, and we didn't lose sleep over it!

An advantage of working with $\mathbb{C}$ is that while in $\mathbb{R}$ there is no solution $x \in \mathbb{R}$ to the equation $x^2 + 1 = 0$, now with complex numbers we have with $x := (0,1) \in \mathbb{C}$ that $x^2 + 1 = (0,1) \cdot (0,1) + (1,0) = (-1,0) + (1,0) = (0,0) = 0$, where we have made the usual identification of the real numbers 1 and 0 with their corresponding complex numbers $(1,0)$ and $(0,0)$. If we give a special symbol, say i, to the number $(0,1)$, then the above says that $i^2 + 1 = 0$. Henceforth, for the complex number (x,y), where x, y are real, we write $x + yi$, since

$$(x,y) = \underbrace{(x,0)}_{\equiv x} + \underbrace{(y,0)}_{\equiv y} \cdot \underbrace{(0,1)}_{\equiv i} = x + yi.$$

As complex multiplication is commutative, $yi = iy$, and so $x + yi = x + iy$.

For $z \in \mathbb{C}\backslash\{0\}$, we will denote its inverse z^{-1} by $\frac{1}{z}$, and if $w \in \mathbb{C}$, then $\frac{w}{z} := w\frac{1}{z}$.

Exercise 1.2. Let $\theta \in (-\frac{\pi}{2}, \frac{\pi}{2})$. Express $\frac{1+i\tan\theta}{1-i\tan\theta}$ in the form $x + yi$, where $x, y \in \mathbb{R}$.

Historical development of complex numbers. Contrary to popular belief, historically, it wasn't the need for solving *quadratic* equations, but rather *cubic* equations, that led mathematicians to take complex numbers seriously. The gist of this is as follows. Around the 16th century, one viewed solving equations like $x^2 + bx + c = 0$ as the geometric problem of finding the intersection point of the parabola $y = x^2$ with the line $y = -bx - c$. Based on this geometric interpretation, it was easy to dismiss the lack of solvability in reals of a quadratic such as $x^2 + 1 = 0$, since that just reflected the geometric fact that parabola $y = x^2$ did not meet the line $y = -1$. See the picture on the left below.

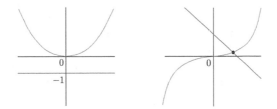

Meanwhile, Cardano (1501-1576) gave a formula for solving the cubic equation $x^3 = 3px + 2q$, namely,

$$x = \sqrt[3]{q + \sqrt{q^2 - p^3}} + \sqrt[3]{q - \sqrt{q^2 - p^3}}.$$

For example, one can check that for the equation $x^3 = 6x + 6$, with $p = 2$ and $q = 3$, this yields one solution to be $x = \sqrt[3]{4} + \sqrt[3]{2}$. Using the Intermediate Value Theorem, it can be shown that the cubic $y = x^3$ *always* intersects the line $y = 3px + 2q$. See the picture on the right above. For the equation $x^3 = 15x + 4$, when $p = 5$ and $q = 2$, we have $q^2 - p^3 = -121 < 0$, and so Cardano's formula fails with real arithmetic. But we do have a real root, namely $x = 4$: $4^3 = 64 = 60 + 4 = 15 \cdot 4 + 4$. Three decades after the appearance of Cardano's work, Bombelli suggested that with the use of *complex* arithmetic, Cardano's formula gives the desired real root. So we may ask if $x = \sqrt[3]{2 + 11i} + \sqrt[3]{2 - 11i} \overset{?}{=} 4$. As $(2 + i)^3 = 2 + 11i$ and $(2 - i)^3 = 2 - 11i$, the above does work, with appropriate choices of cube roots. Thus Bombelli's work established that even for *real* problems, complex arithmetic can be relevant. From then on, complex numbers entered mainstream mathematics.

Exercise 1.3. A field $\mathbb{F}$ is called *ordered* if there is a subset $P \subset \mathbb{F}$, called the *set of positive elements of* $\mathbb{F}$, satisfying the following:

(P1) For all $x, y \in P$, $x + y \in P$.

(P2) For all $x, y \in P$, $x \cdot y \in P$.

(P3) For each $x \in \mathbb{F}$, exactly one of the following is true: 1° $x = 0$. 2° $x \in P$. 3° $-x \in P$.

The field $\mathbb{R}$ of real numbers is ordered, since $P := (0, \infty)$ is a set of positive elements of $\mathbb{R}$. Given a set P of positive elements in an ordered field $\mathbb{F}$, elements of $\mathbb{F}$ can be compared via the 'order relation' $>$ in $\mathbb{F}$, defined by setting $y > x$ for $x, y \in \mathbb{F}$ if $y - x \in P$.

Show that $\mathbb{C}$ is not an ordered field. *Hint:* Consider $x := i$, and first look at $x \cdot x$.

1.2. Geometric representation of complex numbers

Since $\mathbb{C} = \mathbb{R}^2$, we can identify complex numbers with points in the plane.

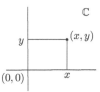

Exercise 1.4. Locate 0, 1, $-\frac{3}{2}$, i, $-\sqrt{2}i$, $\cos\frac{\pi}{3} + i\sin\frac{\pi}{3}$ in the complex plane.

So we can identify $\mathbb{C}$ as a *set* with the set of points in the plane $\mathbb{R}^2$. Do the field operations in $\mathbb{C}$ have some geometric meaning in the plane? We'll now see that addition in $\mathbb{C}$ corresponds to 'vector addition' in the plane, and multiplication in $\mathbb{C}$ too has a special geometric meaning in the plane, as explained below.

Geometric meaning of complex addition. With the identification of complex numbers with points in the plane, complex addition corresponds to addition of vectors in $\mathbb{R}^2$. By addition of vectors, we mean the usual way of combining two vectors, that is, by completing the parallelogram formed by the line segments joining $(0,0)$ to each of the two complex numbers as sides, and then taking the endpoint of the diagonal from $(0,0)$ as the sum of the two given complex numbers.

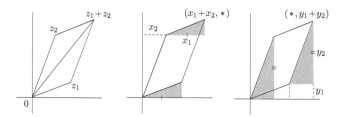

The middle picture shows that addition of z_1 and z_2 as vectors in the plane yields the correct x-coordinate of their sum as complex numbers by looking at the two congruent shaded triangles. Similarly, from the rightmost picture, we see that the y-coordinate is also correct.

Geometric meaning of complex multiplication. We will now see the special geometric meaning of complex multiplication. In order to do this, it will be convenient to use polar coordinates. Let the point $(x,y) \in \mathbb{R}^2\backslash\{(0,0)\}$ have polar coordinates $r > 0$ and $\theta \in (-\pi, \pi]$. This means that the Euclidean distance of (x,y) to $(0,0)$ is r (> 0), and the ray joining $(0,0)$ to (x,y) makes an angle (in the counterclockwise fashion) θ with the positive real axis (the x-axis). If $(x,y) = (0,0)$, then we set $r = 0$ and $\theta = 0$.

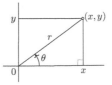

In the right-angled triangle shown above, $x = r\cos\theta$, $y = r\sin\theta$. Using polar coordinates, we have $x + yi = r\cos\theta + (r\sin\theta)i = r(\cos\theta + i\sin\theta)$. Now we give the geometric interpretation of complex multiplication. For two complex numbers expressed in polar coordinates as $z_1 = r_1(\cos\theta_1 + i\sin\theta_1)$, $z_2 = r_2(\cos\theta_2 + i\sin\theta_2)$,

$$z_1 \cdot z_2 = r_1(\cos\theta_1 + i\sin\theta_1) \cdot r_2(\cos\theta_2 + i\sin\theta_2)$$
$$= r_1 r_2 \big((\cos\theta_1)\cos\theta_2 - (\sin\theta_1)\sin\theta_2 + i((\cos\theta_1)\sin\theta_2 + (\cos\theta_2)\sin\theta_1)\big)$$
$$= r_1 r_2 (\cos(\theta_1 + \theta_2) + i\sin(\theta_1 + \theta_2)),$$

using the trigonometric identities for angle addition. Thus $z_1 \cdot z_2$ has polar coordinates $(r_1 r_2, \theta_1 + \theta_2)$. The angles z_1 and z_2 make with the positive real axis are *added* in order to get the angle $z_1 \cdot z_2$ makes with the positive real axis, and the distances to the origin are *multiplied* to get the distance of $z_1 \cdot z_2$ to the origin.

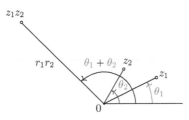

As a special case, consider multiplication by $\cos\alpha + i\sin\alpha$, which is at a distance of 1 from the origin. If $z \in \mathbb{C}$, then $z \cdot (\cos\alpha + i\sin\alpha)$ is obtained by rotating the line joining 0 to z anticlockwise through an angle of α. In particular, multiplying z by $i = 0 + i\cdot 1 = \cos\frac{\pi}{2} + i\sin\frac{\pi}{2}$ produces a counterclockwise rotation of $90°$.

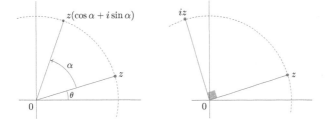

Exercise 1.5. In the complex plane, let the points corresponding to $0, z, w, zw, 1$ be labelled O, P, Q, R, S. Show that the triangle $\triangle POS$ is similar to $\triangle ROQ$.

De Moivre's formula and nth roots. *de Moivre's formula* states that for all $n \in \mathbb{N}$, and $\theta \in \mathbb{R}$, $(\cos\theta + i\sin\theta)^n = \cos(n\theta) + i\sin(n\theta)$.

Exercise 1.6. Recover the identity $\cos(3\theta) = 4(\cos\theta)^3 - 3\cos\theta$ using de Moivre's formula.

Exercise 1.7. Without expanding, express $(1 + i)^{10}$ in the form $x + iy$, $x, y \in \mathbb{R}$.

Exercise 1.8. Considering $(2 + i)(3 + i)$, show that $\frac{\pi}{4} = \tan^{-1}\frac{1}{2} + \tan^{-1}\frac{1}{3}$.

Exercise 1.9. *Gaussian integers* are complex numbers of the form $m + in$, where m, n are integers. Thus they are integral lattice points in the complex plane. Show that it is impossible to draw an equilateral triangle such that all vertices are Gaussian integers. *Hint:* Rotation of one of the sides should give the other. Recall that $\sqrt{3} \notin \mathbb{Q}$.

De Moivre's formula gives a way of finding the *nth roots of a complex number z,* i.e., complex numbers w satisfying $w^n = z$.

Proposition 1.2. *Let $z = r(\cos\theta + i\sin\theta)$, $r > 0$, $\theta \in [0, 2\pi)$, $n \in \mathbb{N}$, $w \in \mathbb{C}$. Then $w^n = z$ if and only if $w \in \{ \sqrt[n]{r}(\cos(\frac{\theta}{n} + k\frac{2\pi}{n}) + i\sin(\frac{\theta}{n} + k\frac{2\pi}{n})) : k = 0, 1, \cdots, n-1\}$.*

Proof. ('If part':) If $w \in \{ \sqrt[n]{r}(\cos(\frac{\theta}{n} + k\frac{2\pi}{n}) + i\sin(\frac{\theta}{n} + k\frac{2\pi}{n})) : k = 0, \cdots, n-1\}$, then $w^n = (\sqrt[n]{r})^n(\cos(n(\frac{\theta}{n} + k\frac{2\pi}{n})) + i\sin(n(\frac{\theta}{n} + k\frac{2\pi}{n}))) = r(\cos(\theta + 2\pi k) + \sin(\theta + 2\pi k)) = z$.

('Only if part':) Let $w = \rho(\cos\alpha + i\sin\alpha)$, where $\rho \geq 0$ and $\alpha \in \mathbb{R}$. As $w^n = z$, we have $\rho^n(\cos(n\alpha) + i\sin(n\alpha)) = w^n = z = r(\cos\theta + i\sin\theta)$. Equating the distance to the origin, $\rho^n = r$. Hence $\rho = \sqrt[n]{r}$, as $\rho, r \geq 0$. The angle w^n makes with the positive real axis is $n\alpha \in \{\cdots, \theta - 4\pi, \theta - 2\pi, \theta, \theta + 2\pi, \theta + 4\pi, \theta + 6\pi, \cdots\}$, since the angle made by a nonzero z with the positive real axis is unique only up to integral multiples of 2π, i.e., instead of θ, we can use $\theta + 2\pi k$ for any $k \in \mathbb{Z}$.

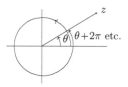

So $\alpha \in \{\frac{\theta}{n} + \frac{2\pi}{n}k : k \in \mathbb{Z}\}$. For $\alpha \in \{\frac{\theta}{n}, \frac{\theta}{n} + \frac{2\pi}{n}, \frac{\theta}{n} + 2\frac{2\pi}{n}, \cdots, \frac{\theta}{n} + (n-1)\frac{2\pi}{n}\}$, we get distinct w. (Indeed, for any integer k, the division algorithm for integers gives $q, r \in \mathbb{Z}$ such that $k = nq + r$ and $0 \leq r < n$, so that $\frac{2\pi}{n}k = \frac{2\pi}{n}r + 2\pi q$.) $\qquad\square$

In particular, if $z = 1$, we get the nth roots of unity, which are located at the vertices of an n-sided regular polygon inscribed in a circle. For example, the following picture depicts the six 6th roots of unity.

Exercise 1.10. Find all $w \in \mathbb{C}$ such that $w^4 = -1$. Depict these in the complex plane.

Exercise 1.11. Find all $z \in \mathbb{C}$ such that $z^6 - z^3 - 2 = 0$.

Exercise 1.12. If a, b, c are *real* numbers such that $a^2 + b^2 + c^2 = ab + bc + ca$, then they must be equal. Indeed, doubling both sides and rearranging gives $(a-b)^2 + (b-c)^2 + (c-a)^2 = 0$, and since each summand is nonnegative, each must then be 0. Show that if a, b, c are *complex* numbers such that $a^2 + b^2 + c^2 = ab + bc + ca$, then they must lie on the vertices of an equilateral triangle in the complex plane. Explain the real case result in light of this. *Hint:* Compute $((b-a)\omega + (b-c))((b-a)\omega^2 + (b-c))$, ω being a nonreal cube root of unity.

Exercise 1.13. The Binomial Theorem says that if $a, b \in \mathbb{C}$, $n \in \mathbb{N}$, and $\binom{n}{k} := \frac{n!}{k!(n-k)!}$ for $k = 0, 1, \cdots, n$, then $(a + b)^n = \sum_{k=0}^{n} \binom{n}{k} a^k b^{n-k}$. A proof can be given using induction. Show that $\binom{3n}{0} + \binom{3n}{3} + \binom{3n}{6} + \cdots + \binom{3n}{3n} = \frac{2^{3n} + 2(-1)^n}{3}$. *Hint:* If ω denotes a nonreal cube root of unity, find $(1 + 1)^{3n} + (1 + \omega)^{3n} + (1 + \omega^2)^{3n}$.

Exercise 1.14. Show, using the geometry of complex numbers, that the line segments joining the centres of opposite external squares described on sides of an arbitrary convex quadrilateral are perpendicular and have equal lengths.

Exercise 1.15. Let p be a polynomial of degree $d \in \mathbb{N}$, i.e., there are $c_0, \cdots, c_d \in \mathbb{C}$, with $c_d \neq 0$, such that for all $z \in \mathbb{C}$, $p(z) = c_0 + c_1 z + \cdots + c_d z^d$. Let $z_0 \in \mathbb{C}$. Show that there exists a polynomial q of degree $d - 1$ such that for all $z \in \mathbb{C}$, $p(z) = q(z)(z - z_0) + p(z_0)$. In particular, if $p(z_0) = 0$, then $p(z) = q(z)(z - z_0)$. Conclude that if $n \in \mathbb{N}$, and $z_1, \cdots, z_n$ are the n distinct n^{th} roots of unity, then $z^n - 1 = (z - z_1) \cdots (z - z_n)$.

Exercise 1.16. Show that if $d \in \mathbb{N} \cup \{0\}$, and $c_0, \cdots, c_d \in \mathbb{C}$ are such that for all $z \in \mathbb{C}$, $c_0 + c_1 z + \cdots + c_d z^d = 0$, then $c_0 = \cdots = c_d = 0$.

Exercise 1.17 (Division Algorithm). Let h be a polynomial of degree $d \in \mathbb{N}$. Show that for any polynomial p, there exist unique polynomials q (the quotient) and r (the remainder) such that for all $z \in \mathbb{C}$, $p(z) = q(z)h(z) + r(z)$, where the degree of r is strictly less than d. (We define the degree of the zero polynomial to be 0.)

Absolute value and complex conjugate. The *absolute value* or *modulus* $|z|$ of the complex number $z = x + iy$, where $x, y \in \mathbb{R}$, is defined by $|z| = \sqrt{x^2 + y^2}$. By Pythagoras's Theorem, $|z|$ is the distance of the complex number z to 0 in the complex plane. See the picture on the left below. By expressing $z_1, z_2 \in \mathbb{C}$ in terms of polar coordinates, or by a direct calculation, it is clear that $|z_1 z_2| = |z_1||z_2|$. If $z \neq 0$, then $1 = |1| = |z \frac{1}{z}| = |z||\frac{1}{z}|$ and so $|\frac{1}{z}| = \frac{1}{|z|}$, and for $w \in \mathbb{C}$, $|\frac{w}{z}| = \frac{|w|}{|z|}$.

Exercise 1.18. Verify $|z_1 z_2| = |z_1||z_2|$ by expressing z_1, z_2 using Cartesian coordinates.

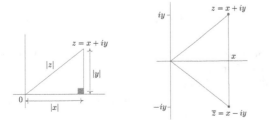

The *complex conjugate* $\bar{z}$ of $z = x + iy$ where $x, y \in \mathbb{R}$, is $\bar{z} := x - iy$. In the complex plane, $\bar{z}$ is the reflection in the real axis of z. With this geometric interpretation, it can be seen that for all $z_1, z_2 \in \mathbb{C}$, $\overline{z_1 + z_2} = \overline{z_1} + \overline{z_2}$ and $\overline{z_1 z_2} = \overline{z_1}\,\overline{z_2}$. The following are easy to check: $\overline{\overline{z}} = z$, $z\bar{z} = |z|^2$, $\operatorname{Re} z = \frac{z+\bar{z}}{2}$, $\operatorname{Im} z = \frac{z-\bar{z}}{2i}$.

Exercise 1.19. Verify that the four equalities above.

Exercise 1.20. Prove that for all $z \in \mathbb{C}$, $|z| = |\bar{z}|$, $|\operatorname{Re} z| \leqslant |z|$, and $|\operatorname{Im} z| \leqslant |z|$. Give geometric interpretations of each.

Exercise 1.21. If $a, z \in \mathbb{C}$ satisfy $|a| < 1$ and $|z| \leqslant 1$, then prove that $\left|\frac{z-a}{1-\bar{a}z}\right| \leqslant 1$.

Exercise 1.22. Let $z, w \in \mathbb{C}$ be distinct complex numbers. Show that the 'convex combination of z, w', namely, $L := \{(1-t)z + tw : t \in [0, 1]\}$ is the line segment joining z and w. *Hint*: $(1-t)z + tw = z + t(w - z)$.

Exercise 1.23. Let p be a polynomial given by $p(z) = c_0 + c_1 z + \cdots + c_d z^d$, where $d \in \mathbb{N}$, $c_0, c_1, \ldots, c_d \in \mathbb{R}$ and $c_d \neq 0$. Show that if $w \in \mathbb{C}$ such that $p(w) = 0$, then $p(\bar{w}) = 0$.

Exercise 1.24. Show that the area of the triangle formed by $0, a, b \in \mathbb{C}$ is $\left|\frac{\operatorname{Im}(a\bar{b})}{2}\right|$.

Exercise 1.25. On a visit to Cornell University, Richard Feynman gave a colloquium on the history of the Feynman-Kac formula. After the talk, Kai Lai Chung entertained Feynman to dinner, at which much of the discussion centred on the following result:

> **Theorem.** For a triangle in the plane, if each vertex is joined to the point one-third along the opposite side (measured anticlockwise), then the area of the inner triangle formed by these lines is exactly one-seventh of the area of the initial triangle.

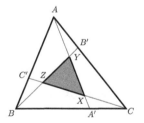

Feynman could not believe that the ratio of the areas of the triangles was $\frac{1}{7}$ since this had nothing to do with the number three. He spent most of the evening trying to disprove it, but finally proved it in the special case when the triangle was equilateral. Use the result of the previous exercise to prove it in general.

Exercise 1.26. Prove for all $z_1, z_2, z_3 \in \mathbb{C}$, $i \det \begin{bmatrix} 1 & z_1 & \overline{z_1} \\ 1 & z_2 & \overline{z_2} \\ 1 & z_3 & \overline{z_3} \end{bmatrix}$ is real.

Exercise 1.27. Let $z, w \in \mathbb{C} \backslash \{0\}$. Show that $|z + w| = \left|\frac{|w|}{|z|}z + \frac{|z|}{|w|}w\right|$. What is the geometric interpretation?

Exercise 1.28. Show that $S^1 = \{z \in \mathbb{C} : |z| = 1\}$ is an Abelian group with respect to complex multiplication. For any $n \in \mathbb{N}$, show that S^1 has a subgroup of order n. Does S^1 have an infinite subgroup which is countable?

Exercise 1.29. Let $n \in \mathbb{N}$ be such that $n \geqslant 2$, and $z_1, \cdots, z_n$ be the n distinct n^{th} roots of unity. Prove that $z_1 + \cdots + z_n = 0$.

1.3. Topology of $\mathbb{C}$

The concepts in ordinary calculus in the setting of $\mathbb{R}$, like convergence of sequences, or continuity and differentiability of functions, all rely on the notion of closeness of points in $\mathbb{R}$. For example, when we talk about the convergence of a real sequence $(a_n)_{n\in\mathbb{N}}$ to its limit $L \in \mathbb{R}$, we mean that given any positive ϵ, there is a large enough index N such that beyond that index, the corresponding terms a_n all have a *distance* to L which is at most ϵ. This 'distance of a_n to L' is taken as $|a_n - L|$, and this is the length of the line segment joining the numbers a_n and L on the real number line. In order to do calculus with *complex* numbers, we need a notion of distance $d(z_1, z_2)$ between complex numbers z_1 and z_2.

1.3.1. Metric on $\mathbb{C}$. Since $\mathbb{C}$ is just $\mathbb{R}^2$, we use the usual Euclidean distance in $\mathbb{R}^2$ as the metric in $\mathbb{C}$. Thus, for complex numbers $z_1 = x_1 + iy_1$ and $z_2 = x_2 + iy_2$, we have $d(z_1, z_2) = \sqrt{(x_1 - x_2)^2 + (y_1 - y_2)^2} = |z_1 - z_2|$. By Pythagoras's Theorem, this is the length of the line segment joining the points $(x_1, y_1), (x_2, y_2)$ in $\mathbb{R}^2$.

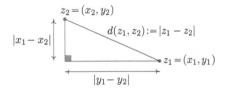

Using the geometric meaning of addition of complex numbers, and the well-known result from Euclidean geometry that the sum of the lengths of any two sides of a triangle is at least as big as the length of the third side, we obtain the following *triangle inequality* for the absolute value: $|z_1 + z_2| \leqslant |z_1| + |z_2|$ for $z_1, z_2 \in \mathbb{C}$. The triangle inequality can also be verified analytically by using the Cauchy-Schwarz inequality for $x_1, x_2, y_1, y_2 \in \mathbb{R}$: $(x_1^2 + y_1^2)(x_2^2 + y_2^2) \geqslant (x_1 x_2 + y_1 y_2)^2$, or noting $|z_1 + z_2|^2 = (z_1 + z_2)(\overline{z_1} + \overline{z_2}) = |z_1|^2 + |z_2|^2 + 2\,\mathrm{Re}(z_1\overline{z_2}) \leqslant |z_1|^2 + |z_2|^2 + 2|z_1||z_2|$.

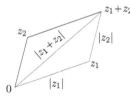

Exercise 1.30. Show that for all $z_1, z_2 \in \mathbb{C}$, $|z_1 - z_2| \geqslant ||z_1| - |z_2||$.

Exercise 1.31. Let $a \in \mathbb{C}$, and $r > 0$. Show that $C := \{z \in \mathbb{C} : |z|^2 - 2\,\mathrm{Re}(\overline{a}z) + |a|^2 = r^2\}$ is a circle centred at a with radius r.

Exercise 1.32. Show that for all $z_1, z_2 \in \mathbb{C}$, $|z_1 + z_2|^2 + |z_1 - z_2|^2 = 2(|z_1|^2 + |z_2|^2)$. What is the geometric interpretation of this equality?

Exercise 1.33. Consider a circle C of radius $r > 0$ and centre 0. For any point z on C, determine $|z + r|^2 + |z - r|^2$, and use the converse of the Pythagoras Theorem to conclude that the angle subtended by the diameter at the point z is a right angle.

Exercise 1.34. Sketch the following sets in the complex plane:

(1) $\{z \in \mathbb{C} : |z - (1 - i)| = 2\}$.

(2) $\{z \in \mathbb{C} : |z - (1 - i)| < 2\}$.

(3) $\{z \in \mathbb{C} : 1 < |z - (1 - i)| < 2\}$.

(4) $\{z \in \mathbb{C} : \mathrm{Re}(z - (1 - i)) = 3\}$.

(5) $\{z \in \mathbb{C} : |\mathrm{Im}(z - (1 - i))| < 3\}$.

(6) $\{z \in \mathbb{C} : |z - (1 - i)| = |z - (1 + i)|\}$.

(7) $\{z \in \mathbb{C} : |z - (1 - i)| + |z - (1 + i)| = 2\}$.

(8) $\{z \in \mathbb{C} : |z - (1 - i)| + |z - (1 + i)| < 3\}$.

Exercise 1.35 (Cosine Formula). Let z, w be distinct nonzero complex numbers. Show that if A, O, B are the points corresponding to $z, 0, w$, then $\cos \angle AOB = \frac{AO^2 + BO^2 - AB^2}{2(AO)(BO)}$.

Exercise 1.36 (Riemann sphere). Let $S^2 := \{(x, y, z) \in \mathbb{R}^3 : x^2 + y^2 + z^2 = 1\}$ be the unit sphere in $\mathbb{R}^3$, and denote the 'north pole' by $\mathbf{n} := (0, 0, 1)$. For each $p \in S^2 \backslash \{\mathbf{n}\}$, define the 'stereographic' projection of $p = (x, y, z)$ to be the point $\varphi(p) = (u, v)$ where the ray starting from $\mathbf{n}$, and passing through p, meets the (x, y)-plane (thought of as the complex plane). See the picture below. From the two similar triangles shown there, we obtain $(1 - z) : 1 = x : u$ and $(1 - z) : 1 = y : v$. Thus we obtain a map φ, given by $S^2 \backslash \{\mathbf{n}\} \ni (x, y, z) \mapsto \varphi(x, y, z) := \frac{1}{1-z}(x, y) \in \mathbb{R}^2$. From the geometric construction of φ, it is clear that $\varphi : S^2 \backslash \{\mathbf{n}\} \to \mathbb{C}$ is a bijection.

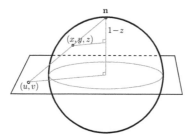

Show that $\mathbb{R}^2 \ni (u, v) \mapsto \left(\frac{2u}{u^2 + v^2 + 1}, \frac{2v}{u^2 + v^2 + 1}, \frac{u^2 + v^2 - 1}{u^2 + v^2 + 1} \right) \in S^2 \backslash \{\mathbf{n}\}$ is the inverse of φ.

Hint: $u^2 + v^2 = \frac{x^2 + y^2}{(1-z)^2} = \frac{1 - z^2}{(1-z)^2}$.

As φ and φ^{-1} are continuous, the complex plane is homeomorphic to $S^2 \backslash \{\mathbf{n}\}$ (where the latter is given the Euclidean metric d_2 on $\mathbb{R}^3$). Via the bijective correspondence φ, we can equip the complex plane with another metric d_c, called the *chordal metric*, defined by $d_c(w, w') = d_2(\varphi^{-1} w, \varphi^{-1} w')$ for all $w, w' \in \mathbb{C}$. Show that $d_c(w, w') = \frac{2|w - w'|}{\sqrt{(1 + |w|^2)(1 + |w'|^2)}} \leqslant 2$.

Hint: Use the Cosine Formula in the triangles formed by $\mathbf{n}, \varphi^{-1} w, \varphi^{-1} w'$ and by $\mathbf{n}, w, w'$. If $w, w' \in \mathbb{C} \backslash \{0\}$, then show that $d_c(w, w') = d_c(\frac{1}{w}, \frac{1}{w'})$.

1.3.2. Open discs, open sets, closed sets and compact sets. To talk about sets of points near a given point, we introduce the following. An *open ball/disc* $D(z_0, r)$ *with centre* z_0 *and radius* $r > 0$ is the set $D(z_0, r) := \{z \in \mathbb{C} : |z - z_0| < r\}$.

A subset U of $\mathbb{C}$ is called *open* if for every $z \in U$, there exists an $r_z > 0$ such that $D(z, r_z) \subset U$. In other words, no matter what point we choose in U, there is always some 'room' around that point comprising points only from U. For example, it can be checked that an open disc $D(z_0, r)$ is an open set. So using the adjective *open* in the name for $D(z_0, r)$ makes sense. Here are some more examples of open sets. The right half-plane $\mathbb{H} := \{z \in \mathbb{C} : \operatorname{Re} z > 0\}$, and for $0 < r < R$, the annulus $\mathbb{A} := \{z \in \mathbb{C} : r < |z| < R\}$, are both open sets. It is also convenient to give a special name to sets whose complement is open, and these are called *closed sets*. One can also give a characterisation of closed sets in terms of sequential convergence (discussed in the following subsection): A set $F \subset \mathbb{C}$ is closed if and only if for every sequence $(z_n)_{n \in \mathbb{N}}$ in F which is convergent in $\mathbb{C}$ to L, the limit $L \in F$. A subset S of $\mathbb{C}$ is called *bounded* if there exists an $M > 0$ such that for all $z \in S$, $|z| \leqslant M$. Thus S is contained in a big enough disc in the complex plane. A subset $K \subset \mathbb{C}$ is called *compact* if it is both closed and bounded. We will often use the known fact from real analysis that a real-valued continuous function on a compact set possesses a maximiser and a minimiser.

1.3.3. Convergence and continuity. A sequence $(z_n)_{n \in \mathbb{N}}$ of complex numbers is said to be *convergent with limit* $L \in \mathbb{C}$ if for every $\epsilon > 0$, there exists an index $N \in \mathbb{N}$ such that for every $n > N$, we have $|z_n - L| < \epsilon$. Using the triangle inequality, it can be seen that for a convergent sequence the limit is unique, and we write $\lim\limits_{n \to \infty} z_n = L$, or that '$z_n \to L$ as $n \to \infty$'.

Example 1.1. If $z \in \mathbb{C}$ and $|z| < 1$, then $(z^n)_{n \in \mathbb{N}}$ converges to 0: $|z^n - 0| = ||z|^n - 0|$, and as $|z| < 1$, the real sequence $(|z|^n)_{n \in \mathbb{N}}$ converges[1] to 0. $\diamond$

Exercise 1.37. Let p be the polynomial given by $p(z) = c_0 + c_1 z + \cdots + c_d z^d$ for all $z \in \mathbb{C}$, where $d \in \mathbb{N}$, $c_0, \cdots, c_d \in \mathbb{C}$, $c_d \neq 0$. Show that there exist $m, M, R > 0$ such that for all $z \in \mathbb{C}$ such that $|z| > R$, $M|z|^d \geqslant |p(z)| \geqslant m|z|^d$.

Exercise 1.38. Show that a sequence $(z_n)_{n \in \mathbb{N}}$ in $\mathbb{C}$ is convergent to $L \in \mathbb{C}$ if and only if $(\operatorname{Re} z_n)_{n \in \mathbb{N}}$ and $(\operatorname{Im} z_n)_{n \in \mathbb{N}}$ are convergent in $\mathbb{R}$ to $\operatorname{Re} L$ and $\operatorname{Im} L$, respectively.

Exercise 1.39. Show that a sequence $(z_n)_{n \in \mathbb{N}}$ in $\mathbb{C}$ is convergent to $L \in \mathbb{C}$ if and only if $(\overline{z_n})_{n \in \mathbb{N}}$ converges to $\overline{L}$.

---

[1]This can be seen as follows. Let $h = \frac{1}{|z|} - 1 > 0$. By the Binomial Theorem, $(1+h)^n = 1 + nh + \cdots + h^n > nh$. So for all $n \in \mathbb{N}$, $\frac{1}{nh} > \frac{1}{(1+h)^n} = |z|^n \geqslant 0$, and the claim follows from the Sandwich Theorem.

Exercise 1.40. Prove that $\mathbb{C}$ is complete, that is, every Cauchy sequence in $\mathbb{C}$ converges in $\mathbb{C}$. (A sequence $(z_n)_{n\in\mathbb{N}}$ is called a *Cauchy sequence* if for every $\epsilon > 0$, there is an index $N \in \mathbb{N}$ such that for all indices $m, n > N$, we have $|z_n - z_m| < \epsilon$.)

Exercise 1.41. A sequence $(z_n)_{n\in\mathbb{N}}$ in $\mathbb{C}$ is said to *tend to* ∞ if for all $R > 0$, there exists an $N \in \mathbb{N}$ such that for all $n \in \mathbb{N}$ satisfying $n > N$, we have $|z_n| > R$.

(1) Show that $(ni^n)_{n\in\mathbb{N}}$ tends to ∞.

(2) Recall Exercise 1.36. Show $(z_n)_{n\in\mathbb{N}}$ tends to ∞ if and only if $\lim_{n\to\infty} d_2(\varphi^{-1}z_n, \mathbf{n}) = 0$.

It is sometimes convenient to adjoin a point called *infinity*, denoted by ∞, to the complex plane to form the *extended complex plane* $\widehat{\mathbb{C}} = \mathbb{C} \cup \{\infty\}$, and extend $\varphi : S^2\setminus\{\mathbf{n}\} \to \mathbb{C}$ to a bijection $\varphi : S^2 \to \widehat{\mathbb{C}}$ by setting $\varphi(\mathbf{n}) = \infty$. If we use this bijection φ and the Euclidean metric on S^2 to obtain a metric on $\widehat{\mathbb{C}}$ (which is the chordal distance d_c we had seen earlier in Exercise 1.36, together with the extra definition $d_c(z, \infty) := d_2(\varphi^{-1}z, \mathbf{n}) = 2/\sqrt{1 + |z|^2}$ for all $z \in \mathbb{C}$), then it can be shown that the metric space $(\widehat{\mathbb{C}}, d_c)$ is complete and compact. This space is called the *one-point compactification* of $\mathbb{C}$.

Let S be a subset of $\mathbb{C}$, $z_0 \in S$ and $f : S \to Y$, where $Y = \mathbb{C}$ or $Y = \mathbb{R}$. Then f is said to be *continuous at* z_0 if for every $\epsilon > 0$, there exists a $\delta > 0$ such that whenever $z \in S$ satisfies $|z - z_0| < \delta$, there holds that $|f(z) - f(z_0)| < \epsilon$. The function f is said to be *continuous* (on S) if for every $z \in S$, f is continuous at z. One can also give a characterisation of continuity at a point in terms of convergent sequences: f is continuous at $z_0 \in S$ if and only if for every sequence $(z_n)_{n\in\mathbb{N}}$ in S convergent to z_0, the sequence $(f(z_n))_{n\in\mathbb{N}}$ is convergent to $f(z_0)$.

Example 1.2. Complex conjugation, i.e., the map $z \mapsto \overline{z} : \mathbb{C} \to \mathbb{C}$, is continuous. Indeed, $|\overline{z} - \overline{z_0}| = |\overline{z - z_0}| = |z - z_0|$ for all $z, z_0 \in \mathbb{C}$. Thus complex conjugation is continuous at each $z_0 \in \mathbb{C}$. As $z_0 \in \mathbb{C}$ was arbitrary, $\overline{}$ is a continuous mapping. This is geometrically obvious, since complex conjugation is just reflection in the real axis, and so the image stays close to the reflected point if we are close to the point. Since $\overline{(\overline{z})} = z$ for all $z \in \mathbb{C}$, complex conjugation is its own inverse. So complex conjugation is invertible with a continuous inverse. Hence complex conjugation is a homeomorphism (that is, a continuous bijective mapping with a continuous inverse) from $\mathbb{C}$ to $\mathbb{C}$. $\diamond$

Exercise 1.42. Prove that the maps $z \mapsto \operatorname{Re} z$ and $z \mapsto |z|$ from $\mathbb{C}$ to $\mathbb{R}$ are continuous.

Exercise 1.43. Let C be any circle of radius r, and $P_1, \cdots, P_n$ be the vertices of a regular n-gon inscribed in C. Show that the product of the diagonal lengths passing through P_1 is nr^{n-1}, i.e., $(P_1P_2)(P_1P_3)\cdots(P_1P_n) = nr^{n-1}$. *Hint:* Consider the unit circle first. There is no loss of generality in assuming $P_1, \cdots, P_n$ are the nth roots of unity, say $1, z_1, \cdots, z_{n-1}$. Use continuity to show $(z - z_1)\cdots(z - z_{n-1}) = z^{n-1} + \cdots + z + 1$ for all $z \in \mathbb{C}$.

Exercise 1.44. Let $S \subset \mathbb{C}$, $z_0 \in S$, f is a real/complex-valued function on S. Show that f is continuous at z_0 if and only if for every sequence $(z_n)_{n\in\mathbb{N}}$ in S convergent to z_0, $(f(z_n))_{n\in\mathbb{N}}$ is convergent to $f(z_0)$.

Exercise 1.45. Let $S \subset \mathbb{C}$, $z_0 \in S$ and $f : S \to \mathbb{C}$. For each $z \in S$, let $u(z) := \operatorname{Re} f(z)$ and $v(z) := \operatorname{Im} f(z)$. Then $u, v : S \to \mathbb{R}$. Show that f is continuous at z_0 if and only if u and v are both continuous at z_0.

1.3.4. Domains. In the rest of the book, the notion of a *path-connected open set* will play an important role. We will prove results about our central object of study, namely functions $f : D \to \mathbb{C}$ that are complex differentiable at every point of the set D ($\subset \mathbb{C}$), and for the validity of many of these theorems, we will need D to be a 'nice' subset of $\mathbb{C}$. Sets which satisfy this 'niceness' assumption, stipulated precisely below, will be what we call a 'domain'. An open subset of $\mathbb{C}$ which is path-connected is called a *domain*. We already know what 'open' means. Now let us explain what we mean by 'path-connectedness'. A *path* (or *curve*) in $\mathbb{C}$ is a continuous function $\gamma : [a, b] \to \mathbb{C}$.

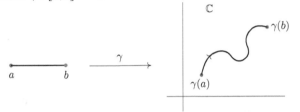

A *stepwise path* is a path $\gamma : [a, b] \to \mathbb{C}$ such that there exists some integer $n \geqslant 0$ and points $t_0, \cdots, t_{n+1}$ such that $t_0 = a < t_1 < \cdots < t_n < t_{n+1} = b$, and for each $k = 0, 1, \cdots, n$, the restriction $\gamma : [t_k, t_{k+1}] \to \mathbb{C}$ is a path with either a constant real part or a constant imaginary part. See the picture on the left below.

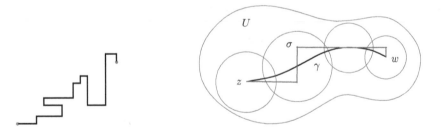

An open set $U \subset \mathbb{C}$ is called *path-connected* if for every $z, w \in U$, there is a stepwise path $\gamma : [a, b] \to \mathbb{C}$ such that $\gamma(a) = z$, $\gamma(b) = w$, and for all $t \in [a, b]$, $\gamma(t) \in U$.

Remark 1.1. In the definition of path-connected open sets, the restriction that the paths should be *stepwise*, can be relaxed: That is, if we look at those open sets in which any two points can be joined merely by a path, then this class of open sets coincides with our path-connected sets[2]. As this is an unnecessary diversion for the path we follow, we will use our definition with stepwise paths.   ✳

---

[2]The idea behind this is as follows. Let U be an open set, and let $\gamma : [a, b] \to \mathbb{C}$ be a path such that for all $t \in [a, b]$, $\gamma(t) \in U$. We want a stepwise path σ joining $\gamma(a)$ and $\gamma(b)$. As U is open, for each $t \in [0, 1]$, $\gamma(t) \in U$ belongs to an open disc $D_t \subset U$. The compact set $K := \{\gamma(t) : t \in [a, b]\}$ has the open cover $\{D_t : t \in [a, b]\}$, which has a finite subcover $\{D_{t_1}, \cdots, D_{t_n}\}$. It is then believable that we can construct a stepwise path by staying within the region formed by these discs. See the picture on the right above. A detailed proof can be found e.g. in [**20**, Ch. 2].

Example 1.3. The entire plane $\mathbb{C}$ is clearly a domain. The open unit disc, namely $\mathbb{D} := \{z \in \mathbb{C} : |z| < 1\}$, is a domain. More generally, if $z_0 \in \mathbb{C}$ and $r > 0$, the disc $D(z_0, r) := \{z \in \mathbb{C} : |z - z_0| < r\}$ is a domain. For $0 < r < R$, the annulus $\mathbb{A} := \{z \in \mathbb{C} : r < |z - z_0| < R\}$ is a domain. Extreme cases of annuli, namely,

- the 'punctured disc' $D_*(z_0, R) := \{z \in \mathbb{C} : 0 < |z - z_0| < R\} = D(z_0, R)\backslash\{z_0\}$,
- the exterior of a disc, $D_e(z_0, r) := \{z \in \mathbb{C} : r < |z - z_0|\}$, and
- the 'punctured plane' $\{z \in \mathbb{C} : 0 < |z - z_0|\} = \mathbb{C}\backslash\{z_0\}$

are domains. The right half-plane $\mathbb{H} := \{z \in \mathbb{C} : \text{Re}(z) > 0\}$ is a domain.

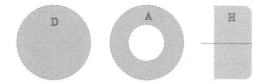

The set $S := \{z \in \mathbb{C} : |z| \neq 1\} =: \mathbb{C}\backslash\mathbb{T}$ is not a domain. Although S is open, it is not path-connected. Indeed, there is no path joining 0 to 2, and staying within S: If there was such a path γ, then by the Intermediate Value Theorem applied to the continuous function $[a, b] \ni t \mapsto |\gamma(t)| \in \mathbb{R}$, as $|\gamma(a)| = 0 < 1 < 2 = |\gamma(b)|$, there exists a $t_* \in [a, b]$ such that $|\gamma(t_*)| = 1$, giving $\gamma(t_*) \notin S$, a contradiction. $\diamond$

Exercise 1.46. Show that the set $\{z \in \mathbb{C} : (\text{Re}\, z)(\text{Im}\, z) > 1\}$ is open, but not a domain.

Exercise 1.47. If D is a domain, then show that $\tilde{D} := \{z \in \mathbb{C} : \overline{z} \in D\}$ is a domain.

1.4. The exponential function and kith

In this last section, we discuss some basic complex functions.

$$z \mapsto \exp z \qquad \text{(exponential function)}$$
$$z \mapsto \sin z, \cos z \;\; \text{(trigonometric functions)}$$
$$z \mapsto \text{Log}\, z \qquad \text{(logarithm function)}.$$

Later we will see that the first two are complex differentiable in the entire complex plane, while the logarithm is complex differentiable in the domain formed by deleting the ray corresponding to the real interval $(-\infty, 0]$. Of course, we will also see that polynomials and rational functions are complex differentiable (the latter in the domain where they are defined). But the above are frequently arising important maps. They are the counterparts to the familiar special/non-rational functions from calculus, to which they reduce to, when restricted to the real axis. In other words, when we restrict our functions to the argument $z = x \in \mathbb{R}$, then we get the usual real-valued functions

$$x \mapsto e^x,$$
$$x \mapsto \sin x, \cos x,$$
$$x \mapsto \log x.$$

So our definitions provide *extensions* of the usual real-valued counterparts.

We will see that these extensions have new and interesting properties in the complex domain that are not possessed by them when the argument is only allowed to be real. Let us begin with the exponential function.

1.4.1. The exponential $\exp z$.

Definition 1.1 (Complex exponential). For $z = x + iy \in \mathbb{C}$, where x, y are real, we define the *complex exponential* $\exp z$ *of* z by $\exp z = e^x(\cos y + i \sin y)$.

When $y = 0$, the right hand side is simply e^x. So our definition extends the usual exponential function $(\mathbb{R} \ni) x \mapsto e^x (\in \mathbb{R})$. Why do we define the complex extension in this way? After all, $z \mapsto e^{\mathrm{Re}\,z}$ also gives an extension of the real exponential. So why not use this simpler definition instead? We define our exp function in the way we do, because we will see that this is the *unique* extension of the real exponential to the whole complex plane having the property that the extension is complex differentiable everywhere; see Example 4.10 on page 92. In fact, just as for the real counterpart, where $\frac{d}{dx}e^x = e^x$ for all $x \in \mathbb{R}$, we will see that $\frac{d}{dz}\exp z = \exp z$ for all $z \in \mathbb{C}$. So eventually we will learn that our complicated definition is actually quite natural. Right now, let us check the following elementary properties.

Proposition 1.3.

(1) $\exp 0 = e^0(\cos 0 + i \sin 0) = (1)(1 + i0) = 1.$

(2) *For all* $z_1, z_2 \in \mathbb{C}$, $\exp(z_1 + z_2) = (\exp z_1)(\exp z_2).$

(3) *For all* $z \in \mathbb{C}$, $\exp z \neq 0$, *and* $(\exp z)^{-1} = \exp(-z).$

(4) *For all* $z \in \mathbb{C}$, $\exp(z + 2\pi i) = \exp z.$

(5) *For all* $z \in \mathbb{C}$, $|\exp z| = e^{\mathrm{Re}\,z}.$

Proof. (2) If $z_1 = x_1 + iy_1$ and $z_2 = x_2 + iy_2$, where $x_1, x_2, y_1, y_2 \in \mathbb{R}$, then

$$
\begin{aligned}
\exp(z_1 + z_2) &= \exp((x_1 + x_2) + i(y_1 + y_2)) \\
&= e^{x_1 + x_2}(\cos(y_1 + y_2) + i \sin(y_1 + y_2)) \\
&= e^{x_1}e^{x_2}(\cos y_1 \cos y_2 - \sin y_1 \sin y_2 + i(\sin y_1 \cos y_2 + \cos y_1 \sin y_2)) \\
&= e^{x_1}(\cos y_1 + i \sin y_1)e^{x_2}(\cos y_2 + i \sin y_2) = (\exp z_1)(\exp z_2).
\end{aligned}
$$

(3) From the previous part, we see that $1 = \exp 0 = \exp(z - z) = (\exp z)(\exp(-z))$, showing that $\exp z \neq 0$, and $(\exp z)^{-1} = \exp(-z)$. (Thus exp maps $\mathbb{C}$ to the 'punctured' plane $\mathbb{C}\backslash\{0\}$.)

(4) We have

$$
\begin{aligned}
\exp(z + 2\pi i) &= (\exp z)(\exp(2\pi i)) \\
&= (\exp z)e^0(\cos(2\pi) + i \sin(2\pi)) \\
&= (\exp z)1(1 + i0) = \exp z.
\end{aligned}
$$

(This shows that exp is 'periodic in the y-direction' with a period of 2π.)

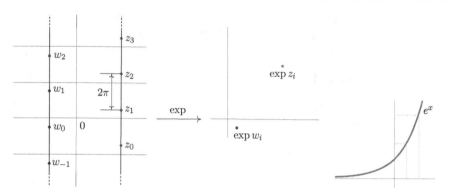

This phenomenon is not present in the x-direction, where as in the real setting, the function $x \mapsto \exp(x + iy_0)$ (with $y_0 \in \mathbb{R}$ fixed) is one-to-one.

(5) For $x, y \in \mathbb{R}$, $|e^x \cos y + ie^x \sin y| = \sqrt{e^{2x}((\cos y)^2 + (\sin y)^2)} = \sqrt{e^{2x}(1)} = e^x$. It follows that $|\exp(x + iy)| = e^x$. Thus exp maps vertical lines in the complex plane, i.e., all points having a common real part, into circles, i.e., points having the same absolute value (which is the distance to the origin). $\square$

Proposition 1.3(3) implies exp is *not* one-to-one; rather it is periodic with period $2\pi i$. The picture below shows the effect of the map $z \mapsto \exp z$ on horizontal (fixed imaginary part y) and vertical lines (fixed real part x).

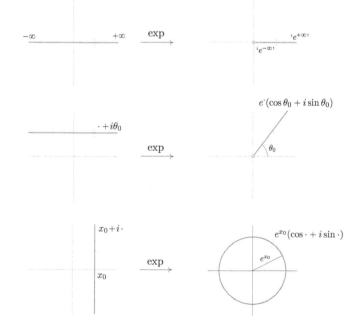

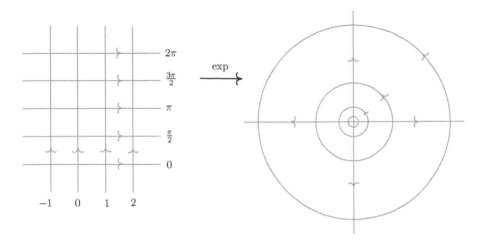

In the picture, we see that exp preserves the angle between the curves considered in its domain: The horizontal and vertical lines, which are mutually perpendicular, are mapped to circles and radial rays, which are also mutually perpendicular. We will see that this is no coincidence, and this property of preservation of angles between curves in the domain together with their 'orientation' is something which is possessed by all complex differentiable functions in domains, and is called 'conformality'; see page 44.

For $z = iy$, where $y \in \mathbb{R}$, $\exp(iy) = \cos y + i \sin y$. This is called *Euler's formula*. If $z \in \mathbb{C}$ has polar coordinates (r, θ), then $z = r(\cos \theta + i \sin \theta) = r \exp(i\theta)$.

Exercise 1.48. Compute $\exp z$ for the following values of z: $3 + \pi i$, $i \frac{9\pi}{2}$.

Exercise 1.49. Find all $z \in \mathbb{C}$ that satisfy $\exp z = \pi i$.

Exercise 1.50. Sketch the path $[0, 2\pi] \ni t \mapsto \exp(it) \in \mathbb{C}$.

Exercise 1.51. Describe $D = \{z \in \mathbb{C} : |\exp(-iz)| < 1\}$. Is D a domain?

Exercise 1.52. Describe the image of the line $y = x$ in $\mathbb{R}^2 = \mathbb{C}$, under the exponential map $z = x + iy \mapsto \exp z$. Proceed as follows: Start with the parametric form $x = t$, $y = t$, and get an expression for the image curve in parametric form. Sketch this curve, explaining what happens as $t \to \infty$, and what happens as $t \to -\infty$.

Exercise 1.53. Find the absolute value and the real and imaginary parts of $\exp(z^2)$ and of $\exp \frac{1}{z}$ in terms of the real and imaginary parts x, y of $z = x + iy$.

Exercise 1.54. Show that for all $z \in \mathbb{C}$, $\overline{\exp z} = \exp \bar{z}$.

Exercise 1.55. Let $a \in \mathbb{R}$. Given $\epsilon > 0$, consider the square region S_ϵ in the complex plane, defined by $S_\epsilon := \{(x, y) \in \mathbb{C} : |x - a| < \epsilon, |y| < \epsilon\}$. Let $\exp S_\epsilon$ denote the image of S_ϵ under exp. Sketch S_ϵ and $\exp S_\epsilon$ in the complex plane. Let $\rho(\epsilon)$ denote the ratio of the area of $\exp S_\epsilon$ to the area of S_ϵ. Determine the limiting value of $\rho(\epsilon)$ as $\epsilon \to 0$.

1.4.2. Trigonometric functions. We now extend the real trigonometric functions of a real variable to complex trigonometric functions of a complex variable. By the Euler formula, for $x \in \mathbb{R}$, $\exp(ix) = \cos x + i\sin x$ and $\exp(-ix) = \cos x - i\sin x$, which give

$$\cos x = \frac{\exp(ix) + \exp(-ix)}{2} \quad \text{and} \quad \sin x = \frac{\exp(ix) - \exp(-ix)}{2i}.$$

This prompts the following definitions: For $z \in \mathbb{C}$, we *define*

$$\cos z = \frac{\exp(iz) + \exp(-iz)}{2} \quad \text{and} \quad \sin z = \frac{\exp(iz) - \exp(-iz)}{2i}.$$

Clearly these definitions give extensions of the usual real trigonometric functions because when we put $z = x \in \mathbb{R}$, we get $\cos z = \cos x$ and $\sin z = \sin x$, as shown above (using Euler's formula). Several trigonometric identities continue to hold in the complex setting. For instance, $\cos(z_1 + z_2) = (\cos z_1)(\cos z_2) - (\sin z_1)(\sin z_2)$ for all $z_1, z_2 \in \mathbb{C}$: Indeed, using properties of exp,

$$(\cos z_1)(\cos z_2) - (\sin z_1)(\sin z_2)$$
$$= \frac{(\exp(iz_1)+\exp(-iz_1))}{2}\frac{(\exp(iz_2)+\exp(-iz_2))}{2} - \frac{(\exp(iz_1)-\exp(-iz_1))}{2i}\frac{(\exp(iz_2)-\exp(-iz_2))}{2i}$$
$$= \frac{2\exp(i(z_1+z_2))+2\exp(-i(z_1+z_2))}{4} = \cos(z_1 + z_2).$$

Exercise 1.56. Show that $\sin(z_1 + z_2) = (\sin z_1)(\cos z_2) + (\cos z_1)(\sin z_2)$ for all $z_1, z_2 \in \mathbb{C}$.

For all $z \in \mathbb{C}$, $(\sin z)^2 + (\cos z)^2 = 1$, since

$$(\sin z)^2 + (\cos z)^2 = \left(\frac{\exp(iz)-\exp(-iz)}{2i}\right)^2 + \left(\frac{\exp(iz)+\exp(-iz)}{2}\right)^2$$
$$= \frac{\exp(2iz)-2+\exp(-2iz)}{-4} + \frac{\exp(2iz)+2+\exp(-2iz)}{4} = 1.$$

However, as opposed to the real trigonometric functions which satisfy $|\sin x| \leqslant 1$ and $|\cos x| \leqslant 1$ for real x, $z \mapsto \sin z$ and $z \mapsto \cos z$ are *not* bounded. Indeed, for $z = iy$, where y is real, we have

$$\cos(iy) = \frac{\exp(i(iy))+\exp(-i(iy))}{2} = \frac{\exp(-y)+\exp y}{2} = \frac{e^{-y}+e^{y}}{2},$$

and so $\cos(iy) \to +\infty$ as $y \to \pm\infty$. Similarly, as $\sin(iy) = \frac{e^{-y}-e^{y}}{2i}$, $|\sin(iy)| \to \infty$ as $y \to \pm\infty$.

We will see later that $z \mapsto \cos z, \sin z$ are complex differentiable everywhere in the complex plane.

Exercise 1.57. For $y \in \mathbb{R}$, let $\cosh y := \frac{e^{y}+e^{-y}}{2}$ and $\sinh y := \frac{e^{y}-e^{-y}}{2}$. Show that $(\cosh y)^2 - (\sinh y)^2 = 1$ for all $y \in \mathbb{R}$. For $z = x + iy$, where x, y are real, show that $\cos z = (\cos x)(\cosh y) - i(\sin x)(\sinh y)$ and $|\cos z|^2 = (\cosh y)^2 - (\sin x)^2$.

Exercise 1.58. We know that the equation $\cos x = 3$ has no real solution x. However, show that there are complex z that satisfy $\cos z = 3$, and find them all.

Exercise 1.59. Determine all $z \in \mathbb{C}$ such that $\cos z + i\sin z = e$.

1.4.3. Logarithm function. In the real setting, given a positive y, $\log y \in \mathbb{R}$ is the unique real number such that $e^{\log y} = y$. Thus $\log : (0, \infty) \to \mathbb{R}$ serves as the inverse of the function $x \mapsto e^x : \mathbb{R} \to (0, \infty)$.

In the complex case, we know that $\exp : \mathbb{C} \to \mathbb{C}\backslash\{0\}$, and we now wonder if there is a 'complex logarithm function' mapping $\mathbb{C}\backslash\{0\}$ to $\mathbb{C}$ that serves as an inverse to the complex exponential function. Given a $z \neq 0$, we seek a complex number w such that $\exp w = z$, and we would like to call w the 'complex logarithm of z'. However, the exponential function $\exp$ is 2π-periodic in the y-direction, and so the moment we find *one* w such that $e^w = z$, we know that there are *infinitely many* others, since $\exp(w + 2\pi i n) = \exp w = z$ for all $n \in \mathbb{Z}$. Given this infinite choice, which w is the complex logarithm of z? We remedy this problem of nonuniqueness by just choosing a w that lies in a *fixed* particular horizontal strip of width 2π. We will show that all possible nonzero complex numbers can be obtained as the $\exp$ of something lying in the fixed strip, and now for the purpose of defining the complex logarithm, we choose (somewhat arbitrarily), the strip $\mathbb{S} := \{z \in \mathbb{C} : -\pi < \mathrm{Im}(z) \leqslant \pi\}$. The strip $\mathbb{S} := \mathbb{R} \times (-\pi, \pi]$ is mapped by $\exp$ onto $\mathbb{C}\backslash\{0\}$.

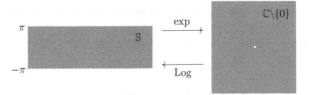

This will give, as we shall show below, a unique w in the strip such that $\exp w = z$, and we will call this unique w the 'principal logarithm of z', denoted by $\mathrm{Log}\, z$. To do this, we first introduce the notion of the principal argument of a nonzero complex number.

Definition 1.2 (Principal argument).
If $z \in \mathbb{C}\backslash\{0\}$, then the *principal argument of z* is the unique number $\mathrm{Arg}\, z \in (-\pi, \pi]$ such that $z = |z|(\cos(\mathrm{Arg}\, z) + i \sin(\mathrm{Arg}\, z))$. If $z = 0$, then we define $\mathrm{Arg}\, z = 0$.

To see the claimed uniqueness above, we note that if $\alpha, \beta \in (-\pi, \pi]$ are such that $|z|(\cos\alpha + i\sin\alpha) = |z|(\cos\beta + i\sin\beta)$, then $\cos\alpha = \cos\beta$ and $\sin\alpha = \sin\beta$, showing $\alpha - \beta = 2\pi n$ for some $n \in \mathbb{Z}$. But as $\alpha, \beta \in (-\pi, \pi]$, it follows that $|\alpha - \beta| < 2\pi$, so that $2\pi|n| = |\alpha - \beta| < 2\pi$, giving $|n| = 0$, and hence $n = 0$.

For example, $\mathrm{Arg}\, 3 = 0$, $\mathrm{Arg}(-1) = \pi$, $\mathrm{Arg}\, i = \frac{\pi}{2}$, $\mathrm{Arg}(-i) = -\frac{\pi}{2}$. If we start at a point on the positive real axis in the complex plane, and go around anticlockwise in a circle, then there is a sudden jump in the value of the principal argument as we cross the negative real axis: On the negative real axis, the value of the principal argument is π, while just below the negative real axis, the principal argument is close to $-\pi$.

Exercise 1.60. Depict $\{z \in \mathbb{C} : z \neq 0,\ \frac{\pi}{4} < |\operatorname{Arg} z| < \frac{\pi}{3}\}$ in the complex plane.

Now we are ready to define the principal logarithm of nonzero complex numbers.

Definition 1.3 (Principal Logarithm).
If $z \in \mathbb{C}\backslash\{0\}$, then the *principal logarithm* $\operatorname{Log} z$ *of* z is $\operatorname{Log} z = \log|z| + i\operatorname{Arg} z$.

For example, $\operatorname{Log}(-i) = \log|-i| + i\operatorname{Arg}(-i) = \log 1 - \frac{\pi}{2}i = 0 - \frac{\pi}{2}i = -\frac{\pi}{2}i$. We now show that $\exp : \mathbb{S} \to \mathbb{C}\backslash\{0\}$ and $\operatorname{Log} : \mathbb{C}\backslash\{0\} \to \mathbb{S}$ are inverses of each other. In particular, $\exp : \mathbb{S} \to \mathbb{C}\backslash\{0\}$ and $\operatorname{Log} : \mathbb{C}\backslash\{0\} \to \mathbb{S}$ are both bijections. Since $\exp(\operatorname{Log} w) = e^{\log|w|}(\cos(\operatorname{Arg} w) + i\sin(\operatorname{Arg} w)) = |w|(\cos(\operatorname{Arg} w) + i\sin(\operatorname{Arg} w)) = w$ for all $w \in \mathbb{C}\backslash\{0\}$, we have $\exp \circ \operatorname{Log} = \operatorname{Id}_{\mathbb{C}\backslash\{0\}}$, where $\operatorname{Id}_{\mathbb{C}\backslash\{0\}}$ denotes the identity map on $\mathbb{C}\backslash\{0\}$, $\mathbb{C}\backslash\{0\} \ni w \mapsto w$.

If $z \in \mathbb{S}$, and $z = x + iy$, where $x \in \mathbb{R}$ and $y \in (-\pi, \pi]$, then we have that $\operatorname{Log}(\exp z) = \operatorname{Log}(e^x(\cos y + i\sin y)) = \log e^x + iy = x + iy = z$. Here we used the fact that $\operatorname{Arg}(e^x(\cos y + i\sin y)) = y$, since $y \in (-\pi, \pi]$. Hence $\operatorname{Log} \circ \exp = \operatorname{Id}_{\mathbb{S}}$, where $\operatorname{Id}_{\mathbb{S}}$ is the identity map on $\mathbb{S}$, $\mathbb{S} \ni z \mapsto z$.

Of course, had we chosen to define the principal argument θ of a nonzero $z = |z|\exp(i\theta)$ to lie in a different interval $(a, 2\pi + a]$ or $[a, 2\pi + a)$ for some other a, we would have obtained a different well-defined notion of the logarithm (which would also be equally legitimate). But in this book, the principal logarithm of z will *always* mean $\log|z| + i\operatorname{Arg} z$, with the principal argument $\operatorname{Arg} z \in (-\pi, \pi]$.

Continuity of Log **on** $\mathbb{C}\backslash(-\infty, 0]$. If $I \subset \mathbb{R}$ is an interval, then the set $I \times \{0\} \subset \mathbb{C}$ will often be denoted simply by I. Intuitively, $\operatorname{Arg} : \mathbb{C}\backslash\{0\} \to (-\pi, \pi]$ is not continuous at any point of the negative real axis $(-\infty, 0)$. Thus Log will not be continuous at these points either (since $\operatorname{Log} z = |z| + i\operatorname{Arg} z$, and $z \mapsto |z|$ is continuous everywhere). We show below that at each point of $(-\infty, 0) \subset \mathbb{C}$, Log is not continuous.

Let $z \in (-\infty, 0)$. Consider the sequence $(|z|\exp(-\pi i + \frac{i}{n}))_{n \in \mathbb{N}}$ in $\mathbb{C}\backslash\{0\}$, which converges to $z = -|z|$:

$$\lim_{n \to \infty} |z|\exp(i(-\pi + \tfrac{1}{n})) = \lim_{n \to \infty} -|z|(\cos\tfrac{1}{n} + i\sin\tfrac{1}{n}) = -|z|(1 + i0) = -|z|.$$

Also, $\operatorname{Log}(|z|\exp(i(-\pi + \frac{1}{n}))) = \log|z| + i(-\pi + \frac{1}{n})$. Thus

$$\lim_{n \to \infty} \operatorname{Log}(|z|\exp(i(-\pi + \tfrac{1}{n}))) = \log|z| - i\pi, \text{ whereas}$$
$$\operatorname{Log}(\lim_{n \to \infty} |z|\exp(i(-\pi + \tfrac{1}{n}))) = \operatorname{Log}(-|z|) = \log|z| + i\pi.$$

Using Exercise 1.44, it follows that Log is not continuous at $z \in (-\infty, 0)$.

On the other hand, Log *is* continuous on the smaller set $\mathbb{C}\backslash(-\infty, 0]$. This is because the principal argument Arg is continuous on $\mathbb{C}\backslash(-\infty, 0]$, as argued below. Let z_0 be any complex number not in $(-\infty, 0]$. Then $\text{Arg}\,z_0 \in (-\pi, \pi)$. There is some room around z_0 not touching the negative real axis, and we can find a small enough r such that the disc $D(z_0, r)$ does not touch the ray $(-\infty, 0]$. Thus, given an $\epsilon > 0$, by shrinking r further if necessary (so that the disc $D(z_0, r)$ is contained in the angular wedge formed by the rays emanating from the origin making angles of $\text{Arg}\,z_0 - \epsilon$ and $\text{Arg}\,z_0 + \epsilon$ with the positive real axis), we can ensure that the points z in $D(z_0, r)$ satisfy $|\text{Arg}\,z - \text{Arg}\,z_0| < \epsilon$.

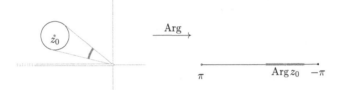

As both $z \mapsto \log|z|$ and Arg are continuous in $\mathbb{C}\backslash(-\infty, 0]$, it follows that Log is continuous there too. Using the continuity of Log on $\mathbb{C}\backslash(-\infty, 0]$, we will see later on that Log is complex differentiable in $\mathbb{C}\backslash(-\infty, 0]$.

The principal value of a^b for $a \in \mathbb{C}\backslash\{0\}$ and $b \in \mathbb{C}$. Recall that if $a \in \mathbb{C}\backslash\{0\}$, and $n \in \mathbb{N}$, then

$$a^n := a \cdots a \quad (n \text{ times})$$
$$= (\exp \text{Log}\,a) \cdots (\exp \text{Log}\,a) = \exp(\text{Log}\,a + \cdots + \text{Log}\,a)$$
$$= \exp(n\,\text{Log}\,a).$$

Recall also that if $a > 0$ and $b \in \mathbb{R}$, then

$$a^b := e^{b \log a}$$
$$= \exp(b \log a) \quad (\text{since } b \log a \in \mathbb{R})$$
$$= \exp(b\,\text{Log}\,a) \quad (\text{since } a > 0).$$

These considerations motivate defining $a^b := \exp(b\,\text{Log}\,a)$ for those complex number a, b for which the right-hand side make sense, that is, $a \in \mathbb{C}\backslash\{0\}$ and $b \in \mathbb{C}$.

Definition 1.4. For $a \in \mathbb{C}\backslash\{0\}$, $b \in \mathbb{C}$, the *principal value of a^b* is $a^b := \exp(b\,\text{Log}\,a)$.

For example, the principal value of i^i is
$$\exp(i\,\text{Log}\,i) = \exp(i(\log|i| + i\text{Arg}\,i)) = \exp(i(0 + i\tfrac{\pi}{2})) = \exp(-\tfrac{\pi}{2}) = e^{-\frac{\pi}{2}}.$$
Taking $a := e$, for all complex numbers $b := z$, we have
$$e^z = \exp(z\,\text{Log}\,e) = \exp(z(\log e + i\text{Arg}\,e)) = \exp(z(1 + i0)) = \exp z.$$
Hence if $z \in \mathbb{C}$, then swapping the notations $\exp z$ and e^z is allowed.

Exercise 1.61. Find $\mathrm{Log}\,(1+i)$.

Exercise 1.62. Find $\mathrm{Log}(-1)$ and $\mathrm{Log}\,1$. Show that $\mathrm{Log}(z^2)$ isn't always equal to $2\,\mathrm{Log}\,z$. If $n \in \mathbb{N}$, then characterise the set of all $z \in \mathbb{C}\backslash\{0\}$ for which we have $\mathrm{Log}(z^n) = n\,\mathrm{Log}\,z$.

Exercise 1.63. Find the image of $\{w \in \mathbb{C} : 1 < |w| < e\}$ under the principal logarithm.

Exercise 1.64. Find the real and imaginary parts of the principal value of $(1+i)^{1-i}$.

Exercise 1.65. If $w \in \mathbb{C}\backslash\{0\}$ and $n \in \mathbb{N}$, then show that $w^{-n} = (\frac{1}{w})^n$. Show that for all $m \in \mathbb{Z}$, and $t \in \mathbb{R}$, $(e^{it})^m = e^{imt}$.

Exercise 1.66 (Möbius transformations). Let $a, b, c, d \in \mathbb{C}$ be such that $ad - bc \neq 0$. We define the complex-valued function f by $f(z) = \frac{az+b}{cz+d}$ for all $z \in \mathbb{C}$ such that $cz + d \neq 0$. The constants c and d cannot both be zero since $ad - bc \neq 0$. If $c = 0$, then f is defined on $\mathbb{C}$, while if $c \neq 0$, then f is defined in $\mathbb{C}\backslash\{-\frac{d}{c}\}$. It will be convenient to extend f to the extended complex plane $\widehat{\mathbb{C}} = \mathbb{C} \cup \{\infty\}$. If $c = 0$, then we set $f(\infty) = \infty$, while if $c \neq 0$, then we set $f(-\frac{d}{c}) = \infty$ and $f(\infty) = \frac{a}{c}$. Such an $f : \widehat{\mathbb{C}} \to \widehat{\mathbb{C}}$ is called a *Möbius transformation*.

(1) Show that f is a bijection, with the inverse given by the Möbius transformation g, where $g(w) = \frac{-dw+b}{cw-a}$. Note that $(-d)(-a) - bc = ad - bc \neq 0$.

(2) Prove that the composition of Möbius transformations is also a Möbius transformation. For simiplicity, just consider the case of transformations $f(z) = \frac{az+b}{cz+d}$, $g(z) = \frac{a'z+b'}{c'z+d'}$, where c, c' are both nonzero. (The set of all Möbius transformations forms a group with composition of maps taken as the group operation.)

(3) Describe the fate of circles and lines under the following Möbius transformations:

(Translation) $f(z) = z + b$	(Dilation) $f(z) = az$ $(a > 0)$
(Rotation) $f(z) = e^{i\theta}z$ $(\theta \in \mathbb{R})$	(Inversion) $f(z) = \frac{1}{z}$.

For the image under inversion of a line not passing through 0, it can be assumed (why?) that the line is $L = \{z_0 + tiz_0 : t \in \mathbb{R}\}$, where $z_0 \neq 0$. For such a line, show that the image lies on a circle with centre $\frac{1}{2z_0}$ and radius $\frac{1}{2|z_0|}$.

(If $c \neq 0$, then $\frac{az+b}{cz+d} = \frac{a}{c} + \frac{bc-ad}{c(cz+d)}$. So every Möbius transformation is the composition of the special ones described above. Imagining straight lines as circles of infinite radius, define a *sircle* be a line or a circle. Möbius transformations map sircles to sircles.)

Chapter 2

Complex differentiability

In this chapter we will learn three main things:

- The definition of complex differentiability, i.e., for $f : U \to \mathbb{C}$, where U is an open subset $\mathbb{C}$, and $z_0 \in U$, we will learn the meaning of the statement 'f is complex differentiable at z_0 with complex derivative $f'(z_0)$'.

- The Cauchy-Riemann equations, $\frac{\partial u}{\partial x} = \frac{\partial v}{\partial y}$ and $\frac{\partial u}{\partial y} = -\frac{\partial v}{\partial x}$, are partial differential equations that are satisfied by the the real and imaginary parts u, v of a complex differentiable function $f : U \to \mathbb{C}$ wherever it is complex differentiable.

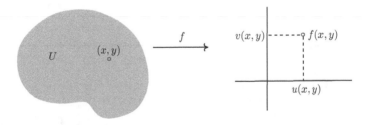

Vice versa, if the Cauchy-Riemann equations are satisfied at $z_0 \in U$ and u, v are real differentiable there, then $f = u + iv$ is complex differentiable at z_0.

- The geometric meaning of the complex derivative $f'(z_0)$: infinitesimally the map f is an amplification by $|f(z_0)|$ together with a twist (a counterclockwise rotation) through $\text{Arg}(f'(z_0))$.

The central result in this chapter is that a function is complex differentiable at a point if and only if its real and imaginary parts are real differentiable, and they satisfy the Cauchy-Riemann equations.

2.1. Complex differentiability

We can define the notion of the limit of a complex-valued function just as in ordinary calculus. Let U be an open subset of $\mathbb{C}$, $h : U \backslash \{z_0\} \to \mathbb{C}$, and $z_0 \in U$. If there exists an $L \in \mathbb{C}$ such that for all $\epsilon > 0$, there exists a $\delta > 0$ such that whenever $z \in U$ satisfies $0 < |z - z_0| < \delta$, we have $|h(z) - L| < \epsilon$, then we say that *the limit of h as $z \to z_0$ exists, and equals L.* We then write $\lim_{z \to z_0} h(z) = L$.

Exercise 2.1. Let U be an open subset of $\mathbb{C}$, $z_0 \in U$, and $h_1, h_2 : U \backslash \{z_0\} \to \mathbb{C}$ be such that $\lim_{z \to z_0} h_i(z)$ exist for $i = 1, 2$. Let $h_1 + h_2$, respectively $h_1 h_2$ denote the pointwise sum, respectively product, of h_1 and h_2. Show that $\lim_{z \to z_0} (h_1 + h_2)(z) = \lim_{z \to z_0} h_1(z) + \lim_{z \to z_0} h_2(z)$, and $\lim_{z \to z_0} (h_1 h_2)(z) = \left(\lim_{z \to z_0} h_1(z) \right) \left(\lim_{z \to z_0} h_2(z) \right)$.

For an open set $U \subset \mathbb{C}$ and an $f : U \to \mathbb{C}$, it is clear that f is continuous at $z_0 \in U$ if and only if $\lim_{z \to z_0} f(z) = f(z_0)$.

Exercise 2.2. Let $U \subset \mathbb{C}$ be open, $z_0 \in U$, $f : U \backslash \{z_0\} \to \mathbb{C}$ be such that $f(z) \neq 0$ for all $z \in U \backslash \{z_0\}$ and $\lim_{z \to z_0} f(z) \neq 0$. Define $\frac{1}{f} : U \to \mathbb{C}$ by $(\frac{1}{f})(z) = \frac{1}{f(z)}$ for all $z \in U \backslash \{z_0\}$. Show that $\lim_{z \to z_0} (\frac{1}{f})(z) = \frac{1}{\lim\limits_{z \to z_0} f(z)}$.

Definition 2.1 (Complex differentiable, holomorphic, entire).
Let U be an open subset of $\mathbb{C}$, $f : U \to \mathbb{C}$ and $z_0 \in U$. Then f is said to be *complex differentiable at z_0* if there exists a complex number L such that $\lim_{z \to z_0} \frac{f(z) - f(z_0)}{z - z_0} = L$, i.e., for every $\epsilon > 0$, there exists a $\delta > 0$ such that whenever $z \in U$ satisfies $0 < |z - z_0| < \delta$, we have $|\frac{f(z) - f(z_0)}{z - z_0} - L| < \epsilon$. This (unique limit) L is denoted by $f'(z_0)$ or $\frac{df}{dz}(z_0)$, and is called the *complex derivative of f at z_0.*

We say $f : U \to \mathbb{C}$ is *holomorphic*[1] in U if for all $z \in U$, f is complex differentiable at z. We denote by $\mathcal{O}(U)$ the set of all holomorphic functions in U.

A function that is holomorphic in $\mathbb{C}$ is called *entire*, that is, the domain of f is understood to be the whole of $\mathbb{C}$ and moreover, f is holomorphic in $\mathbb{C}$.

Example 2.1. Consider the function $f : \mathbb{C} \to \mathbb{C}$ defined by $f(z) = z^2$ for all $z \in \mathbb{C}$. We will now show that f is entire. We have $\frac{f(z) - f(z_0)}{z - z_0} = \frac{z^2 - z_0^2}{z - z_0} = z + z_0 \approx 2z_0$ for z near z_0, and so we *guess* that $f'(z_0) = 2z_0$. Let us show this now. For $z \neq z_0$, we have $|\frac{f(z) - f(z_0)}{z - z_0} - 2z_0| = |\frac{z^2 - z_0^2}{z - z_0} - 2z_0| = |z + z_0 - 2z_0| = |z - z_0|$. So the left-hand side can be made as small as we please when z is close enough to z_0. Let $\epsilon > 0$. Set $\delta := \epsilon > 0$. Then whenever $z \in \mathbb{C}$ satisfies $0 < |z - z_0| < \delta$, we have $|\frac{f(z) - f(z_0)}{z - z_0} - 2z_0| = |z - z_0| < \delta = \epsilon$. Hence $f'(z_0) = 2z_0$. As $z_0 \in \mathbb{C}$ was arbitrary, f is holomorphic in $\mathbb{C}$, i.e., f is entire. We have $f'(z) = \frac{df}{dz}(z) = 2z$ for all $z \in \mathbb{C}$, and we often we simply write $\frac{d}{dz} z^2 = 2z$ $(z \in \mathbb{C})$. $\diamond$

---

[1]'Holomorphic' is derived from the Greek 'holos' meaning 'entire', and 'morphe' meaning 'form'.

Here's an example of a map which at each $z \in \mathbb{C}$ is *not* complex differentiable at z.

Example 2.2. Let $g : \mathbb{C} \to \mathbb{C}$ be defined by $g(z) = \bar{z}$ ($z \in \mathbb{C}$). We show that g is complex differentiable nowhere. Suppose g is complex differentiable at $z_0 \in \mathbb{C}$. Let $\epsilon := \frac{1}{2} > 0$. Then there exists $\delta > 0$ such that whenever $z \in \mathbb{C}$ satisfies $0 < |z - z_0| < \delta$, we have

$$\left| \frac{g(z) - g(z_0)}{z - z_0} - g'(z_0) \right| = \left| \frac{\bar{z} - \bar{z_0}}{z - z_0} - g'(z_0) \right| < \epsilon.$$

So whenever z is in the punctured disc of radius δ with centre z_0, we are guaranteed that this inequality holds. We will now make special choices of the z as the midpoints on the horizontal and vertical radii shown in the figure, and show that the special cases of the inequality above yield that $g'(z_0)$ must lie in discs of radius $\frac{1}{2}$ with centres at -1 and 1. But these discs have an empty intersection, and this will be our contradiction. We give the details below.

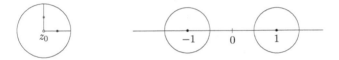

Taking $z = z_0 + \frac{\delta}{2}$, we have $0 < |z - z_0| < \delta$, and so

$$\left| \frac{\bar{z} - \bar{z_0}}{z - z_0} - g'(z_0) \right| = \left| \frac{\frac{\delta}{2}}{\frac{\delta}{2}} - g'(z_0) \right| = |1 - g'(z_0)| < \epsilon. \tag{2.1}$$

Also, taking $z = z_0 + i\frac{\delta}{2}$, we have $0 < |z - z_0| < \delta$, and so

$$\left| \frac{\bar{z} - \bar{z_0}}{z - z_0} - g'(z_0) \right| = \left| \frac{-i\frac{\delta}{2}}{i\frac{\delta}{2}} - g'(z_0) \right| = |-1 - g'(z_0)| = |1 + g'(z_0)| < \epsilon. \tag{2.2}$$

It follows from (2.1) and (2.2) that

$$2 = |1 - g'(z_0) + 1 + g'(z_0)| \leqslant |1 - g'(z_0)| + |1 + g'(z_0)| < \epsilon + \epsilon = 2\epsilon = 2\frac{1}{2} = 1,$$

a contradiction. So g is not complex differentiable at z_0. $\qquad \diamond$

Exercise 2.3. Show that $f : \mathbb{C} \to \mathbb{C}$ defined by $f(z) = |z|^2$ for $z \in \mathbb{C}$, is complex differentiable at 0 and that $f'(0) = 0$. We will see later (in Exercise 2.12) that f is not complex differentiable at any nonzero complex number.

Exercise 2.4. Let D be a domain, and $f : D \to \mathbb{C}$ be holomorphic in D. By Exercise 1.47 $\tilde{D} := \{z \in \mathbb{C} : \bar{z} \in D\}$ is a domain as well. Define $\tilde{f} : \tilde{D} \to \mathbb{C}$ by $\tilde{f}(z) = \overline{f(\bar{z})}$ for all $z \in \tilde{D}$. Prove that $\tilde{f}$ is holomorphic in $\tilde{D}$.

Exercise 2.5. Let $f(z) = \frac{1}{z}$ for all $z \in \mathbb{C}\backslash\{0\}$. Prove that $f'(z) = -\frac{1}{z^2}$ for all $z \in \mathbb{C}\backslash\{0\}$.

The following reformulation of complex differentiability is useful to prove elementary facts about complex differentiation. Roughly speaking, the result says that for a complex differentiable function f with complex derivative L at z_0, $f(z) - f(z_0) - L(z - z_0)$ goes to 0 'faster than $z - z_0$'.

Lemma 2.1. *Let U be an open set in $\mathbb{C}$, $z_0 \in U$, $f : U \to \mathbb{C}$, and $L \in \mathbb{C}$. Then the following are equivalent:*

(1) *f is complex differentiable at z_0 with $f'(z_0) = L$.*

(2) *There exist $r > 0$, and $h : D(z_0, r)(\subset U) \to \mathbb{C}$, such that*
  (a) *$f(z) = f(z_0) + (L + h(z))(z - z_0)$ for all $z \in D(z_0, r)$ and*
  (b) *$\lim_{z \to z_0} h(z) = 0$.*

Proof. $(2) \Rightarrow (1)$: Upon rearranging, we have $\lim_{z \to z_0} \left(\frac{f(z) - f(z_0)}{z - z_0} - L \right) = \lim_{z \to z_0} h(z) = 0$. So f is complex differentiable at z_0, and $f'(z_0) = L$.

$(1) \Rightarrow (2)$: Let f be complex differentiable at z_0, and $L := f'(z_0)$. As U is open, and since f is complex differentiable at z_0 with $f'(z_0) = L$, there exists a $\delta_1 > 0$ such that $D(z_0, \delta_1) \subset U$ and whenever $z \in D(z_0, \delta_1) \backslash \{z_0\}$, $|\frac{f(z) - f(z_0)}{z - z_0} - L| < 1$. Set $r := \delta_1$, and define $h : D(z_0, r) \to \mathbb{C}$ by

$$h(z) = \begin{cases} \frac{f(z) - f(z_0)}{z - z_0} - L & \text{if } z \neq z_0, \\ 0 & \text{if } z = z_0. \end{cases}$$

Then $f(z) = f(z_0) + (L + h(z))(z - z_0)$ for all $z \in D(z_0, r)$. If $\epsilon > 0$, then there is a $\delta > 0$ (which can be chosen smaller than r) such that whenever $0 < |z - z_0| < \delta$, we have $(|h(z) - 0| =) \, |\frac{f(z) - f(z_0)}{z - z_0} - L| < \epsilon$. So $\lim_{z \to z_0} h(z) = 0$. □

Exercise 2.6. Let $U \subset \mathbb{C}$ be open. Use Lemma 2.1 to show that if $f : U \to \mathbb{C}$ is complex differentiable at $z_0 \in U$, then f is continuous at z_0.

Proposition 2.2. *Let U be an open subset of $\mathbb{C}$. Let $f, g : U \to \mathbb{C}$ be complex differentiable functions at $z_0 \in U$. Then:*

(1) *$f + g$ is complex differentiable at z_0 and $(f + g)'(z_0) = f'(z_0) + g'(z_0)$.*
  (Here $f + g : U \to \mathbb{C}$ is defined by $(f + g)(z) = f(z) + g(z)$ for all $z \in U$.)

(2) *If $\alpha \in \mathbb{C}$, then $\alpha \cdot f$ is complex differentiable and $(\alpha \cdot f)'(z_0) = \alpha f'(z_0)$.*
  (Here $\alpha \cdot f : U \to \mathbb{C}$ is defined by $(\alpha \cdot f)(z) = \alpha f(z)$ for all $z \in U$.)

(3) *fg is complex differentiable at z_0, and $(fg)'(z_0) = f'(z_0)g(z_0) + f(z_0)g'(z_0)$.*
  (Here $fg : U \to \mathbb{C}$ is defined by $(fg)(z) = f(z)g(z)$ for all $z \in U$.)

Remark 2.1. Let U be an open subset of $\mathbb{C}$. It follows from the above that $\mathcal{O}(U)$ is a complex vector space with pointwise operations. The third statement above shows that also the pointwise product of two holomorphic functions is holomorphic, and so $\mathcal{O}(U)$ also has the structure of a complex algebra (that is, it is a complex vector space with a bilinear product $\mathcal{O}(U) \times \mathcal{O}(U) \ni (f, g) \mapsto fg \in \mathcal{O}(U)$).   ✳

Example 2.3. It is easy to see that if $f(z) := z$ $(z \in \mathbb{C})$, then $f'(z) = 1$. Moreover, if $g(z) := 1$ $(z \in \mathbb{C})$, then $g'(z) = 0$. Using the rule for complex differentiation of a pointwise product of holomorphic functions, induction gives that for all $n \in \mathbb{N}$, $z \mapsto z^n$ is entire, and $\frac{d}{dz} z^n = n z^{n-1}$. In particular, all polynomials are entire. $\diamond$

Exercise 2.7. Use Lemma 2.1 to prove Proposition 2.2.

Exercise 2.8. Is $\mathcal{O}(\mathbb{C})$ a finite-dimensional vector space (with pointwise operations)?

Exercise 2.9. Let $U \subset \mathbb{C}$ be open, and let $f \in \mathcal{O}(U)$ be such that $f(z) \neq 0$ for $z \in U$. Define $\frac{1}{f} : U \to \mathbb{C}$ by $(\frac{1}{f})(z) = \frac{1}{f(z)}$ for all $z \in U$. Prove $(\frac{1}{f})'(z) = -\frac{f'(z)}{(f(z))^2}$ for all $z \in U$.

Exercise 2.10. Show that in $\mathbb{C} \backslash \{0\}$, for each $m \in \mathbb{Z}$, $\frac{d}{dz} z^m = m z^{m-1}$.

Just as we have the chain rule for the derivative of composition of functions in the real setting, there is an analogous chain rule for holomorphic maps.

Proposition 2.3 (Chain rule).
If • U_f, U_g *are open subsets of* $\mathbb{C}$,
  • $f : U_f \to \mathbb{C}$ *is complex differentiable at* $z_0 \in U_f$, $f(U_f) \subset U_g$,
  • $g : U_g \to \mathbb{C}$ *is complex differentiable at* $f(z_0)$,
then their composition[2] *is complex differentiable at* z_0 *and* $(g \circ f)'(z_0) = g'(f(z_0)) f'(z_0)$.

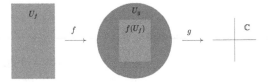

Proof. By the complex differentiability of f at z_0, and of g at $f(z_0)$, there exist $r_f, r_g > 0$, and functions h_f, h_g such that $f(z) - f(z_0) = (f'(z_0) + h_f(z))(z - z_0)$ for all $z \in D(z_0, r_f) \subset U_f$, $g(w) - g(f(z_0)) = (g'(f(z_0)) + h_g(w))(w - f(z_0))$ for all $w \in D(f(z_0), r_g) \subset U_g$, $\lim_{z \to z_0} h_f(z) = 0$, and $\lim_{w \to f(z_0)} h_g(w) = 0$. Set

$$\tilde{h}_g(w) := \begin{cases} h_g(w) & \text{if } w \in D(f(z_0), r_g) \backslash \{f(z_0)\}, \\ 0 & \text{if } w = f(z_0). \end{cases}$$

Then $g(w) - g(f(z_0)) = (g'(f(z_0)) + \tilde{h}_g(w))(w - f(z_0))$ for all $w \in D(f(z_0), r_g) \subset U_g$, and $\lim_{w \to f(z_0)} \tilde{h}_g(w) = 0$. As f is continuous at z_0, there exists a $\delta \in (0, r_f)$ such that if $0 < |z - z_0| < \delta$, then $f(z) \in D(f(z_0), r_g)$. So for all $z \in D(z_0, \delta)$, we have

$$(g \circ f)(z) - (g \circ f)(z_0) = (g'(f(z_0)) + \tilde{h}_g(f(z)))(f'(z_0) + h_f(z))(z - z_0)$$
$$= (g'(f(z_0)) f'(z_0) + \varphi(z))(z - z_0),$$

where $\varphi(z) = f'(z_0) \tilde{h}_g(f(z)) + g'(f(z_0)) h_f(z) + \tilde{h}_g(f(z)) h_f(z)$ for all $z \in D(z_0, \delta)$.

---

[2]The composition $g \circ f : U_f \to \mathbb{C}$ is defined by $(g \circ f)(z) = g(f(z))$ for all $z \in U_f$.

We claim that $\lim_{z \to z_0} \varphi(z) = 0$. As $\lim_{z \to z_0} h_f(z) = 0$, it is enough to show $\lim_{z \to z_0} \tilde{h}_g(f(z)) = 0$.

Let $\epsilon > 0$. Since $\tilde{h}_g(w) \to 0$ as $w \to f(z_0)$, there exists a $\tilde{\delta} \in (0, r_g)$ such that for all $w \in D(f(z_0), \tilde{\delta}) \backslash \{f(z_0)\}$, $|\tilde{h}_g(w)| = |h_g(w)| < \epsilon$. Clearly if $w = f(z_0)$, then $|\tilde{h}_g(w)| = |0| = 0 < \epsilon$ too. So for all $w \in D(f(z_0), \tilde{\delta})$, $|\tilde{h}_g(w)| < \epsilon$. By the continuity of f at z_0, there exists a $\delta' \in (0, \delta)$ such that whenever $z \in D(z_0, \delta')$, we have $|f(z) - f(z_0)| < \tilde{\delta}$, i.e., $f(z) \in D(f(z_0), \delta)$. So for all $z \in D(z_0, \delta')$, $|\tilde{h}_g(f(z))| < \epsilon$. The claim follows from Lemma 2.1.   □

Example 2.4. From Exercise 2.5, $\frac{d}{dz} \frac{1}{z} = -\frac{1}{z^2}$ in $\mathbb{C} \backslash \{0\}$.

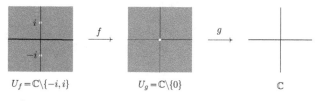

$$U_f = \mathbb{C} \backslash \{-i, i\} \qquad\qquad U_g = \mathbb{C} \backslash \{0\} \qquad\qquad\qquad \mathbb{C}$$

With $f(z) = 1 + z^2$ for $z \in U_f = \mathbb{C} \backslash \{-i, i\}$, and $g(w) = \frac{1}{w}$ for $w \in U_g = \mathbb{C} \backslash \{0\}$, we have $f(U_f) \subset U_g$. By the Chain rule, $\frac{d}{dz} \frac{1}{1+z^2} = -\frac{1}{(1+z^2)^2} 2z = -\frac{2z}{(1+z^2)^2}$ in $\mathbb{C} \backslash \{-i, i\}$. ◇

Exercise 2.11. Assuming that exp is entire and that $\exp' z = \exp z$ (shown later), prove that $z \mapsto \exp(-\frac{1+z}{1-z})$ is holomorphic in $\mathbb{D} := \{z \in \mathbb{C} : |z| < 1\}$, and find its derivative.

2.2. Cauchy-Riemann equations

We now prove the main result in this chapter, which says roughly that a function $f = u + iv$ is holomorphic if and only if its real and imaginary parts u, v (viewed as real-valued functions living in an open subset of $\mathbb{R}^2$) satisfy a pair of partial differential equations, called the Cauchy-Riemann equations.

Let U be an open subset of $\mathbb{C}$, and let $f : U \to \mathbb{C}$ be a function. Then taking any point $(x, y) \in U$, we have $f(x + iy) \in \mathbb{C}$, and we can look at the real part $u(x, y)$ of $f(x + iy)$, and the imaginary part $v(x, y)$ of $f(x + iy)$.

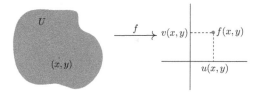

If one changes the point (x, y), then $f(x + iy)$ changes, and so do $u(x, y)$ and $v(x, y)$. In this manner, associated with f, we obtain two *real-valued* functions

$$u : U \to \mathbb{R}, \qquad U \ni (x, y) \mapsto \mathrm{Re}(f(x + iy)) =: u(x, y),$$
$$v : U \to \mathbb{R}, \qquad U \ni (x, y) \mapsto \mathrm{Im}(f(x + iy)) =: v(x, y).$$

Our first result in this section is the necessity of the Cauchy-Riemann equations for complex differentiability, and we will prove this result in Theorem 2.4 below. The result says that if f is complex differentiable at $z_0 = (x_0, y_0) \in U$, then

$$\frac{\partial u}{\partial x}(x_0, y_0) = \frac{\partial v}{\partial y}(x_0, y_0) \text{ and } \frac{\partial u}{\partial y}(x_0, y_0) = -\frac{\partial v}{\partial x}(x_0, y_0),$$

and these two equations are called the *Cauchy-Riemann equations*. If these equations are not satisfied by the real and imaginary part of some complex-valued function at some z_0, then the function cannot be complex differentiable at z_0. Let us revisit Example 2.2. We saw (by 'brute force', using the ϵ-δ definition of complex differentiability), that $z \mapsto \bar{z}$ is not complex differentiable anywhere in the complex plane. Now we will use the Cauchy-Riemann equations to show the same result.

Example 2.5. Define $g : \mathbb{C} \to \mathbb{C}$ by $g(z) = \bar{z}$ for all $z \in \mathbb{C}$. Writing $z = x + iy$, where $x, y \in \mathbb{R}$, we have that $u(x,y) = \text{Re}(g(x + iy)) = \text{Re}(x - iy) = x$ and $v(x,y) = \text{Im}(g(x + iy)) = \text{Im}(x - iy) = -y$. Thus $\frac{\partial u}{\partial x}(x,y) = 1 \neq -1 = \frac{\partial v}{\partial y}(x,y)$, showing that the Cauchy-Riemann equations cannot hold at any point in $\mathbb{C}$. So we recover our previous observation that g is complex differentiable nowhere. $\diamond$

Before proving the necessity of the Cauchy-Riemann equations for complex differentiability, let us also mention the second important result we will show in this section, namely the sufficiency of the Cauchy-Riemann equations for complex differentiability. Suppose that U is an open subset of $\mathbb{C}$, $z_0 = (x_0, y_0) \in U$, the real and imaginary parts u, v of $f : U \to \mathbb{C}$ are real differentiable at (x_0, y_0), and the Cauchy-Riemann equations are satisfied by u and v at (x_0, y_0). Then f is complex differentiable at z_0, and moreover, $f'(z_0) = \frac{\partial u}{\partial x}(x_0, y_0) + i\frac{\partial v}{\partial x}(x_0, y_0)$. This important result will enable us to establish the holomorphicity of important functions without having to go through the rigmarole of verifying the ϵ-δ definition. As an example, we revisit Example 2.1.

Example 2.6. Define $f : \mathbb{C} \to \mathbb{C}$ by $f(z) = z^2$ for all $z \in \mathbb{C}$. Writing $z = x + iy$, where $x, y \in \mathbb{R}$, we have that $u(x,y) = \text{Re}(f(x+iy)) = \text{Re}(x^2 - y^2 + 2xyi) = x^2 - y^2$, and $v(x,y) = \text{Im}(f(x + iy)) = \text{Im}(x^2 - y^2 + 2xyi) = 2xy$. As u, v are polynomials, they are real differentiable, and moreover the partial derivatives satisfy the Cauchy-Riemann equations: $\frac{\partial u}{\partial x}(x,y) = 2x = \frac{\partial v}{\partial y}(x,y)$, and $\frac{\partial u}{\partial y}(x,y) = -2y = -\frac{\partial v}{\partial x}(x,y)$. We recover that f is entire and $f'(z) = \frac{\partial u}{\partial x}(x,y) + i\frac{\partial v}{\partial x}(x,y) = 2x + 2yi = 2z$ in $\mathbb{C}$. $\diamond$

Theorem 2.4 (C-R necessity). *Let $U \subset \mathbb{C}$ be open, $z_0 = (x_0, y_0) \in U$, $f : U \to \mathbb{C}$. Define $u, v : U \to \mathbb{R}$ by $u(x,y) = \text{Re } f(x+iy)$ and $v(x,y) = \text{Im } f(x+iy)$, $(x,y) \in U$. If f is complex differentiable at z_0, then u, v are real differentiable at (x_0, y_0), and*

$$\frac{\partial u}{\partial x}(x_0, y_0) = \frac{\partial v}{\partial y}(x_0, y_0) \text{ and } \frac{\partial u}{\partial y}(x_0, y_0) = -\frac{\partial v}{\partial x}(x_0, y_0).$$

Moreover, $f'(z_0) = \frac{\partial u}{\partial x}(x_0, y_0) + i\frac{\partial v}{\partial x}(x_0, y_0)$.

Proof. (The idea of the proof is that we let (x, y) tend to (x_0, y_0) first keeping y fixed at y_0, and then keeping x fixed at x_0, and look at what this gives us.)

Let $\epsilon > 0$. Then there is $\delta > 0$ for all $z \in U$ satisfying $0 < |z - z_0| < \delta$,

$$|\tfrac{f(z)-f(z_0)}{z-z_0} - f'(z_0)| < \epsilon. \qquad (\star)$$

Step 1. We will show that $\frac{\partial u}{\partial x}(x_0, y_0)$ exists and equals $\operatorname{Re} f'(z_0)$.

Let $z := x + iy_0$, where $x \in \mathbb{R}$ is such that $0 < |x - x_0| < \delta$. Then $z - z_0 = x - x_0$, and so $0 < |z - z_0| = |x - x_0| < \delta$. Thus

$$
\begin{aligned}
|\tfrac{u(x,y_0)-u(x_0,y_0)}{x-x_0} - \operatorname{Re} f'(z_0)| &= |\operatorname{Re}(\tfrac{f(x+iy_0)-f(x_0+iy_0)}{x-x_0}) - \operatorname{Re} f'(z_0)| \\
&= |\operatorname{Re}(\tfrac{f(z)-f(z_0)}{z-z_0}) - \operatorname{Re} f'(z_0)| \\
&\leqslant |\tfrac{f(z)-f(z_0)}{z-z_0} - f'(z_0)| < \epsilon,
\end{aligned}
$$

using $(\star)$. Thus the partial derivative $\frac{\partial u}{\partial x}(x_0, y_0) = \lim\limits_{x \to x_0} \frac{u(x,y_0)-u(x_0,y_0)}{x-x_0} = \operatorname{Re} f'(z_0)$.

Step 2. We show that $\frac{\partial v}{\partial x}(x_0, y_0) = \operatorname{Im} f'(z_0)$.

Proceeding in a manner similar to Step 1, we have with the same notation that

$$
\begin{aligned}
|\tfrac{v(x,y_0)-v(x_0,y_0)}{x-x_0} - \operatorname{Im} f'(z_0)| &= |\operatorname{Im}(\tfrac{f(x+iy_0)-f(x_0+iy_0)}{x-x_0}) - \operatorname{Im} f'(z_0)| \\
&= |\operatorname{Im}(\tfrac{f(z)-f(z_0)}{z-z_0}) - \operatorname{Im} f'(z_0)| \\
&\leqslant |\tfrac{f(z)-f(z_0)}{z-z_0} - f'(z_0)| < \epsilon.
\end{aligned}
$$

So $\frac{\partial v}{\partial x}(x_0, y_0) = \lim\limits_{x \to x_0} \frac{v(x,y_0)-v(x_0,y_0)}{x-x_0} = \operatorname{Im} f'(z_0)$. Thus $f'(z_0) = \frac{\partial u}{\partial x}(x_0, y_0) + i\frac{\partial v}{\partial x}(x_0, y_0)$.

Step 3. We show that $\frac{\partial u}{\partial y}(x_0, y_0) = -\operatorname{Im} f'(z_0)$.

Now let $z := x_0 + iy$, where $y \in \mathbb{R}$ is such that $0 < |y - y_0| < \delta$. Then $z - z_0 = i(y - y_0)$ and so $0 < |z - z_0| = |y - y_0| < \delta$. For $a, b \in \mathbb{R}$, $\operatorname{Re}(a + ib) = \operatorname{Im}(i(a + ib))$. Hence

$$
\begin{aligned}
|\tfrac{u(x_0,y)-u(x_0,y_0)}{y-y_0} + \operatorname{Im} f'(z_0)| &= |\operatorname{Im}(\tfrac{i(f(z)-f(z_0))}{y-y_0}) + \operatorname{Im} f'(z_0))| \\
&= |\operatorname{Im}(-\tfrac{f(z)-f(z_0)}{z-z_0} + f'(z_0))| \\
&\leqslant |\tfrac{f(z)-f(z_0)}{z-z_0} - f'(z_0)| < \epsilon.
\end{aligned}
$$

Thus the partial derivative $\frac{\partial u}{\partial y}(x_0, y_0) = \lim\limits_{y \to y_0} \frac{u(x_0,y)-u(x_0,y_0)}{y-y_0} = -\operatorname{Im} f'(z_0)$.

Recall from Step 2 that $\frac{\partial v}{\partial x}(x_0, y_0) = \operatorname{Im} f'(z_0)$, and so, we have established one of the two Cauchy-Riemann equations, namely $\frac{\partial u}{\partial y}(x_0, y_0) = -\frac{\partial v}{\partial x}(x_0, y_0)$.

Step 4. We show that $\frac{\partial v}{\partial y}(x_0, y_0) = \mathrm{Re}\, f'(z_0)$.

We proceed as in Step 3. For $a, b \in \mathbb{R}$, $\mathrm{Im}(a + ib) = -\mathrm{Re}(i(a + ib))$. Hence

$$\left| \tfrac{v(x_0,y) - v(x_0,y_0)}{y - y_0} - \mathrm{Re}\, f'(z_0) \right| = \left| -\mathrm{Re}(i \tfrac{f(z) - f(z_0)}{y - y_0}) - \mathrm{Re}\, f'(z_0) \right|$$

$$\leqslant \left| -i \tfrac{f(z) - f(z_0)}{y - y_0} - f'(z_0) \right| = \left| \tfrac{f(z) - f(z_0)}{z - z_0} - f'(z_0) \right| < \epsilon.$$

Thus $\frac{\partial v}{\partial y}(x_0, y_0) = \lim\limits_{y \to y_0} \frac{v(x_0,y) - v(x_0,y_0)}{y - y_0} = \mathrm{Re}\, f'(z_0)$. From Steps 1 and 4, we obtain $\frac{\partial u}{\partial x}(x_0, y_0) = \frac{\partial v}{\partial y}(x_0, y_0)$. So we have got both Cauchy-Riemann equations.

Finally, we show that u, v are real differentiable (as real-valued functions of two real variables) at (x_0, y_0). For $z = (x, y)$ satisfying $0 < |z - z_0| < \delta$,

$$\frac{\left| u(x,y) - u(x_0,y_0) - \left[\frac{\partial u}{\partial x}(x_0,y_0) \ \ \frac{\partial u}{\partial y}(x_0,y_0) \right] \left[\begin{smallmatrix} x - x_0 \\ y - y_0 \end{smallmatrix} \right] \right|}{\| (x,y) - (x_0,y_0) \|_2}$$

$$= \frac{\left| u(x,y) - u(x_0,y_0) - (\frac{\partial u}{\partial x}(x_0,y_0))(x - x_0) + (\frac{\partial v}{\partial x}(x_0,y_0))(y - y_0) \right|}{\| (x,y) - (x_0,y_0) \|_2}$$

$$= \frac{\left| \mathrm{Re}(f(z) - f(z_0) - f'(z_0)(z - z_0)) \right|}{|z - z_0|} < \epsilon.$$

So u is real differentiable at (x_0, y_0). Similarly, v is real differentiable at (x_0, y_0). □

We will see later on that in fact the real and imaginary parts of a holomorphic function are infinitely many times real differentiable.

Exercise 2.12. Consider Exercise 2.3 again. Show that f is not complex differentiable at any point of the open set $\mathbb{C} \backslash \{0\}$.

One also has the following converse to Theorem 2.4. This is a very useful result to check the holomorphicity of functions.

Theorem 2.5 (C-R sufficiency). *Let $U \subset \mathbb{C}$ be open, $z_0 = (x_0, y_0) \in U$, $f : U \to \mathbb{C}$. Define $u, v : U \to \mathbb{R}$ by $u(x, y) = \mathrm{Re}\, f(x + iy)$ and $v(x, y) = \mathrm{Im}\, f(x + iy)$, $(x, y) \in U$. If u, v are real differentiable at (x_0, y_0),*

$$\tfrac{\partial u}{\partial x}(x_0, y_0) = \tfrac{\partial v}{\partial y}(x_0, y_0), \text{ and } \tfrac{\partial u}{\partial y}(x_0, y_0) = -\tfrac{\partial v}{\partial x}(x_0, y_0),$$

then f is complex differentiable at z_0, and $f'(z_0) = \frac{\partial u}{\partial x}(x_0, y_0) + i \frac{\partial v}{\partial x}(x_0, y_0)$.

Proof. Let $A := \frac{\partial u}{\partial x}(x_0, y_0) = \frac{\partial v}{\partial y}(x_0, y_0)$, $B := -\frac{\partial u}{\partial y}(x_0, y_0) = \frac{\partial v}{\partial x}(x_0, y_0)$. Let $\epsilon > 0$. As u is real differentiable at (x_0, y_0), there exists $\delta_1 > 0$ such that if $(x, y) \in U$ satisfies $0 < \| (x, y) - (x_0, y_0) \|_2 < \delta_1$, then

$$\frac{\left| u(x,y) - u(x_0,y_0) - [A \ -B] \left[\begin{smallmatrix} x - x_0 \\ y - y_0 \end{smallmatrix} \right] \right|}{\| (x,y) - (x_0,y_0) \|_2} = \frac{\left| u(x,y) - u(x_0,y_0) - \left[\frac{\partial u}{\partial x}(x_0,y_0) \ \frac{\partial u}{\partial y}(x_0,y_0) \right] \left[\begin{smallmatrix} x - x_0 \\ y - y_0 \end{smallmatrix} \right] \right|}{\| (x,y) - (x_0,y_0) \|_2} < \frac{\epsilon}{2}.$$

Similarly, let $\delta_2 > 0$ be such that if $(x, y) \in U$ satisfies $0 < \| (x, y) - (x_0, y_0) \|_2 < \delta_2$,

$$\frac{\left| v(x,y) - v(x_0,y_0) - [B \ A] \left[\begin{smallmatrix} x - x_0 \\ y - y_0 \end{smallmatrix} \right] \right|}{\| (x,y) - (x_0,y_0) \|_2} = \frac{\left| v(x,y) - v(x_0,y_0) - \left[\frac{\partial v}{\partial x}(x_0,y_0) \ \frac{\partial v}{\partial y}(x_0,y_0) \right] \left[\begin{smallmatrix} x - x_0 \\ y - y_0 \end{smallmatrix} \right] \right|}{\| (x,y) - (x_0,y_0) \|_2} < \frac{\epsilon}{2}.$$

Let $z_0 := x_0 + iy_0$. Set $\delta := \min\{\delta_1, \delta_2\} > 0$. If $z = x + iy$ $(x, y \in \mathbb{R})$, satisfies $0 < |z - z_0| < \delta$, then we have

$$\left| \frac{f(z) - f(z_0)}{z - z_0} - \left(\frac{\partial u}{\partial x}(x_0, y_0) + i\frac{\partial v}{\partial x}(x_0, y_0) \right) \right| = \frac{|f(z) - f(z_0) - (A + iB)(z - z_0)|}{|z - z_0|}$$

$$\leqslant \frac{|\mathrm{Re}(f(z) - f(z_0) - (A + iB)(z - z_0))|}{|z - z_0|} + \frac{|\mathrm{Im}(f(z) - f(z_0) - (A + iB)(z - z_0))|}{|z - z_0|}$$

$$= \frac{|u(x, y) - u(x_0, y_0) - (A(x - x_0) - B(y - y_0))|}{|(x - x_0) + i(y - y_0)|} + \frac{|v(x, y) - v(x_0, y_0) - (B(x - x_0) + A(y - y_0))|}{|(x - x_0) + i(y - y_0)|}$$

$$= \frac{|u(x, y) - u(x_0, y_0) - [A \ -B]\begin{bmatrix} x - x_0 \\ y - y_0 \end{bmatrix}|}{\|(x, y) - (x_0, y_0)\|_2} + \frac{|v(x, y) - v(x_0, y_0) - [B \ A]\begin{bmatrix} x - x_0 \\ y - y_0 \end{bmatrix}|}{\|(x, y) - (x_0, y_0)\|_2}$$

$$< \frac{\epsilon}{2} + \frac{\epsilon}{2} = \epsilon.$$

Thus f is complex differentiable at z_0 and $f'(z_0) = \frac{\partial u}{\partial x}(x_0, y_0) + i\frac{\partial v}{\partial x}(x_0, y_0)$. $\qquad \square$

If $u, v : U \to \mathbb{R}$ have the partial derivatives $\frac{\partial u}{\partial x}, \frac{\partial u}{\partial y}, \frac{\partial v}{\partial x}, \frac{\partial v}{\partial y}$ in U, which moreover are continuous on U, then u, v are real differentiable in U.

Example 2.7 (exp, sin, cos are entire). For $x, y \in \mathbb{R}$,

$$u(x, y) = \mathrm{Re}\exp(x + iy) = \mathrm{Re}(e^x(\cos y + i\sin y)) = e^x \cos y,$$

$$v(x, y) = \mathrm{Im}\exp(x + iy) = \mathrm{Im}(e^x(\cos y + i\sin y)) = e^x \sin y.$$

Thus $\frac{\partial u}{\partial x}(x, y) = e^x \cos y = \frac{\partial v}{\partial y}(x, y)$, and $\frac{\partial u}{\partial y}(x, y) = -e^x \sin y = -\frac{\partial v}{\partial x}(x, y)$. So the partial derivatives are continuous, and the Cauchy-Riemann equations hold in $\mathbb{C}$. Thus exp is entire, and $\exp' z = \frac{\partial u}{\partial x}(x, y) + i\frac{\partial v}{\partial x}(x, y) = e^x \cos y + ie^x \sin y = \exp z$. By Propositions 2.3 and 2.2, also $\sin z = \frac{e^{iz} - e^{-iz}}{2i}$ and $\cos z = \frac{e^{iz} + e^{-iz}}{2}$ are entire, $\frac{d}{dz}\sin z = \frac{ie^{iz} - (-i)e^{-iz}}{2i} = \frac{e^{iz} + e^{-iz}}{2} = \cos z$, and $\frac{d}{dz}\cos z = \frac{ie^{iz} + (-i)e^{-iz}}{2} = -\sin z$. $\quad \Diamond$

Example 2.8. Let u, v be defined by $u(x, y) = \frac{x}{x^2 + y^2}$, $v(x, y) = \frac{-y}{x^2 + y^2}$ for all $(x, y) \in \mathbb{R}^2 \backslash \{(0, 0)\}$. Then we have

$$\frac{\partial u}{\partial x} = \frac{1(x^2 + y^2) - x2x}{(x^2 + y^2)^2} = \frac{y^2 - x^2}{(x^2 + y^2)^2},$$

$$\frac{\partial u}{\partial y} = \frac{-x2y}{(x^2 + y^2)^2} = \frac{-2xy}{(x^2 + y^2)^2},$$

$$\frac{\partial v}{\partial x} = \frac{2xy}{(x^2 + y^2)^2}, \text{ and } \frac{\partial v}{\partial y} = \frac{y^2 - x^2}{(x^2 + y^2)^2}.$$

Clearly $(x, y) \mapsto (x^2 + y^2)^2, y^2 - x^2, \pm 2xy$ are continuous in $\mathbb{R}^2$ and $(x^2 + y^2)^2$ is pointwise nonzero in $\mathbb{R}^2 \backslash \{(0, 0)\}$. So each of the above partial derivatives is continuous in $\mathbb{R}^2 \backslash \{(0, 0)\}$. Thus u, v are real differentiable in $\mathbb{R}^2 \backslash \{(0, 0)\}$. Also the Cauchy-Riemann equations hold. Hence $f := u + iv$ is holomorphic in $\mathbb{C} \backslash \{0\}$.

In fact, f is inversion, $\mathbb{C} \backslash \{0\} \ni z \mapsto \frac{1}{z}$: For $(x, y) \in \mathbb{R}^2 \backslash \{(0, 0)\}$, setting $z = x + iy$, we have $f(z) = u(x, y) + iv(x, y) = \frac{x}{x^2 + y^2} + i\frac{-y}{x^2 + y^2} = \frac{x - iy}{x^2 + y^2} = \frac{\bar{z}}{|z|^2} = \frac{\bar{z}}{z\bar{z}} = \frac{1}{z}$. $\quad \Diamond$

Example 2.9 (Holomorphicity of Log in $\mathbb{C}\backslash(-\infty,0]$). The principal logarithm is holomorphic in the open set[3] $\mathbb{C}\backslash(-\infty,0]$. This can be done using the Cauchy-Riemann equations; see Exercise 2.13. A different argument is given below.

The principal logarithm is defined in the bigger set $\mathbb{C}\backslash\{0\}$, but it is not continuous in this bigger set (because at each negative real number, it is discontinuous). In the smaller set $\mathbb{C}\backslash(-\infty,0]$, the principal logarithm is continuous. We will now use this continuity to show that Log is in fact holomorphic in $\mathbb{C}\backslash(-\infty,0]$, and that $\frac{d}{dz}\mathrm{Log}\,z = \frac{1}{z}$ for $z \in \mathbb{C}\backslash(-\infty,0]$. If $z, z_0 \in \mathbb{C}\backslash(-\infty,0]$ are distinct, then $\mathrm{Log}\,z \neq \mathrm{Log}\,z_0$ (otherwise $z = \exp(\mathrm{Log}\,z) = \exp(\mathrm{Log}\,z_0) = z_0$, a contradiction). Let $\epsilon > 0$. Set $\epsilon_1 := \min\{\frac{|z_0|}{2}, \frac{|z_0|^2}{2}\epsilon\} > 0$. Since exp is complex differentiable at $w_0 := \mathrm{Log}\,z_0$, there is a $\delta_1 > 0$ such that whenever w satisfies $0 < |w - w_0| = |w - \mathrm{Log}\,z_0| < \delta_1$, $|\frac{\exp w - \exp w_0}{w - w_0} - \exp w_0| = |\frac{\exp w - z_0}{w - \mathrm{Log}\,z_0} - z_0| < \epsilon_1$. But by the continuity and injectivity of Log in $\mathbb{C}\backslash(-\infty,0]$, there exists a $\delta > 0$ such that whenever $0 < |z - z_0| < \delta$, we have $0 < |\mathrm{Log}\,z - \mathrm{Log}\,z_0| < \delta_1$. Thus with $w := \mathrm{Log}\,z$, and $0 < |z - z_0| < \delta$, we have $0 < |w - w_0| < \delta_1$, and so $|\frac{z - z_0}{\mathrm{Log}\,z - \mathrm{Log}\,z_0} - z_0| < \epsilon_1$. By the reverse triangle inequality, $|z_0| - |\frac{z - z_0}{\mathrm{Log}\,z - \mathrm{Log}\,z_0}| \leqslant |\frac{z - z_0}{\mathrm{Log}\,z - \mathrm{Log}\,z_0} - z_0| < \epsilon_1$, and so $|\frac{z - z_0}{\mathrm{Log}\,z - \mathrm{Log}\,z_0}| \geqslant |z_0| - \epsilon_1 \geqslant \frac{|z_0|}{2}$. Thus whenever $0 < |z - z_0| < \delta$, we have

$$\left|\frac{\mathrm{Log}\,z - \mathrm{Log}\,z_0}{z - z_0} - \frac{1}{z_0}\right| = \left|\left(z_0 - \frac{z - z_0}{\mathrm{Log}\,z - \mathrm{Log}\,z_0}\right)\frac{1}{\frac{z - z_0}{\mathrm{Log}\,z - \mathrm{Log}\,z_0}}\frac{1}{z_0}\right|$$

$$= \left|z_0 - \frac{z - z_0}{\mathrm{Log}\,z - \mathrm{Log}\,z_0}\right|\frac{1}{\left|\frac{z - z_0}{\mathrm{Log}\,z - \mathrm{Log}\,z_0}\right|}\frac{1}{|z_0|} < \epsilon_1 \frac{1}{\frac{|z_0|}{2}}\frac{1}{|z_0|} \leqslant \epsilon.$$

Thus Log is holomorphic in $\mathbb{C}\backslash(-\infty,0]$ and moreover, $\frac{d}{dz}\mathrm{Log}\,z = \frac{1}{z}$. $\diamond$

Exercise 2.13. Recall that $\cos^{-1} : (-1,1) \to (0,\pi)$ has the derivative $\frac{d}{dy}\cos^{-1}y = \frac{-1}{\sqrt{1-y^2}}$, and that $\sin^{-1} : (-1,1) \to (-\frac{\pi}{2}, \frac{\pi}{2})$ has the derivative $\frac{d}{dy}\sin^{-1}y = \frac{1}{\sqrt{1-y^2}}$.

For $z = x + iy \in \mathbb{C}\backslash(-\infty,0]$, where $x, y \in \mathbb{R}$, $\mathrm{Arg}\,z = \begin{cases} \cos^{-1}\dfrac{x}{\sqrt{x^2+y^2}} & \text{if } y > 0, \\ \sin^{-1}\dfrac{y}{\sqrt{x^2+y^2}} & \text{if } x > 0, \\ -\cos^{-1}\dfrac{x}{\sqrt{x^2+y^2}} & \text{if } y < 0. \end{cases}$

Use the Cauchy-Riemann equations to show that Log is holomorphic in $\mathbb{C}\backslash(-\infty,0]$, and find its complex derivative.

Exercise 2.14. Let $U, V \subset \mathbb{C}$ be open, $f : U \to V$, $g : V \to U$ be continuous maps such that $f \circ g = \mathrm{id}_V$ and $g \circ f = \mathrm{id}_U$. (For a set X, id_X denotes the identity map $X \ni x \mapsto x$.) Show that if f is complex differentiable at $z_0 \in U$ and $f'(z_0) \neq 0$, then g is complex differentiable at $f(z_0)$ and $g'(f(z_0)) = \frac{1}{f'(z_0)}$. Use this to show $\mathrm{Log}'z = \frac{1}{z}$, $z \in \mathbb{C}\backslash(-\infty,0]$.

The Cauchy-Riemann equations can also be used to prove some interesting facts, for example the following one, which highlights the 'rigidity' of holomorphic functions alluded to earlier. We will use this later to prove the 'Maximum Modulus Theorem' (Theorem 4.14).

---

[3]Here the 'interval notation' $(-\infty,0]$' actually means the set $(-\infty,0] \times \{0\}$ in the complex plane, i.e., the ray in the complex plane comprising the nonpositive reals.

Example 2.10 ($f \in \mathcal{O}(D(z_0, r))$ with constant $|f|$ is constant). Let $z_0 \in \mathbb{C}$ and $r > 0$. We will show using the Cauchy-Riemann equations that if $f \in \mathcal{O}(D(z_0, r))$ and there exists a $c \in \mathbb{R}$ such that $|f(z)| = c$ for all $z \in D(z_0, r)$, then f is constant.

With $u := \operatorname{Re} f$ and $v := \operatorname{Im} f$, $c^2 = |f|^2 = u^2 + v^2$. Differentiating, $u\frac{\partial u}{\partial x} + v\frac{\partial v}{\partial x} = 0$ and $u\frac{\partial u}{\partial y} + v\frac{\partial v}{\partial y} = 0$. Using $\frac{\partial v}{\partial x} = -\frac{\partial u}{\partial y}$ in the first equation and $\frac{\partial v}{\partial y} = \frac{\partial u}{\partial x}$ in the second,

$$u\frac{\partial u}{\partial x} - v\frac{\partial u}{\partial y} = 0, \qquad (*)$$
$$u\frac{\partial u}{\partial y} + v\frac{\partial u}{\partial x} = 0. \qquad (**)$$

For eliminating $\frac{\partial u}{\partial y}$, $u(*) + v(**)$ gives: $(u^2 + v^2)\frac{\partial u}{\partial x} = 0$, i.e., $c^2\frac{\partial u}{\partial x} = 0$.

For eliminating $\frac{\partial u}{\partial x}$, $-v(*) + u(**)$ gives: $(u^2 + v^2)\frac{\partial u}{\partial y} = 0$, i.e., $c^2\frac{\partial u}{\partial y} = 0$.

(If $c = 0$, then $u^2 + v^2 = c^2 = 0$, and so $u = v = 0$, giving $f = 0$ in $D(z_0, r)$.)

If $c \neq 0$, then from the above $\frac{\partial u}{\partial x} = \frac{\partial u}{\partial y} = 0$. The Cauchy-Riemann equations yield $\frac{\partial v}{\partial x} = \frac{\partial v}{\partial y} = 0$. By the Mean Value Theorem, for $x \neq x_0$, and $y \neq y_0$, there exists a ξ in the open interval with endpoints x, x_0, and there exists an η in the open interval with endpoints y, y_0 such that $\frac{u(x, y_0) - u(x_0, y_0)}{x - x_0} = \frac{\partial u}{\partial x}(\xi, y_0) = 0$, and $\frac{u(x, y) - u(x, y_0)}{y - y_0} = \frac{\partial u}{\partial y}(x, \eta) = 0$. So $u(x, y) = u(x_0, y_0)$ for all $(x, y) \in D(z_0, r)$. Thus u is constant in $D(z_0, r)$.

Similarly, v is constant in D. Consequently, $f = u + iv$ is constant in D.   ◇

Exercise 2.15. Show that $z \mapsto z^3$ is entire using the Cauchy-Riemann equations.

Exercise 2.16. Show that $z \mapsto \operatorname{Re} z$ is complex differentiable nowhere.

Exercise 2.17. Let $D \subset \mathbb{C}$ be a domain. Show that if $f : D \to \mathbb{C}$ is holomorphic in D, with the property that $f(z) \in \mathbb{R}$ for all $z \in D$, then f is constant in D.

Exercise 2.18. Let $D \subset \mathbb{C}$ be a domain. Show that if $f : D \to \mathbb{C}$ is holomorphic in D, with the property that $f'(z) = 0$ for all $z \in D$, then f is constant in D.

Exercise 2.19. Let $f : \mathbb{C} \to \mathbb{C}$ be entire. Set $u := \operatorname{Re} f$, $v := \operatorname{Im} f$. Show that if there exists a real differentiable $h : \mathbb{R} \to \mathbb{R}$ such that $u = h \circ v$, then f is constant in $\mathbb{C}$.

Exercise 2.20. Let $k \in \mathbb{R}$ be a fixed, and let f be defined by $f(z) = x^2 - y^2 + kxyi$ for $z = x + iy$, $x, y \in \mathbb{R}$. Show that f is entire if and only if $k = 2$.

2.3. Geometric meaning of the complex derivative

Recall the geometric meaning of the derivative $f'(x_0)$ of a function $f : \mathbb{R} \to \mathbb{R}$ at a point $x_0 \in \mathbb{R}$: $f'(x_0)$ is the slope of the tangent to the graph of f at x_0.

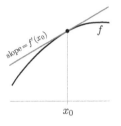

$\lim_{x \to x_0} \frac{f(x)-f(x_0)}{x-x_0} = f'(x_0)$ implies $\frac{f(x)-f(x_0)}{x-x_0} \approx f'(x_0)$ for x near x_0, i.e.,

$$f(x) - f(x_0) \approx f'(x_0)(x - x_0).$$

This means that locally around x_0, $f(x) - f(x_0)$ looks like the action of the *linear* map $h \mapsto f'(x_0)h : \mathbb{R} \to \mathbb{R}$ on $x - x_0$. Visually this means that near x_0, there is very little difference between the (tangent) line with slope $f'(x_0)$ passing through $(x_0, f(x_0))$ and the graph of f. That is, if we zoom into the graph of the function around the point $(x_0, f(x_0))$, then the graph looks like a straight line.

We pose an analogous question for a complex-valued function map $f : U \to \mathbb{C}$ defined on an open set U, that happens to be complex differentiable at a point z_0:

> What is the geometric meaning of the complex number $f'(z_0)$?

We cannot draw a graph of f, because z as well as $f(z)$ belong to $\mathbb{C} = \mathbb{R}^2$, and so $(z, f(z))$ would be a point in $\mathbb{R}^2 \times \mathbb{R}^2 = \mathbb{R}^4$. But we can draw a copy of U in the plane on the left-hand side, and a copy of $\mathbb{C}$ on the right-hand side, with f mapping points from U on the left to points on the right, as shown below.

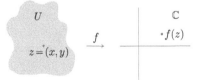

We will show that the complex number $f'(z_0)$ describes the action of the complex differentiable function locally infinitesimally around z_0 by an anticlockwise rotation through the angle $\operatorname{Arg} f'(z_0)$, along with a scaling/magnification by $|f'(z_0)|$.

$\lim_{z \to z_0} \frac{f(z)-f(z_0)}{z-z_0} = f'(z_0)$, implies $\frac{f(z)-f(z_0)}{z-z_0} \approx f'(z_0)$ for z near z_0, i.e.,

$$f(z) - f(z_0) \approx f'(z_0)(z - z_0).$$

But from the geometric meaning of complex multiplication, when we multiply
$z - z_0$ by $f'(z_0)$, $z - z_0$ gets rotated anticlockwise through the angle $\operatorname{Arg} f'(z_0)$,
and the length of $z - z_0$ gets multiplied by the length of $f'(z_0)$, namely we get a
magnification by the factor $|f'(z_0)|$. See the picture below.

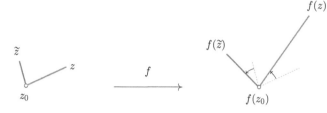

Suppose that $f'(z_0) = \sqrt{3} + i$, so that $|f'(z_0)| = 2$ and $\operatorname{Arg}(f'(z_0)) = \frac{\pi}{6}$. First look
at $z - z_0$ shown in the domain U as the solid line segment between z and z_0. On
the right-hand side, we have shown a translated version of this line segment as a
dashed line, emanating from $f(z_0)$. In order to find out where $f(z)$ is, we just use
the fact that $f(z) - f(z_0)$ is approximately equal to $f'(z_0)$ multiplied by $z - z_0$. So
the solid line joining $f(z_0)$ to $f(z)$ on the right-hand side is obtained by rotating
the rightmost dashed line anticlockwise through an angle of $\operatorname{Arg} f'(z_0)$ (assumed to
be 30° in this picture), and magnifying the length of the dashed line by $|f'(z_0)| = 2$.
If want to find the image of another point $\tilde{z}$ near z_0, we repeat the same procedure.
Namely we first look at the line segment joining z_0 to $\tilde{z}$, which is the solid line on
the left. We have shown a translated version of this as a dashed line, emanating
from $f(z_0)$, in the picture on right-hand side. To find the location of $f(\tilde{z})$, we
first rotate the leftmost dashed line anticlockwise through the argument of $f'(z_0)$,
that is 30°, and magnify the length of the leftmost dashed line by $|f'(z_0)| = 2$. In
this manner, we obtain the solid line on the right-hand side joining $f(z_0)$ to $f(\tilde{z})$
(approximately!). So, locally, the action of f is as follows. Imagine the domain
as a rubber sheet, and look at a point z_0 on this rubber sheet. Tear out a small
portion of this rubber sheet around z_0. Then f takes the point z_0 on this rubber
sheet to a point $f(z_0)$ somewhere in the complex plane. If we want to know how
the rest of the points on our little torn rubber sheet are mapped by f, one follows
this procedure. We place our rubber sheet such that z_0 on our rubber sheet is lying
over the point $f(z_0)$ in the complex plane. (Imagine pinning it on the plane with
the pin passing through the point marked z_0 on our little torn rubber sheet.) Then
we stretch out our rubber sheet about the point z_0 by a factor of $|f'(z_0)|$, and then
rotate this stretched rubber sheet anticlockwise by an angle of $\operatorname{Arg} f'(z_0)$ around
the point z_0 on the rubber sheet. The images of the points on the original little
torn rubber sheet neighbourhood of z_0 in the domain are then approximately given
by their corresponding new positions on the stretched and rotated rubber sheet
pinned at $f(z_0)$.

In order to stress this geometric interpretation, let us revisit Example 2.1 yet
again, where we considered the squaring map $z \mapsto z^2$.

Example 2.11. Suppose we assume complex differentiability of the squaring map f, $z \mapsto z^2$, at a point $z_0 \in \mathbb{C}$. Let us then show that the complex derivative of the squaring map at z_0 must be $2z_0$ by figuring out geometrically the approximate local amplification and rotation produced by the squaring map at z_0.

We first ask: What is the local rotation produced? To find this out, consider a point z close to z_0 along the ray joining 0 to z_0. See the picture below, which shows the effect of the squaring map: The angle is doubled, and the distance to 0 is squared. Thus z^2 lies on the ray joining 0 to z_0^2, which makes an angle $2\mathrm{Arg}\, z_0$ with the positive real axis. Hence the line segment joining z_0^2 to z^2 is obtained by rotating the line segment joining z_0 to z anticlockwise through the angle $\mathrm{Arg}\, z_0$. Consequently, $\mathrm{Arg}\, f'(z_0) = \mathrm{Arg}\, z_0$.

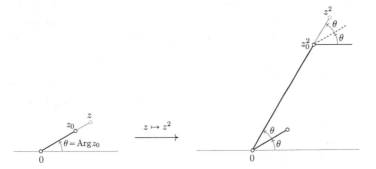

Next we ask: What is the local amplification produced? To find this out, consider a point z close to z_0 which is at the same distance from 0 as z_0, but it makes a slightly bigger angle $\theta + d\theta$ with the positive real axis. See the following picture. Since $d\theta$ is tiny, the length $|z - z_0|$ is approximately $|z_0|\, d\theta$, while the length $|z^2 - z_0^2|$ is approximately $|z_0|^2\, 2d\theta$. So the magnification factor $|f'(z_0)|$ is equal to $\frac{|z_0|^2\, 2d\theta}{|z_0|\, d\theta} = 2|z_0|$.

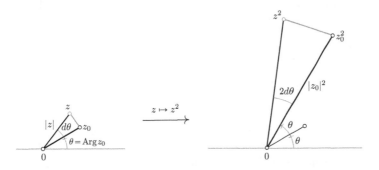

So $f'(z_0) = |f'(z_0)|(\cos \mathrm{Arg}\, f'(z_0) + i \sin \mathrm{Arg}\, f'(z_0)) = 2|z_0|(\cos \mathrm{Arg}\, z_0 + i \sin \mathrm{Arg}\, z_0) = 2z_0$. Thus by investigating the local behaviour of the squaring map f near z_0, we could find out the complex derivative $f'(z_0)$.   ◇

Example 2.12 (Complex conjugation is complex differentiable nowhere).

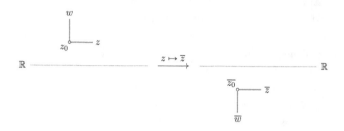

Suppose that $z \mapsto \overline{z}$ is complex differentiable at z_0. Then the local behaviour of the map around z_0 should be a rotation followed by an amplification. Consider the point z near z_0 obtained by a tiny horizontal translation. From the picture, by looking at the images $\overline{z_0}$ and $\overline{z}$, we see that since $\overline{z} - \overline{z_0} = z - z_0$, no rotation is produced. On the other hand, if we look at w which is near z_0 obtained by a tiny vertical displacement, then from the picture, we see that $\overline{w} - \overline{z_0} = -(z - z_0)$, and so there is a rotation through $180°$. But this means that locally the map is not a rotation (because if it were, *all* infinitesimal vectors emanating at z_0 would be rotated by a same fixed amount). $\diamond$

Exercise 2.21. We know that the power function $z \mapsto z^n$, $n \in \mathbb{N}$, is entire. Find its complex derivative by investigating its local behaviour.

Exercise 2.22. We know that the inversion map $f \in \mathcal{O}(\mathbb{C}\backslash\{0\})$, where $f(z) = \frac{1}{z}$, $z \in \mathbb{C}\backslash\{0\}$. Find the complex derivative of f at $z \in \mathbb{C}\backslash\{0\}$ by investigating its local behaviour.

Exercise 2.23. We know that the exponential function $z \mapsto \exp z$ is entire. Find its complex derivative by investigating its local behaviour. *Hint:* Displace a point z_0 vertically (respectively horizontally) to find the local amplification (respectively rotation).

Exercise 2.24. We know that $\mathrm{Log} \in \mathcal{O}(\mathbb{C}\backslash(-\infty, 0])$. Find the complex derivative of Log at a point $w \in \mathbb{C}\backslash(-\infty, 0]$ by investigating its local behaviour.

Exercise 2.25. Give a visual argument to show that the map $z \mapsto \mathrm{Re}\,z$ is not complex differentiable anywhere in $\mathbb{C}$.

Exercise 2.26. In Exercise 1.55, show that the limiting value we calculated there equals the number $|\exp'(a + i0)|^2$. Is this expected?

Conformality. The picture on page 23 shows the action of the entire mapping exp. Just as in the domain, the images of the vertical and horizontal lines under exp are mutually perpendicular. This is a manifestation of the 'conformality' property of holomorphic functions, that is, of the preservation, under the mapping action, of the angles between curves in the domain. Let us now see why holomorphic functions possess this property, based on what we have learnt about the local action of complex differentiable functions.

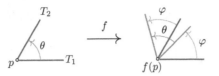

Let $f : U \to \mathbb{C}$ be holomorphic in U. Imagine two smooth curves intersecting at a point $p \in U$. Since the curves are smooth, they have tangents at p, say T_1 and T_2. Near the point p, there is very little difference between the curve and its tangent line at p, so we may assume that the curves are replaced by their tangent lines. These tangent lines make a certain angle. Now let us look at what f does to these lines. Each of the curves is mapped to new curves in $\mathbb{C}$ by f intersecting at $f(p)$, and these new curves are smooth too, possessing tangent lines at $f(p)$. But since the infinitesimal local action of f around p is rotation counterclockwise by $\theta := \operatorname{Arg} f'(p)$ followed by magnification, the new tangent lines are obtained by just rotating counterclockwise the old tangent lines, and magnifying the image. Thus it is obvious that the angle will be the same. Hence the conformality of holomorphic maps is not a mystery anymore.

Complex versus real differentiability. Let $U \subset \mathbb{C}$ be open, and $f : U \to \mathbb{C}$ be complex differentiable at $z_0 = (x_0, y_0) \in U$. If $u := \operatorname{Re} f$, $v := \operatorname{Im} f$, then $u, v : U \to \mathbb{R}$ are real differentiable at (x_0, y_0), and the Cauchy-Riemann equations hold, i.e., $a := \frac{\partial u}{\partial x}(x_0, y_0) = \frac{\partial v}{\partial y}(x_0, y_0)$ and $b := -\frac{\partial u}{\partial y}(x_0, y_0) = \frac{\partial v}{\partial x}(x_0, y_0)$. Thus $(x, y) \overset{f}{\mapsto} (u(x, y), v(x, y)) : U \to \mathbb{R}^2$ is real differentiable, and its real derivative is the linear transformation $\mathbb{R}^2 \ni \mathbf{x} \mapsto A\mathbf{x} \in \mathbb{R}^2$, where the matrix A is given by

$$A := \begin{bmatrix} \frac{\partial u}{\partial x}(x_0, y_0) & \frac{\partial u}{\partial y}(x_0, y_0) \\ \frac{\partial v}{\partial x}(x_0, y_0) & \frac{\partial v}{\partial y}(x_0, y_0) \end{bmatrix} = \begin{bmatrix} a & -b \\ b & a \end{bmatrix}.$$

Set $r := |f'(z_0)| = |\frac{\partial u}{\partial x}(x_0, y_0) + i\frac{\partial v}{\partial x}(x_0, y_0)| = |a + ib| = \sqrt{a^2 + b^2}$. If $\theta := \operatorname{Arg} f'(z_0)$, then $a = \frac{\partial u}{\partial x}(x_0, y_0) = \operatorname{Re} f'(z_0) = r \cos\theta$, and similarly, $b = r \sin\theta$. Thus the real derivative is the linear transformation

$$A = \begin{bmatrix} a & -b \\ b & a \end{bmatrix} = r \begin{bmatrix} \cos\theta & -\sin\theta \\ \sin\theta & \cos\theta \end{bmatrix}.$$

But we recognise this as the linear transformation describing a counterclockwise rotation about the origin through θ, followed by a dilation/magnification by r.

Exercise 2.27. Show that if $\theta \in \mathbb{R}$, and $R = \begin{bmatrix} \cos\theta & -\sin\theta \\ \sin\theta & \cos\theta \end{bmatrix}$, then the map $\mathbb{R}^2 \ni \mathbf{x} \mapsto R\mathbf{x} \in \mathbb{R}^2$ is a counterclockwise rotation about the origin through θ.

So, if f is complex differentiable at a point z_0, then it is real differentiable (as a map from $U \subset \mathbb{R}^2$ to $\mathbb{R}^2$), but what distinguishes complex differentiability from mere real differentiability is that the real derivative for a complex differentiable mapping is not just any linear transformation, but a special one: It is a counterclockwise rotation through an angle $\theta \in [0, 2\pi)$, followed by a scaling by $r \geqslant 0$.

Exercise 2.28 (The $\frac{\partial}{\partial \bar{z}}$-operator). The two Cauchy-Riemann equations can be written as a single equation via the operator $\frac{\partial}{\partial \bar{z}}$. Let $U \subset \mathbb{C}$ be open. If $\varphi : U \to \mathbb{R}$ is real differentiable at $z_0 \in U$, then $\frac{\partial \varphi}{\partial z}(z_0) := \frac{1}{2}(\frac{\partial \varphi}{\partial x}(z_0) - i\frac{\partial \varphi}{\partial y}(z_0))$ and $\frac{\partial \varphi}{\partial \bar{z}}(z_0) := \frac{1}{2}(\frac{\partial \varphi}{\partial x}(z_0) + i\frac{\partial \varphi}{\partial y}(z_0))$. Suppose $f : U \to \mathbb{C}$ is such that $u := \operatorname{Re} f$, $v := \operatorname{Im} f$ are real-differentiable at $z_0 \in U$. Then define $\frac{\partial f}{\partial z}(z_0) := \frac{\partial u}{\partial z}(z_0) + i\frac{\partial v}{\partial z}(z_0)$ and $\frac{\partial f}{\partial \bar{z}}(z_0) := \frac{\partial u}{\partial \bar{z}}(z_0) + i\frac{\partial v}{\partial \bar{z}}(z_0)$.

(1) Show that f is complex differentiable at z_0 if and only if $\frac{\partial f}{\partial \bar{z}}(z_0) = 0$.

(2) Show that if f is complex differentiable at z_0, then $\frac{\partial f}{\partial z}(z_0) = f'(z_0)$.

(3) Determine $\frac{\partial \bar{z}}{\partial \bar{z}}$ and $\frac{\partial |z|^2}{\partial \bar{z}}$.

(4) Determine $\frac{\partial \bar{z}^2}{\partial \bar{z}}$ and $\frac{\partial z^2}{\partial \bar{z}}$.

So holomorphic functions are thought of as 'functions of $z, \bar{z}$ that are independent of $\bar{z}$'.

Chapter 3

Cauchy Integral Theorem

Having become familiar with complex differentiation, we now turn to integration. We will learn an important result called the Cauchy Integral Theorem, after introducing the 'contour integral'. The Cauchy Integral Theorem is important because it will lead to a deeper understanding of holomorphic functions. For example, using this, we will show the fact that holomorphic functions are infinitely many times complex differentiable. In this chapter we will learn the following:

- The contour integral and its properties.
- The Fundamental Theorem of Contour Integration.
- The Cauchy Integral Theorem.
- The following consequences of the Cauchy Integral Theorem:
    (a) Existence of a primitive,
    (b) Infinite complex differentiability of holomorphic functions,
    (c) Liouville's Theorem and the Fundamental Theorem of Algebra,
    (d) Morera's Theorem.

3.1. Definition of the contour integral

In ordinary calculus, given a continuous function $f : [a, b] \to \mathbb{R}$, $\int_a^b f(x)dx$ has a clear meaning. Now suppose we wish to generalise this in the complex setting: Given z, w complex numbers, want to give meaning to something like $\int_z^w f(\zeta) \, d\zeta$. Then a first question is: How do we get from z to w? In $\mathbb{R}$, if $a < b$, then there is just one way of going from the real number a to the real number b, and so our starting data in the real case is: $a, b \in \mathbb{R}$, and a continuous function $f : [a, b] \to \mathbb{R}$.

But now z and w are points in the complex plane, and so there are many possible connecting paths along which we could integrate.

So in the complex setting, besides specifying the end points z and w, we also specify the path γ taken to go from z to w, and we will replace the integral $\int_a^b f(x)dx$ in the real case by an expression which looks like this: $\int_\gamma f(z)dz$. We call such an expression a 'contour' integral, for the computation of which we need the following data:

- A domain D ($\subset \mathbb{C}$), and $z, w \in D$.
- A continuous function $f : D \to \mathbb{C}$.
- A smooth path $\gamma : [a, b] \to D$ joining z to w.

Note that we need not merely a path, but a *smooth* path, defined below. A path $\gamma : [a, b] \to D$ is a continuous function. Decompose γ into its real and imaginary parts: $\gamma(t) = x(t) + iy(t)$ for all $t \in [a, b]$, where $x, y : [a, b] \to \mathbb{R}$. Then x, y are continuous functions. The path γ is *smooth* if x, y are continuously real differentiable (that is, the derivatives $x', y' : [a, b] \to \mathbb{R}$ exist, and they are also continuous).

Example 3.1. Define $\gamma : [0, 1] \to \mathbb{C}$ by $\gamma(t) = t(1 + i)$, $t \in [0, 1]$. The real and imaginary parts $x, y : [0, 1] \to \mathbb{R}$ of γ are given by $x(t) = t$, $y(t) = t$ for all $t \in [0, 1]$. Since x, y are continuously real differentiable on $[0, 1]$, γ is a smooth path.

Define the two paths $\gamma_1, \gamma_2 : [0, 2\pi] \to \mathbb{R}$ by $\gamma_1(t) = \exp(it)$ and $\gamma_2(t) = \exp(2it)$, $t \in [0, 2\pi]$. The real and imaginary parts of γ_1, γ_2 are $\cos t, \sin t, \cos(2t), \sin(2t)$, each of which is continuously real differentiable, and so γ_1, γ_2 are smooth paths.

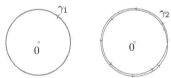

Although the ranges of γ_1 and γ_2 are equal, i.e.,

$\{\gamma_1(t) : t \in [0, 2\pi]\} = \{\gamma_2(t) : t \in [0, 2\pi]\} = \{z \in \mathbb{C} : |z| = 1\}$ (unit circle with center 0),

γ_1 and γ_2 are deemed to be *different* paths, because the functions are not the same: For example, $\gamma_1(\pi) = -1 \neq 1 = \gamma_2(\pi)$. $\diamond$

Remark 3.1. It is convenient to refer to the *range* $\{\gamma(t) : t \in [a, b]\}$ of a path $\gamma : [a, b] \to \mathbb{C}$ as the *path* itself. With this usage, a path becomes a concrete geometric object (as opposed to being a map), such as a circle or a line segment in the complex plane and hence can be easily visualised. The difficulty with this abuse of terminology is that several *different* paths can have the same image, and so it causes ambiguity. $*$

Definition 3.1 (Contour integral). Given

- a domain D,
- a continuous function $f : D \to \mathbb{C}$ (with $u := \operatorname{Re} f$, $v := \operatorname{Im} f$), and
- a smooth path $\gamma : [a, b] \to D$ (with $x := \operatorname{Re} \gamma$, $y := \operatorname{Im} \gamma : [a, b] \to \mathbb{R}$),

we define the *contour integral* $\int_\gamma f(z)dz$ by

$$\int_\gamma f(z)dz := \int_a^b f(\gamma(t))\gamma'(t)dt$$
$$:= \int_a^b \left(u(\gamma(t)) + iv(\gamma(t)) \right) \left(x'(t) + iy'(t) \right) dt$$
$$:= \int_a^b \left(u(\gamma(t))\, x'(t) - v(\gamma(t))\, y'(t) \right) dt + i \int_a^b \left(v(\gamma(t))\, x'(t) + u(\gamma(t))\, y'(t) \right) dt.$$

The last two integrals above are the usual Riemann integrals of real-valued continuous functions. The contour integral can be interpreted geometrically as follows. The term $\gamma'(t)dt = x'(t)dt + iy'(t)dt$ can be viewed as an infinitesimal incremental piece of the contour. We multiply this by the (almost constant) value $f(\gamma(t))$ of f on this incremental piece. Finally, we add up all these contributions along the contour to get the total as the integral $\int_a^b f(\gamma(t))\gamma'(t)dt$.

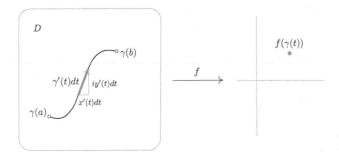

Example 3.2. Let $D = \mathbb{C}$, and $\gamma : [0, 1] \to D$ be the smooth path given by $\gamma(t) = t(1+i)$ for all $t \in [0, 1]$. Then $\gamma'(t) = 1 + i$, $t \in [0, 1]$. If $f = (z \mapsto \overline{z})$, then

$$\int_\gamma f(z)\,dz = \int_0^1 \overline{t(1+i)}(1+i)\,dt = \int_0^1 t(1-i)(1+i)\,dt = \int_0^1 t(1^2 - i^2)\,dt$$
$$= \int_0^1 t(1+1)\,dt = 2\int_0^1 t\,dt = 2\frac{t^2}{2}\Big|_0^1 = 1. \qquad \diamond$$

Exercise 3.1. Consider the three paths $\gamma_1, \gamma_2, \gamma_3 : [0, 2\pi] \to \mathbb{C}$ defined by $\gamma_1(t) = \exp(it)$, $\gamma_2(t) = \exp(2it)$, $\gamma_3(t) = \exp(-it)$, for $t \in [0, 2\pi]$. Show that their images are the same, but the three contour integrals $\int_{\gamma_1} \frac{1}{z}\,dz$, $\int_{\gamma_2} \frac{1}{z}$, $\int_{\gamma_3} \frac{1}{z}\,dz$ are all different.

Exercise 3.2. Let f be holomorphic in a domain and let $\gamma : [a, b] \to D$ be a smooth path. Show that $\frac{d}{dt} f(\gamma(t)) = f'(\gamma(t))\gamma'(t)$ for all $t \in [a, b]$.

Exercise 3.3. Prove or disprove: For every entire function f and every smooth curve $\gamma : [0, 1] \to \mathbb{C}$, $\operatorname{Re}\left(\int_\gamma f(z)\,dz \right) = \int_\gamma \operatorname{Re} f(z)\,dz$.

Equivalent paths give equal integrals. We will often assume that our smooth paths are parametrised by $[0,1]$, rather than some more general interval $[a,b]$. Let us explain why we may assume this. Suppose that $\gamma : [a,b] \to \mathbb{C}$ and $\tilde{\gamma} : [c,d] \to \mathbb{C}$, are two smooth paths, such that there is a continuously differentiable function $\varphi : [c,d] \to [a,b]$ such that $a = \varphi(c)$, $b = \varphi(d)$, and $\tilde{\gamma}(t) = \gamma(\varphi(t))$ for $t \in [c,d]$. We call such smooth paths $\gamma, \tilde{\gamma}$ 'equivalent'. (Imagine going from $\gamma(a) = \tilde{\gamma}(c)$ to $\gamma(b) = \tilde{\gamma}(d)$ along the same route, but with possibly different speeds.)

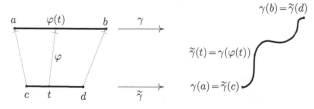

By the chain rule, $\tilde{\gamma}'(t) = \gamma'(\varphi(t))\varphi'(t)$ for all $t \in [c,d]$, and so

$$\int_{\tilde{\gamma}} f(z)\,dz = \int_c^d f(\tilde{\gamma}(t))\tilde{\gamma}'(t)\,dt = \int_c^d f(\gamma(\varphi(t)))\gamma'(\varphi(t))\varphi'(t)\,dt$$

$$\overset{\tau=\varphi(t)}{=} \int_a^b f(\gamma(\tau))\gamma'(\tau)\,d\tau = \int_\gamma f(z)\,dz.$$

In particular, given any $\gamma : [a,b] \to \mathbb{C}$, define $\varphi : [0,1] \to [a,b]$ by $\varphi(t) = (1-t)a + tb$ for all $t \in [0,1]$. Then φ is smooth, and $\varphi(0) = a$, $\varphi(1) = b$. So with $c := 0$, $d := 1$ in the above, and $\tilde{\gamma} : [0,1] \to \mathbb{C}$ defined by $\tilde{\gamma} = \gamma \circ \varphi$, $\int_{\tilde{\gamma}} f(z)\,dz = \int_\gamma f(z)\,dz$. Hence there is no loss of generality (when it comes to statements about contour integrals) in assuming that the smooth path is parametrised by $[0,1]$.

Contour integrals along piecewise smooth paths. We extend the definition above to paths with 'corners'. A path $\gamma : [a,b] \to \mathbb{C}$ is called *piecewise smooth* if there exist points $c_0 := a < c_1 < \cdots < c_n < b =: c_{n+1}$ (for some $n \in \mathbb{N}$) such that the restriction of γ to $[c_k, c_{k+1}]$ is smooth for all $k \in \{0, 1, \cdots, n\}$. For such a piecewise path γ, we define $\int_\gamma f(z)\,dz := \sum_{k=0}^n \int_{c_k}^{c_{k+1}} f(\gamma(t))\gamma'(t)\,dt$.

Example 3.3. Let $\tilde{\gamma}$ be the path given by $\tilde{\gamma}(t) = \begin{cases} t & \text{if } t \in [0,1] \\ 1+(t-1)i & \text{if } t \in (1,2]. \end{cases}$

Then

$$\int_{\tilde{\gamma}} \overline{z}\,dz = \int_0^1 \overline{t}\,1\,dt + \int_1^2 \overline{(1+(t-1)i)}\,i\,dt$$

$$= \int_0^1 t\,dt + \int_1^2 (1-(t-1)i)\,i\,dt$$

$$= \tfrac{1}{2} + \int_1^2 (i+(t-1))\,dt = \tfrac{1}{2} + i + \tfrac{4-1}{2} - (2-1) = 1+i. \qquad \diamond$$

In Examples 3.2 and 3.3, we found that the contour integrals of the (nonholomorphic) function $z \mapsto \bar{z}$ along the paths $\gamma, \tilde{\gamma}$ joining 0 and $1 + i$ are different: $\int_{\gamma} \bar{z} \, dz = 1 \neq 1 + i = \int_{\tilde{\gamma}} \bar{z} \, dz$.

Thus the integral *depends* on the path for the *nonholomorphic* integrand $z \mapsto \bar{z}$. This is not strange, because from the definition of the contour integral of course we expect the value of the contour integral to depend on the route chosen. The main goal in this chapter is to show that the contour integrals of a function f along two paths joining $z, w \in \mathbb{C}$ is the same provided that f is *holomorphic* everywhere in the region between the paths. This result (called the Cauchy Integral Theorem) is fundamental in complex analysis because many further results about holomorphic functions follow from it. First let us check that the result holds in a simple example.

Example 3.4. Let γ, $\tilde{\gamma}$ be the paths considered in Examples 3.2 and 3.3. Instead of the *nonholomorphic* map $z \mapsto \bar{z}$, we take the *entire* function $z \mapsto z$. Then

$\int_{\gamma} z \, dz = \int_0^1 (1 + i)t \, (1 + i) \, dt = \int_0^1 2it \, dt = i$, and

$\int_{\tilde{\gamma}} z \, dz = \int_0^1 t \, 1 \, dt + \int_1^2 (1 + (t-1)i) i \, dt = \int_0^1 t \, dt + \int_1^2 (i - (t-1)) \, dt = \frac{1}{2} + i - \frac{1}{2} = i.$

This time the integrals are the same for γ and $\tilde{\gamma}$: $\int_{\gamma} z \, dz = i = \int_{\tilde{\gamma}} z \, dz$. ◇

Exercise 3.4. Integrate the following over γ, where $\gamma(t) = 2 \exp(it)$ for all $t \in [0, 2\pi]$: $z + \bar{z}$, $z^2 - 2z + 3$, $\mathbb{C} \ni z \mapsto xy$ (where $z = x + iy$, $x, y \in \mathbb{R}$).

Exercise 3.5. Evaluate $\int_{\gamma} \mathrm{Re}\, z \, dz$, where γ is given by:
• The line segment from 0 to $1 + i$.
• The short circular arc with centre i and radius 1 joining 0 to $1 + i$.
• The part of the parabola $y = x^2$ from $x = 0$ to $x = 1$ joining 0 to $1 + i$.

An important integral. Let us now calculate a simple, but important contour integral, which will appear repeatedly. We fix a useful convention: Throughout, a *circular path with centre z_0 and radius $r > 0$ traversed in the anticlockwise direction* will mean the path $C : [0, 2\pi] \to \mathbb{C}$, where $C(t) = z_0 + re^{it}$ for all $t \in [0, 2\pi]$, or the (reparametrised) path $C : [0, 1] \to \mathbb{C}$, where $C(t) = z_0 + re^{2\pi i t}$ for all $t \in [0, 1]$. (Thus the circle is with centre z_0 and radius r traversed *once* counterclockwise.)

Theorem 3.1. *Let C be a circular path with centre z_0 and radius $r > 0$ traversed in the anticlockwise direction. Then* $\int_C (z - z_0)^n \, dz = \begin{cases} 2\pi i & \text{if } n = -1, \\ 0 & \text{if } n \neq -1. \end{cases}$

Proof. We have $C(t) = z_0 + re^{it} = z_0 + r\cos t + ir\sin t$, $t \in [0, 2\pi]$, and so $C'(t) = -r\sin t + ir\cos t = ir(\cos t + i\sin t) = ire^{it}$, $t \in [0, 2\pi]$.

$1°$ If $n = -1$, then $\int_C (z - z_0)^n \, dz = \int_C (z - z_0)^{-1} dz = \int_0^{2\pi} \frac{1}{re^{it}} ire^{it} \, dt = \int_0^{2\pi} i \, dt = 2\pi i$.

$2°$ If $n \neq -1$, then

$$\int_C (z - z_0)^n dz = \int_0^{2\pi} r^n e^{nit} ire^{it} \, dt = \int_0^{2\pi} ir^{n+1} e^{i(n+1)t} dt$$
$$= -r^{n+1} \int_0^{2\pi} \sin((n+1)t) dt + ir^{n+1} \int_0^{2\pi} \cos((n+1)t) dt = 0 + 0 = 0.$$

The claim now also holds for the reparameterised path, $[0, 1] \ni t \mapsto z_0 + re^{2\pi i t}$. $\square$

We will see later that this has significant consequences. For instance, suppose that a function f has an 'expansion in terms of integral powers of z', in an annulus $\mathbb{A} := \{z \in \mathbb{C} : r < |z - z_0| < R\}$ with centre z_0, inner radius r, and outer radius R:

$$f(z) = \sum_{n \in \mathbb{Z}} a_n (z - z_0)^n \text{ for all } z \in \mathbb{A}. \qquad (\star)$$

We will give a precise meaning to the (infinite) 'sum' later, but for now, imagine a finite sum (so that all but finitely many a_ns are zeros). Multiplying both sides by $(z - z_0)^{-(m+1)}$ for some $m \in \mathbb{Z}$, we obtain $\frac{f(z)}{(z-z_0)^{m+1}} = \sum_{n \in \mathbb{Z}} a_n (z - z_0)^{n-m-1}$, and so

$$\frac{1}{2\pi i} \int_C \frac{f(z)}{(z-z_0)^{m+1}} dz = \sum_{n \in \mathbb{Z}} a_n \int_C (z - z_0)^{n-m-1} dz = a_m.$$

Here, we assumed that the sum passes through integration over C, which for finite sums, follows from the definition of the integral, and we will see this in the next section. When the sum is not finite, we will make precise the details later. The upshot of it all is that the coefficients are expressible in terms of a contour integral, and we will see later (Chapter 4) that any function holomorphic in an annulus $\mathbb{A}$ will have such a ('Laurent series') expansion $(\star)$.

Exercise 3.6. Let C be the circular path with centre 0 and radius 1 traversed once in the anticlockwise direction. Show that for $0 \leqslant k \leqslant n$, $\binom{n}{k} = \frac{1}{2\pi i} \int_C \frac{(1+z)^n}{z^{k+1}} dz$.

3.2. Properties of the contour integral

In this section we will show some useful properties of the contour integral. The next result follows in a straightforward manner from the definition of contour integration, and is left as an exercise.

Proposition 3.2. *Let D be a domain in $\mathbb{C}$ and $\gamma : [a, b] \to D$ be a piecewise smooth path. Then the following hold:*

(1) *For all continuous $f, g : D \to \mathbb{C}$, $\int_\gamma (f+g)(z) \, dz = \int_\gamma f(z) \, dz + \int_\gamma g(z) \, dz$.*

(2) *For all continuous $f : D \to \mathbb{C}$ and all $\alpha \in \mathbb{C}$, $\int_\gamma (\alpha f)(z) \, dz = \alpha \int_\gamma f(z) \, dz$.*

Let $C(D;\mathbb{C})$ denote the vector space over $\mathbb{C}$ of all complex-valued continuous functions on D with pointwise operations. Proposition 3.2 implies that each piecewise smooth path γ in D induces a linear transformation from $C(D;\mathbb{C})$ to $\mathbb{C}$: $f \mapsto \int_\gamma f(z)\,dz : C(D;\mathbb{C}) \to \mathbb{C}$.

Exercise 3.7. Prove Proposition 3.2.

Opposite paths. Given a smooth path $\gamma : [a,b] \to D$ in a domain D, its *opposite path*, $-\gamma : [a,b] \to D$, is defined by $(-\gamma)(t) = \gamma(a+b-t)$, $t \in [a,b]$. Then $(-\gamma)(a) = \gamma(b)$ and $(-\gamma)(b) = \gamma(a)$, and so $-\gamma$ starts where γ ends, and ends at the starting point of γ while traversing the same set of points as the range of γ, but in the opposite direction.

Why do we denote the opposite path by $-\gamma$?

Proposition 3.3. *If $\gamma : [a,b] \to D$ be a smooth path in a domain D, and $f : D \to \mathbb{C}$ is a continuous function, then $\int_{-\gamma} f(z)dz = -\int_\gamma f(z)dz$.*

Proof. We have

$$\int_{-\gamma} f(z)\,dz = \int_a^b f((-\gamma)(t))(-\gamma)'(t)\,dt = \int_a^b f(\gamma(a+b-t))(\gamma'(a+b-t))(-1)\,dt$$

$$\overset{\tau=a+b-t}{=} \int_b^a f(\gamma(\tau))\gamma'(\tau)\,d\tau = -\int_a^b f(\gamma(\tau))\gamma'(\tau)\,d\tau = -\int_\gamma f(z)\,dz. \qquad \square$$

Exercise 3.8. If $\gamma : [a,b] \to D$ is a smooth path in a domain D, then show that $-(-\gamma) = \gamma$.

Concatenation of paths. Let $\gamma_1 : [a_1, b_1] \to D$ and $\gamma_2 : [a_2, b_2] \to D$ be two paths in a domain D, such that $\gamma_1(b_1) = \gamma_2(a_2)$ (so that γ_2 starts where γ_1 ends). Define the *concatenation* $\gamma_1 + \gamma_2 : [a_1, b_1 + b_2 - a_2]$ of γ_1 and γ_2 by:

$$(\gamma_1 + \gamma_2)(t) = \begin{cases} \gamma_1(t) & \text{for } a_1 \leq t \leq b_1, \\ \gamma_2(t - b_1 + a_2) & \text{for } b_1 \leq t \leq b_1 + b_2 - a_2. \end{cases}$$

Proposition 3.4. *Let $\gamma_1 : [a_1, b_1] \to D$, $\gamma_2 : [a_2, b_2] \to D$ be two paths in a domain D such that $\gamma_1(b_1) = \gamma_2(a_2)$. Then $\int_{\gamma_1 + \gamma_2} f(z)\,dz = \int_{\gamma_1} f(z)\,dz + \int_{\gamma_2} f(z)\,dz$.*

Proof. We use the substitution $\tau = t - b_1 + a_2$ to obtain the fourth equality below:

$$\int_{\gamma_1 + \gamma_2} f(z)\,dz = \int_{a_1}^{b_1 + b_2 - a_2} f((\gamma_1 + \gamma_2)(t))(\gamma_1 + \gamma_2)'(t)\,dt$$

$$= \int_{a_1}^{b_1} f((\gamma_1 + \gamma_2)(t))(\gamma_1 + \gamma_2)'(t)\,dt + \int_{b_1}^{b_1 + b_2 - a_2} f((\gamma_1 + \gamma_2)(t))(\gamma_1 + \gamma_2)'(t)\,dt$$

$$= \int_{a_1}^{b_1} f(\gamma_1(t))\gamma_1'(t)\,dt + \int_{b_1}^{b_1 + b_2 - a_2} f(\gamma_2(t - b_1 + a_2))\gamma_2'(t - b_1 + a_2)\,dt$$

$$= \int_{\gamma_1} f(z)\,dz + \int_{a_2}^{b_2} f(\gamma_2(\tau))\gamma_2'(\tau)\,d\tau = \int_{\gamma_1} f(z)\,dz + \int_{\gamma_2} f(z)\,dz. \qquad \square$$

Exercise 3.9. If $\gamma : [a, b] \to D$ is a smooth path in a domain D, and $f : D \to \mathbb{C}$ is continuous, then show that $\int_{\gamma + (-\gamma)} f(z)\,dz = 0$.

A useful estimate. We now prove a useful inequality for the size of the contour integral in terms of the size of $|f|$ along the contour, and the length of the contour.

Proposition 3.5 (ML inequality). *Let $\gamma : [a, b] \to D$ be a smooth path in a domain D, and $f : D \to \mathbb{C}$ is a continuous function, then*

$$\left| \int_\gamma f(z)\,dz \right| \leq \Big(\max_{t \in [a,b]} |f(\gamma(t))| \Big)(\text{length of } \gamma).$$

If $x := \operatorname{Re}\gamma$, $y := \operatorname{Im}\gamma : [a, b] \to \mathbb{R}$, then the length of γ is $\int_a^b \sqrt{(x'(t))^2 + (y'(t))^2}\,dt$: The length of the path γ is the sum of the incremental arc lengths ds, where $ds = \sqrt{(x'(t)dt)^2 + (y'(t)dt)^2} = \sqrt{(x'(t))^2 + (y'(t))^2}\,dt$.

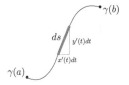

Proof. First, for a path $\varphi : [a, b] \to \mathbb{C}$, we will prove $\left| \int_a^b \varphi(t)\,dt \right| \leq \int_a^b |\varphi(t)|\,dt$. Here $\int_a^b \varphi(t)\,dt := \int_a^b \operatorname{Re}(\varphi(t))\,dt + i \int_a^b \operatorname{Im}(\varphi(t))\,dt$. Let $\int_a^b \varphi(t)\,dt = re^{i\theta}$ for some $r \geq 0$ and $\theta \in (-\pi, \pi]$. Then

$$\left| \int_a^b \varphi(t)\,dt \right| = r = e^{-i\theta} re^{i\theta} = e^{-i\theta} \int_a^b \varphi(t)\,dt = \int_a^b e^{-i\theta} \varphi(t)\,dt$$

$$= \int_a^b \operatorname{Re}(e^{-i\theta}\varphi(t))\,dt + i \int_a^b \operatorname{Im}(e^{-i\theta}\varphi(t))\,dt.$$

But the left-hand side is real, and so the integral of the imaginary part on the right-hand side must be zero. Consequently,

$$\left| \int_a^b \varphi(t)dt \right| = \int_a^b \operatorname{Re}(e^{-i\theta}\varphi(t))dt \leq \int_a^b |\operatorname{Re}(e^{-i\theta}\varphi(t))|dt \leq \int_a^b |e^{-i\theta}\varphi(t)|dt = \int_a^b |\varphi(t)|dt.$$

The proposition now follows, since with $\varphi(t) := f(\gamma(t))\gamma'(t)$ for all $t \in [a, b]$,

$$\left| \int_\gamma f(z)\,dz \right| = \left| \int_a^b f(\gamma(t))\gamma'(t)\,dt \right| \leq \int_a^b |f(\gamma(t))\gamma'(t)|\,dt = \int_a^b |f(\gamma(t))||\gamma'(t)|\,dt$$

$$\leq \Big(\max_{t \in [a,b]} |f(\gamma(t))| \Big) \int_a^b |\gamma'(t)|\,dt.$$

If $\gamma(t) = x(t) + iy(t)$, where x, y are real-valued, then

$$\int_a^b |\gamma'(t)|\,dt = \int_a^b \sqrt{(x'(t))^2 + (y'(t))^2}\,dt = \text{length of } \gamma. \qquad \square$$

By applying the ML inequality to each smooth section of a piecewise smooth path, it follows that the ML inequality also holds for a piecewise smooth path.

Exercise 3.10. For a path $\varphi : [a, b] \to \mathbb{C}$ and $z \in \mathbb{C}$, show that $\int_a^b z\varphi(t)\,dt = z\int_a^b \varphi(t)\,dt$.

Exercise 3.11. Calculate the upper bound given by the ML inequality on the absolute value of the integral $\int_\gamma z^2\,dz$, where γ is the straight line path from 0 to $1+i$. Also, compute the integral and find its absolute value.

Exercise 3.12. Using the calculation done in Exercise 3.6, deduce that $\binom{2n}{n} \leq 4^n$.

Exercise 3.13. Let γ be the straight line path from i to 1. Prove that $|\int_\gamma \frac{1}{z^4}dz| \leq 4\sqrt{2}$.

3.3. Fundamental Theorem of Contour Integration

Let us recall the Fundamental Theorem of Calculus in the real setting:

Theorem 3.6 (Fundamental Theorem of Calculus).
If $f : [a, b] \to \mathbb{R}$ is continuously real differentiable, then $\int_a^b f'(x)dx = f(b) - f(a)$.

This is an important result, because it facilitates the computation of integrals. Indeed, knowing that a function is the derivative of something, it is easy to calculate its integral. For example, since $x = \frac{d}{dx}\frac{x^2}{2}$, we have $\int_a^b x\,dx = \frac{b^2-a^2}{2}$.

Analogously, if $f \in \mathcal{O}(D)$ where D is a domain, and f' is continuous on D, then the calculation of the contour integral $\int_\gamma f'(z)\,dz$ is easy, since similar to the Fundamental Theorem of Calculus in the real setting, we have the following.

Theorem 3.7 (Fundamental Theorem of Contour Integration).
If D is a domain, $f \in \mathcal{O}(D)$ such that f' is continuous on D, and $\gamma : [a,b] \to D$ is a piecewise smooth path, then $\int_\gamma f'(z)\,dz = f(\gamma(b)) - f(\gamma(a))$.

How does this theorem help? One can now calculate some contour integrals very easily (just like in ordinary calculus). Here is an example.

Example 3.5. As $\frac{d}{dz}\frac{z^2}{2} = z$ in $\mathbb{C}$, we have that for *any* piecewise smooth path γ joining 0 to $1 + i$, $\int_\gamma z\,dz = \frac{(1+i)^2}{2} - \frac{0^2}{2} = \frac{1+2i+i^2}{2} = \frac{1+2i-1}{2} = i$. In particular, we recover the answer obtained in Example 3.4. $\diamond$

From the above theorem, if f possesses an 'antiderivative' or 'primitive' F in D (i.e., $F \in \mathcal{O}(D)$ and $F' = f$), then for any path γ joining the points z to w, we have that $\int_\gamma f(z)\,dz = F(w) - F(z)$, so that $\int_\gamma f(z)\,dz$ is independent of the path γ taken to go from z to w.

Example 3.6. There is no function $F : \mathbb{C} \to \mathbb{C}$ such that $F'(z) = \bar{z}$ for all $z \in \mathbb{C}$. Indeed, the calculations in Examples 3.2 and 3.3 show that the contour integral along paths joining 0 to $1 + i$ does depend on the path chosen. $\diamond$

Proof of Theorem 3.7. Let $z = x + iy \in D$, $x, y \in \mathbb{R}$. Define $U, V, u, v : D \to \mathbb{R}$ by $f(x + iy) = U(x, y) + iV(x, y)$ and $f'(x + iy) = u(x, y) + iv(x, y)$. We obtain $u(x, y) + iv(x, y) = f'(x + iy) = \frac{\partial U}{\partial x}(x, y) + i\frac{\partial V}{\partial x}(x, y) = \frac{\partial V}{\partial y}(x, y) - i\frac{\partial U}{\partial y}(x, y)$, where we have used the Cauchy-Riemann equations. Set $\gamma(t) = \mathrm{x}(t) + i\mathrm{y}(t)$ ($t \in [a, b]$), where x, y are real-valued. It is enough to show the theorem for smooth paths. So

$$\frac{d}{dt}U(\mathrm{x}(t), \mathrm{y}(t)) = \frac{\partial U}{\partial x}(\mathrm{x}(t), \mathrm{y}(t))\mathrm{x}'(t) + \frac{\partial U}{\partial y}(\mathrm{x}(t), \mathrm{y}(t))\mathrm{y}'(t)$$
$$= u(\mathrm{x}(t), \mathrm{y}(t))\mathrm{x}'(t) - v(\mathrm{x}(t), \mathrm{y}(t))\mathrm{y}'(t),$$
$$\frac{d}{dt}V(\mathrm{x}(t), \mathrm{y}(t)) = \frac{\partial V}{\partial x}(\mathrm{x}(t), \mathrm{y}(t))\mathrm{x}'(t) + \frac{\partial V}{\partial y}(\mathrm{x}(t), \mathrm{y}(t))\mathrm{y}'(t)$$
$$= v(\mathrm{x}(t), \mathrm{y}(t))\mathrm{x}'(t) + u(\mathrm{x}(t), \mathrm{y}(t))\mathrm{y}'(t),$$

by the chain rule. Thus

$$\int_\gamma f'(z)\,dz = \int_a^b f'(\gamma(t))\gamma'(t)dt = \int_a^b \big(u(\mathrm{x}(t), \mathrm{y}(t)) + iv(\mathrm{x}(t), \mathrm{y}(t))\big)(\mathrm{x}'(t) + i\mathrm{y}'(t))dt$$
$$= \int_a^b \frac{d}{dt}U(\mathrm{x}(t), \mathrm{y}(t))dt + i\int_a^b \frac{d}{dt}V(\mathrm{x}(t), \mathrm{y}(t))dt$$
$$= U(\mathrm{x}(b), \mathrm{y}(b)) - U(\mathrm{x}(a), \mathrm{y}(a)) + i\big(V(\mathrm{x}(b), \mathrm{y}(b)) - V(\mathrm{x}(a), \mathrm{y}(a))\big)$$
$$= f(\gamma(b)) - f(\gamma(a)). \qquad \square$$

Exercise 3.14. Use the Cauchy-Riemann equations to show that $z \mapsto \bar{z}$ has no primitive in $\mathbb{C}$.

Exercise 3.15 (Integration by Parts Formula). Let f, g be holomorphic functions defined in a domain D, such that f', g' are continuous in D. Let γ be a piecewise smooth path in D from $w \in D$ to $z \in D$. Show that $\int_\gamma f(\zeta)g'(\zeta)\,d\zeta = f(z)g(z) - f(w)g(w) - \int_\gamma f'(\zeta)g(\zeta)\,d\zeta$. For $r > 0$, let $S_r(t) := Re^{it}$, $t \in [0, \pi]$. Let $a \in \mathbb{R}\backslash\{0\}$. Show that $\lim_{r\to\infty} |\int_{S_r} \frac{ze^{iz}}{z^2 + a^2}\,dz| = 0$.

Exercise 3.16. Evaluate $\int_\gamma \cos z\,dz$, where γ is any path joining $-i$ to i.

Exercise 3.17. Let $\mathbb{D} = \{z \in \mathbb{C} : |z| < 1\}$. Let $f \in \mathcal{O}(\mathbb{D})$ be such that f' is continuous on $\mathbb{D}$, and for all $z \in \mathbb{D}$, $|f'(z)| \leqslant 1$. Prove that for all $z, w \in \mathbb{D}$, $|f(z) - f(w)| \leqslant |z - w|$.

Exercise 3.18. Use the Fundamental Theorem of Contour Integration to find $\int_0^1 \frac{1}{1 - it}dt$.

A path $\gamma : [a, b] \to \mathbb{C}$ is *closed* if $\gamma(a) = \gamma(b)$.

$\gamma(a) = \gamma(b)$

Corollary 3.8. *If D is a domain, $f \in \mathcal{O}(D)$ is such that f' is continuous on D, and $\gamma : [a, b] \to D$ is a* closed *piecewise smooth path, then $\int_\gamma f'(z)\,dz = 0$.*

Proof. $\int_\gamma f'(z)\,dz = f(\gamma(b)) - f(\gamma(a)) = 0$, since $\gamma(b) = \gamma(a)$. $\qquad \square$

Example 3.7. For $m \in \mathbb{Z}\backslash\{0\}$, $\frac{d}{dz}\frac{z^m}{m} = z^{m-1}$ in $D := \mathbb{C}\backslash\{0\}$. Also, $D \ni z \mapsto z^{m-1}$ is continuous. So for any closed path γ in D, $\int_\gamma z^{m-1}\,dz = 0$.

What if $m = 0$? As $\mathrm{Log}'z = \frac{1}{z}$ for $z \in \tilde{D} := \mathbb{C}\backslash((-\infty, 0] \times \{0\})$, for any closed path $\tilde{\gamma}$ in $\tilde{D}$, $\int_{\tilde{\gamma}} \frac{1}{z}\,dz = 0$. But $\frac{1}{z}$ does not have a primitive in D; see Exercise 3.21. $\qquad \diamond$

Exercise 3.19. Use the Fundamental Theorem of Contour Integration to determine the value of $\int_\gamma \exp z \, dz$, where γ is a path joining 0 and $a + ib$. Equate the answer obtained with the contour integral along the straight line from 0 to $a + ib$, and hence deduce that $\int_0^1 e^{ax} \cos(bx) \, dx = \frac{a(e^a \cos b - 1) + be^a \sin b}{a^2 + b^2}$.

Exercise 3.20. Applying the Fundamental Theorem of Contour Integration to $\exp z$ and integrating round a circular path, show that for all $r > 0$, $\int_0^{2\pi} e^{r \cos \theta} \cos(r \sin \theta + \theta) \, d\theta = 0$.

Exercise 3.21. Show that $\frac{1}{z}$ has no primitive in the punctured plane $\mathbb{C} \backslash \{0\}$.

3.4. The Cauchy Integral Theorem

Theorem 3.9 (Cauchy Integral Theorem).
Let • *D be a domain in $\mathbb{C}$,*
    • *$f : D \to \mathbb{C}$ be holomorphic in D, and*
    • *$\gamma_0, \gamma_1 : [0,1] \to D$ be two closed, piecewise smooth, D-homotopic paths.*
Then $\int_{\gamma_0} f(z) \, dz = \int_{\gamma_1} f(z) \, dz$.

Before we go further, let us try to understand the statement. Firstly, the two paths in D are *closed*. Secondly, what is meant by saying that the two closed paths are 'D-homotopic'? Intuitively, this means the following. See the picture on the left below, where the two paths are depicted in the domain $D = \mathbb{C}$. Imagine a rubber band placed along γ_0. For γ_0 to be D-homotopic to γ_1, it should be possible to deform this rubber band so as to get γ_1, with the condition that each intermediate position of the rubber band lies in D. Clearly this is not possible sometimes, for example if the domain has holes. See for example the picture on the right below, where we expect the two paths in the domain $D = \mathbb{C} \backslash \{0\}$ to be not D-homotopic.

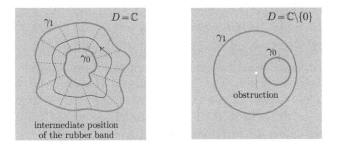

Definition 3.2. Let D be a domain in $\mathbb{C}$ and $\gamma_0, \gamma_1 : [0,1] \to D$ be closed paths. Then γ_0 is D-*homotopic to* γ_1 if there is a continuous $H : [0,1] \times [0,1] \to D$ such that the following hold:
    (H1) For all $t \in [0,1]$, $H(t,0) = \gamma_0(t)$.
    (H2) For all $t \in [0,1]$, $H(t,1) = \gamma_1(t)$.
    (H3) For all $\tau \in [0,1]$, $H(0,\tau) = H(1,\tau)$.

The map H may be thought of as describing a family of closed paths from $[0,1]$ to D, parameterised by the τ-variable, thought of as 'time'. The closed path $H(\cdot, \tau)$ represents the intermediate position of the 'rubber band' at time τ. Initially, for $\tau = 0$, $H(\cdot, 0)$ is the path γ_0, while finally, when the time $\tau = 1$, $H(\cdot, 1)$ is γ_1. This is the content of (H1) and (H2). The requirement (H3) says that at time τ, the intermediate path $H(\cdot, \tau)$ is closed too. Continuity of H means that the rubber band never breaks, and the deformation takes place continuously.

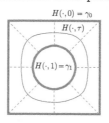

Example 3.8. Let $D = \mathbb{C}$, and $\gamma_0, \gamma_1 : [0,1] \to \mathbb{C}$ be the two circular paths given by $\gamma_0 = 4e^{2\pi i t}$, and $\gamma_1 = 2i + e^{2\pi i t}$, for all $t \in [0,1]$. Then γ_0 is $\mathbb{C}$-homotopic to γ_1: We define H by taking a 'convex combination' of the points $\gamma_0(t)$ and $\gamma_1(t)$.

Define $H : [0,1] \times [0,1] \to \mathbb{C}$ by $H(t, \tau) = (1 - \tau)\gamma_0(t) + \tau\gamma_1(t)$ for all $0 \leqslant t, \tau \leqslant 1$. Then H is continuous, and moreover,

(H1) for all $t \in [0,1]$, $H(t, 0) = \gamma_0(t)$,

(H2) for all $t \in [0,1]$, $H(t, 1) = \gamma_1(t)$, and

(H3) for each $\tau \in [0,1]$, $H(0, \tau) = (1 - \tau)4 + \tau(2i + 1) = H(1, \tau)$.

Hence (H1), (H2), (H3) are satisfied and so γ_0 is $\mathbb{C}$-homotopic to γ_1.

On the other hand, the same two paths are not $\mathbb{C}\backslash\{0\}$-homotopic. Why? If they were $\mathbb{C}\backslash\{0\}$-homotopic, then by the Cauchy Integral Theorem, the contour integral of $\frac{1}{z} \in \mathcal{O}(\mathbb{C}\backslash\{0\})$ for the two paths would be equal. But $\int_{\gamma_0} \frac{1}{z} dz = 2\pi i \neq 0 = \int_{\gamma_1} \frac{1}{z} dz$, where the last equality is due to the Fundamental Theorem of Contour Integration, by observing that $\frac{1}{z}$ has the primitive Log in $\mathbb{C}\backslash((-\infty, 0] \times \{0\})$. $\diamond$

Exercise 3.22. Let $a, b, r > 0$. Define $C, E : [0,1] \to \mathbb{C}\backslash\{0\}$ by $C(t) = re^{2\pi i t}$ and $E(t) = a\cos(2\pi t) + ib\sin(2\pi t)$ for all $t \in [0,1]$. Show that C is $\mathbb{C}\backslash\{0\}$-homotopic to E.

Exercise 3.23. Let D be a domain in $\mathbb{C}$. Show that D-homotopy is an equivalence relation on the set of all closed paths in D. In particular, we can say 'γ_0, γ_1 are D-homotopic' instead of 'γ_0 is D-homotopic to γ_1'.

Exercise 3.24. Let $r > 0$, $C(t) = re^{2\pi i t}$, $\tilde{C}(t) = -2 + e^{2\pi i t}$ for all $t \in [0,1]$. Prove that $C, \tilde{C}$ are not $\mathbb{C}\backslash\{0\}$-homotopic.

Proof.(of Theorem 3.9.) We will make the simplifying assumption that the real and imaginary parts of the homotopy H are twice continuously differentiable. This smoothness condition can be omitted (see §3.10), but the proof is technical. The assumption of twice continuous differentiability is mild, and is used below to exchange the order of partial differentiation: $\frac{\partial^2 H}{\partial \tau \partial t} = \frac{\partial^2 H}{\partial t \partial \tau}$. (See e.g. [1, §8.23].)

Let $\gamma_\tau := H(\cdot, \tau)$ be the path at time τ. Define $I(\tau) := \int_{\gamma_\tau} f(z)\,dz$ for all $\tau \in [0,1]$. (We differentiate under the integral sign with respect to τ to obtain $\frac{dI}{d\tau} \equiv 0$, showing that I is constant. So $\int_{\gamma_0} f(z)\,dz = I(0) = I(1) = \int_{\gamma_1} f(z)\,dz$.) We have

$$
\begin{aligned}
\frac{dI}{d\tau}(\tau) &= \frac{d}{d\tau} \int_{\gamma_\tau} f(z)\,dz = \frac{d}{d\tau} \int_0^1 f(H(t,\tau)) \frac{\partial H}{\partial t}(t,\tau)\,dt \\
&= \int_0^1 \frac{\partial}{\partial \tau} \left(f(H(t,\tau)) \frac{\partial H}{\partial t}(t,\tau) \right) dt \\
&\overset{(*)}{=} \int_0^1 \left(f'(H(t,\tau)) \frac{\partial H}{\partial \tau}(t,\tau) \frac{\partial H}{\partial t}(t,\tau) + f(H(t,\tau)) \frac{\partial^2 H}{\partial \tau \partial t}(t,\tau) \right) dt \\
&= \int_0^1 \left(f'(H(t,\tau)) \frac{\partial H}{\partial t}(t,\tau) \frac{\partial H}{\partial \tau}(t,\tau) + f(H(t,\tau)) \frac{\partial^2 H}{\partial t \partial \tau}(t,\tau) \right) dt \\
&\overset{(\star)}{=} \int_0^1 \frac{d}{dt} \left(f(H(t,\tau)) \frac{\partial H}{\partial \tau}(t,\tau) \right) dt \\
&= f(H(1,\tau)) \frac{\partial H}{\partial \tau}(1,\tau) - f(H(0,\tau)) \frac{\partial H}{\partial \tau}(0,\tau) \quad \text{\small (using the Fundamental Theorem of Calculus)} \\
&= f(H(1,\tau)) \frac{\partial H}{\partial \tau}(1,\tau) - f(H(1,\tau)) \lim_{\sigma \to \tau} \frac{H(0,\sigma) - H(0,\tau)}{\sigma - \tau} \\
&= f(H(1,\tau)) \frac{\partial H}{\partial \tau}(1,\tau) - f(H(1,\tau)) \lim_{\sigma \to \tau} \frac{H(1,\sigma) - H(1,\tau)}{\sigma - \tau} \\
&= f(H(1,\tau)) \frac{\partial H}{\partial \tau}(1,\tau) - f(H(1,\tau)) \frac{\partial H}{\partial \tau}(1,\tau) = 0.
\end{aligned}
$$

So $[0,1] \ni \tau \mapsto I(\tau)$ is constant. In particular, $\int_{\gamma_1} f(z)\,dz = I(1) = I(0) = \int_{\gamma_2} f(z)\,dz$. This completes the proof. But where was the holomorphicity of f used? To get the equalities labelled by $(*)$ and $(\star)$, for e.g., in $(*)$, we used the fact that $\frac{\partial}{\partial \tau}(f(H(t,\tau))) = f'(H(t,\tau)) \frac{\partial H}{\partial \tau}(t,\tau)$. We justify this now.

Claim: If φ is a smooth path, then $\frac{d}{d\tau} f(\varphi(\tau)) = f'(\varphi(\tau)) \varphi'(\tau)$.

Let $u := \operatorname{Re} f$, $v := \operatorname{Im} f$, $x := \operatorname{Re} \varphi$, $y := \operatorname{Im} \varphi$. Then

$$
\begin{aligned}
\frac{d}{d\tau} f(\varphi(\tau)) &= \frac{d}{d\tau} \left(u(x(\tau), y(\tau)) + iv(x(\tau), y(\tau)) \right) \\
&= \frac{\partial u}{\partial x}(\varphi(\tau)) x'(\tau) + \frac{\partial u}{\partial y}(\varphi(\tau)) y'(\tau) + i \left(\frac{\partial v}{\partial x}(\varphi(\tau)) x'(\tau) + \frac{\partial v}{\partial y}(\varphi(\tau)) y'(\tau) \right) \\
&= \frac{\partial u}{\partial x}(\varphi(\tau)) x'(\tau) - \frac{\partial v}{\partial x}(\varphi(\tau)) y'(\tau) + i \left(\frac{\partial v}{\partial x}(\varphi(\tau)) x'(\tau) + \frac{\partial u}{\partial x}(\varphi(\tau)) y'(\tau) \right) \\
&= \left(\frac{\partial u}{\partial x}(\varphi(\tau)) + i \frac{\partial v}{\partial x}(\varphi(\tau)) \right) (x'(\tau) + iy'(\tau)) = f'(\varphi(\tau)) \varphi'(\tau).
\end{aligned}
$$

We also assumed that f' is continuous when differentiating under the integral sign. Again, the result holds without this assumption, and the complete proof of the Cauchy Integral Theorem is given in §3.10. $\qquad \square$

Exercise 3.25. Let $a > 0$, $b > 0$. Define E by $E(t) = a \cos t + ib \sin t$ for all $t \in [0, 2\pi]$. Considering $\int_E \frac{1}{z}\,dz$, show that $\int_0^{2\pi} \frac{1}{a^2(\cos\theta)^2 + b^2(\sin\theta)^2}\,d\theta = \frac{2\pi}{ab}$.

Exercise 3.26. We have seen that if C is the circular path with centre 0 and radius 1 traversed in the anticlockwise direction, then $\int_C \frac{1}{z}\, dz = 2\pi i$. Now consider the path S, comprising the four line segments which are the sides of the square with vertices $\pm 1 \pm i$, traversed anticlockwise. Draw a picture to convince yourself that S is $\mathbb{C}\backslash\{0\}$-homotopic to C. Evaluate parametrically the integral $\int_S \frac{1}{z}\, dz$, and confirm that the answer is indeed $2\pi i$.

Special case: simply connected domains. If D is a domain and $p \in D$, then consider the 'degenerate path' $\gamma_p : [0,1] \to D$ given by $\gamma_p(t) = p$ for all $t \in [0,1]$. As $\gamma_p(0) = p = \gamma_p(1)$, γ_p is closed. If $f : D \to \mathbb{C}$ is continuous, what is $\int_{\gamma_p} f(z)\, dz$? 0, because $\gamma_p' \equiv 0$, which yields $\int_{\gamma_p} f(z)\, dz = \int_0^1 f(\gamma_p(t))\gamma_p'(t)\, dt = 0$.

In light of this, an important special case of the Cauchy Integral Theorem is obtained when a closed path γ is D-homotopic to a *point* (that is, the constant path $\gamma_p(t) = p$ for all $t \in [0,1]$). In this case we say that γ is *D-contractible*. Imagine placing a rubber band along γ, and then shrinking it to a point such that each intermediate position of the rubber band is in D. For a D-contractible path γ (which is D-homotopic to a point $p \in D$), and for any $f \in \mathcal{O}(D)$, the Cauchy Integral Theorem implies $\int_\gamma f(z)dz = \int_{\gamma_p} f(z)dz = 0$. A domain in which *every* closed path is D-contractible is called *simply connected*. We have the following corollary of the Cauchy Integral Theorem.

Corollary 3.10. *If D is a simply connected domain, γ is a closed piecewise smooth path in D, and $f \in \mathcal{O}(D)$, then $\int_\gamma f(z)\, dz = 0$.*

The domains $\mathbb{C}$, $D(z_0, r)$, $\mathbb{C}\backslash((-\infty, 0] \times \{0\})$, $\mathbb{H} := \{z \in \mathbb{C} : \mathrm{Re}\, z > 0\}$ are all simply connected. For example if we take any p in the domains in first two cases, then the homotopy given by $H(t, \tau) := (1-\tau)\gamma(t) + \tau p$ for all $t, \tau \in [0,1]$ does the job. In the case of $D = \mathbb{C}\backslash((-\infty, 0] \times \{0\})$, or $\mathbb{H}$, given any $\gamma : [0,1] \to D$, we first choose any real $p > 0$ for example $p = 1$, and then use the same H as above. In the former case, if $\mathrm{Im}(\gamma(t)) \neq 0$, then $\mathrm{Im}(H(t,\tau)) = (1-\tau)\mathrm{Im}(\gamma(t)) \neq 0$ for all $\tau \in [0,1)$, showing that in this case $H(t,\tau) \in D$, while if $\mathrm{Im}(\gamma(t)) = 0$, then as $\gamma(t) \in D$, we must have $\mathrm{Re}(\gamma(t)) > 0$, which yields $\mathrm{Re}(H(t,\tau)) = (1-\tau)\mathrm{Re}(\gamma(t)) + \tau 1 > 0$, so that again $H(t,\tau) \in D$. None of the above domains have any 'holes' in them. On the other hand, domains with holes are not simply connected. For example, the punctured complex plane $\mathbb{C}\backslash\{0\}$ is not simply connected. For example consider the circular path C with centre 0 and any radius $r > 0$ traversed once in the anticlockwise direction. We have seen that $\int_C \frac{1}{z}\, dz = 2\pi i$. But if C were $\mathbb{C}\backslash\{0\}$-contractible to a point in the punctured plane, we should have $\int_C \frac{1}{z}\, dz = 0$ by the Cauchy Integral Theorem. So this means that C is not $\mathbb{C}\backslash\{0\}$-contractible to a point in $\mathbb{C}\backslash\{0\}$ and so $\mathbb{C}\backslash\{0\}$ is not simply connected. Similarly, it can be shown that the annulus $\{z \in \mathbb{C} : 1 < |z| < 2\}$ is not simply connected. The obstruction of the hole can be thought of as a nail or a pillar emanating from the plane, which prevents a rubber band encircling it from being shrunk to a point in the domain while always staying in the plane.

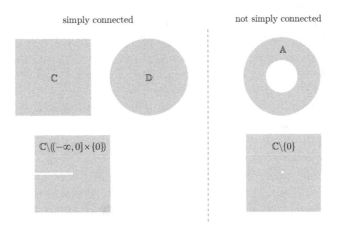

Example 3.9. For any closed path γ, as exp is entire, and as $\mathbb{C}$ is simply connected, $\int_\gamma \exp z \, dz = 0$. For any entire function f, $\int_\gamma f(z) \, dz = 0$, for any closed path γ. $\diamond$

It can happen for a domain which is *not* simply connected, and an $f \in \mathcal{O}(D)$ that $\int_\gamma f(z) \, dz = 0$ for all closed paths γ in D. For example, let $D = \mathbb{C}\backslash\{0\}$, $f := \frac{1}{z^2}$. Then $\int_\gamma \frac{1}{z^2} \, dz = 0$ for all closed paths γ in the punctured plane (as $\frac{1}{z^2}$ possesses a primitive in $\mathbb{C}\backslash\{0\}$): $\frac{d}{dz}(-\frac{1}{z}) = \frac{1}{z^2}$.

Corollary 3.11. *If D is a simply connected domain, $f \in \mathcal{O}(D)$, $\gamma : [a,b] \to D$ and $\tilde{\gamma} : [c,d] \to D$ are two smooth paths such that they have the same start and end points, that is, $\gamma(a) = \tilde{\gamma}(c) =: z$ and $\gamma(b) = \tilde{\gamma}(d) =: w$, then $\int_\gamma f(z) \, dz = \int_{\tilde{\gamma}} f(z) \, dz$.*

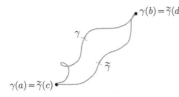

Proof. As $\gamma - \tilde{\gamma}$ is closed, it follows from the Cauchy Integral Theorem that
$$0 = \int_{\gamma - \tilde{\gamma}} f(z) \, dz = \int_\gamma f(z) \, dz - \int_{\tilde{\gamma}} f(z) \, dz. \qquad \square$$

Exercise 3.27. Integrate the following over the circular path given by $|z| = 3$ traversed in the anticlockwise direction: $\text{Log}(z - 4i)$, $\frac{1}{z-1}$, principal value of i^{z-3}.

Exercise 3.28. Define $\gamma_0, \gamma_1 : [0,1] \to \mathbb{C}\backslash\{0\}$ by $\gamma_0(t) = e^{2\pi it}$, $\gamma_1(t) = e^{-2\pi it}$ for all $t \in [0,1]$. Prove or disprove: γ_0 is $\mathbb{C}\backslash\{0\}$-homotopic to γ_1.

Exercise 3.29 (Winding number of a curve).

Suppose $\gamma : [0,1] \to \mathbb{C}$ is a closed smooth path that does not pass through the origin 0. Define the *winding number* $w(\gamma)$ *of* γ (about 0) by $w(\gamma) := \frac{1}{2\pi i} \int_\gamma \frac{1}{z} dz = \frac{1}{2\pi i} \int_0^1 \frac{\gamma'(t)}{\gamma(t)} dt$.

(1) Use the observation $[\exp(2\pi i a) = 1 \Leftrightarrow a \in \mathbb{Z}]$, to show $w(\gamma) \in \mathbb{Z}$ proceeding as follows. We define the function $\varphi : [0,1] \to \mathbb{C}$ by setting $\varphi(t) = \exp \int_0^t \frac{\gamma'(s)}{\gamma(s)} ds$ for all $t \in [0,1]$. To show $w(\gamma) \in \mathbb{Z}$, it suffices to show that $\varphi(1) = 1$. To this end, calculate $\varphi'(t)$, and use this expression to show φ/γ is constant in $[0,1]$. Conclude $\varphi(1) = 1$.

(2) Calculate the winding number of $\Gamma_1 : [0,1] \to \mathbb{C}$ given by $\Gamma_1(t) = \exp(2\pi i t)$ ($t \in [0,1]$).

(3) Prove that if $\gamma_1, \gamma_2 : [0,1] \to \mathbb{C}$ are smooth closed paths not passing through 0, and $\gamma_1 \cdot \gamma_2$ is their pointwise product, then $w(\gamma_1 \cdot \gamma_2) = w(\gamma_1) + w(\gamma_2)$.

(4) Let $m \in \mathbb{N}$. Find the winding number of Γ_m, where $\Gamma_m(t) := \exp(2\pi i m t)$ ($t \in [0,1]$).

(5) Show that the winding number map $\gamma \mapsto w(\gamma)$ is 'locally constant': If $\gamma_0 : [0,1] \to \mathbb{C}\backslash\{0\}$ is a smooth closed path, then there is a $\delta > 0$ such that for every smooth closed path $\gamma : [0,1] \to \mathbb{C}\backslash\{0\}$ such that $\|\gamma - \gamma_0\|_\infty := \max\{|\gamma(t) - \gamma_0(t)| : t \in [0,1]\} < \delta$, $w(\gamma) = w(\gamma_0)$. (In other words, if the set of curves is equipped with the uniform topology, and $\mathbb{Z}$ with the discrete topology, then $\gamma \mapsto w(\gamma)$ is continuous.)

3.4.1. What happens with nonholomorphic functions?

We now highlight the fact that the Cauchy Integral Theorem may fail if one drops the assumption of holomorphicity of f. Let us see what happens when we consider our favourite nonholomorphic function, the complex conjugation map $z \mapsto \overline{z}$. We will show that rather than the integral around the closed loop γ being 0, the contour integral of $\overline{z}$ around γ yields the area enclosed by γ, which is of course very much dependent on γ, and two $\mathbb{C}$-homotopic paths can enclose widely different areas (take two concentric circles with different radii). We will only give a plausibility argument by resorting to a specific picture, e.g. as shown in the following figure.

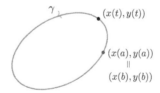

For the smooth path $\gamma : [a,b] \to \mathbb{C}$, with $x := \operatorname{Re}\gamma$ and $y := \operatorname{Im}\gamma$, we have

$$
\begin{aligned}
\int_\gamma \overline{z}\, dz &= \int_a^b (x(t) - iy(t))(x'(t) + iy'(t))\, dt \\
&= \int_a^b \big(x(t)x'(t) + y(t)y'(t) + i(x(t)y'(t) - y(t)x'(t)) \big)\, dt \\
&= \int_a^b \frac{d}{dt} \frac{(x(t))^2 + (y(t))^2}{2}\, dt + i \int_a^b (x(t)y'(t) - y(t)x'(t))\, dt \\
&= \frac{(x(b))^2 - (x(a))^2 + (y(b))^2 - (y(a))^2}{2} + i \int_a^b (x(t)y'(t) - y(t)x'(t))\, dt \\
&= i \int_a^b (x(t)y'(t) - y(t)x'(t))\, dt.
\end{aligned}
$$

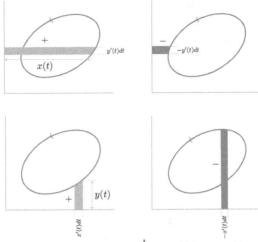

From the top two pictures, we see that $\int_a^b x(t)y'(t)dt = $ (Area enclosed by γ).
From the bottom two pictures, we see that $\int_a^b x'(t)y(t)dt = -$(Area enclosed by γ).
Thus $\int_\gamma \bar{z}\,dz = i\int_a^b(x(t)y'(t) - y(t)x'(t))dt = 2i \cdot$ (Area enclosed by γ).

Exercise 3.30. Suppose a coin of radius r rolls around a fixed bigger coin of radius R. Then the path traced by a point on the rim of the rolling coin is called an *epicycloid*, and it is a closed curve if $R = nr$, for some $n \in \mathbb{N}$. With the centre of the fixed coin at the origin, show that the epicycloid is given by $z(t) = r((n+1)e^{it} - e^{i(n+1)t})$, $t \in [0, 2\pi]$. By evaluating the integral of $\bar{z}$ along the epicycloid, show that the area enclosed by the epicycloid is equal to $\pi r^2(n+1)(n+2)$.

Next, we will show the following consequences of the Cauchy Integral Theorem:

• In simply connected domains every holomorphic function possesses a primitive.

• Holomorphic functions are infinitely many times complex differentiable.

• Bounded entire functions are constants (Liouville's Theorem), and a proof of the Fundamental Theorem of Algebra using Liouville's Theorem.

• Morera's Theorem (which is a sort of a converse to the Cauchy Integral Theorem).

3.5. Existence of a primitive

On a simply connected domain, every holomorphic function f is the derivative of some holomorphic function, that is, f possesses a primitive.

Theorem 3.12. *If D is a simply connected domain and $f \in \mathcal{O}(D)$, then there exists an $F \in \mathcal{O}(D)$ such that $F' = f$ in D.*

Proof. Fix a point $p \in D$. Define $F : D \to \mathbb{C}$ by $F(z) = \int_{\gamma_z} f(\zeta)\, d\zeta$ $(z \in D)$, where γ_z is *any* smooth path in D joining p to z. Then F is well-defined, i.e., $F(z)$ does *not* depend on the path joining p to z. If γ is another smooth path in D that joins p to z, then $\gamma_z - \gamma$ is a closed smooth path in the simply connected domain D. The Cauchy Integral Theorem implies $0 = \int_{\gamma_z - \gamma} f(\zeta)\, d\zeta = \int_{\gamma_z} f(\zeta)\, d\zeta - \int_{\gamma} f(\zeta)\, d\zeta$, so that $\int_{\gamma_z} f(\zeta)\, d\zeta = \int_{\gamma} f(\zeta)\, d\zeta$.

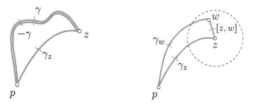

Next, we will show $F \in \mathcal{O}(D)$ and that $F' = f$ in D. Since $f \in \mathcal{O}(D)$, f is continuous in D, and so given a $z \in D$ and an $\epsilon > 0$, there exists a $\delta > 0$ such that if $w \in D$ and $|w - z| < \delta$, then $|f(w) - f(z)| < \epsilon$. If $w \in D$ is such that $0 < |w - z| < \delta$, then $\frac{F(w) - F(z)}{w - z} = \frac{1}{w - z}\left(\int_{\gamma_w} f(\zeta)\, d\zeta - \int_{\gamma_z} f(\zeta)\, d\zeta\right)$. If $[z, w]$ is a straight line path joining z to w, then the concatenation of γ_z with the concatenation of $[z, w]$ with $-\gamma_w$ is a closed path. By the Cauchy Integral Theorem, $0 = \int_{\gamma_z + [z,w] - \gamma_w} f(\zeta)\, d\zeta = \int_{\gamma_z} f(\zeta)\, d\zeta + \int_{[z,w]} f(\zeta)\, d\zeta - \int_{\gamma_w} f(\zeta)\, d\zeta$. The Fundamental Theorem of Contour Integration gives $\int_{[z,w]} 1\, d\zeta = \int_{[z,w]} \zeta'\, d\zeta = w - z$. So $\frac{F(w) - F(z)}{w - z} - f(z) = \frac{1}{w - z}\int_{[z,w]} f(\zeta)\, d\zeta - \frac{1}{w - z}\int_{[z,w]} f(z)\, d\zeta = \frac{1}{w - z}\int_{[z,w]} (f(\zeta) - f(z))\, d\zeta$. Thus

$$\left|\frac{F(w) - F(z)}{w - z} - f(z)\right| = \left|\frac{1}{w - z}\int_{[z,w]} (f(\zeta) - f(z))\, d\zeta\right| = \frac{1}{|w - z|}\left|\int_{[z,w]} (f(\zeta) - f(z))\, d\zeta\right|$$
$$\leq \frac{1}{|w - z|}\left(\max_{\zeta \in [z,w]} |f(\zeta) - f(z)|\right)(\text{length of } [z, w]) \leq \frac{1}{|w - z|}\epsilon |w - z| = \epsilon.$$

Thus $F'(z) = f(z)$. As $z \in D$ was arbitrary, $F \in \mathcal{O}(D)$ and $F' = f$ in D. $\qquad\square$

Remark 3.2. A primitive for a holomorphic function f in a simply connected domain is unique up to a constant. Indeed, if $F, \widetilde{F}$ are both primitives for F, then $F' = f = \widetilde{F}'$ in D, and so $\frac{d}{dz}(F - \widetilde{F}) = F' - \widetilde{F}' = f - f = 0$ in D. By Exercise 2.18, there is a constant C such that $F - \widetilde{F} = C$ in D. So $F = \widetilde{F} + C$ in D. $\qquad *$

Example 3.10. $\exp(-z^2)$ is entire. So there exists an F, which is also entire, such that for all $z \in \mathbb{C}$, $F'(z) = \exp(-z^2)$. But one 'cannot express F in terms of elementary functions'; see e.g. [23]. One primitive is given by $\widetilde{F}(z) = \int_{\gamma_z} e^{-\zeta^2}\, d\zeta$ for $z \in \mathbb{C}$, where γ_z is the straight line path joining 0 to z. Then in particular, for real x, $\widetilde{F}(x) = \int_0^x e^{-\xi^2}\, d\xi$, and it turns out that this (and so any other primitive too) cannot be expressed in terms of elementary functions. $\qquad \diamond$

Exercise 3.31. If D is a domain, and $f \in \mathcal{O}(D)$ is such that there is no $F \in \mathcal{O}(D)$ such that $F' = f$ in D, then D is not simply connected. Give an example of such a D and f.

3.6. The Cauchy Integral Formula

The Cauchy Integral Formula, roughly speaking, says that if γ is a closed path, and f is holomorphic inside γ, then the value of f at any point inside γ is determined by the values of f on γ. This illustrates the 'rigidity' of holomorphic functions. In the next chapter, we will study a more general Cauchy Integral Formula, expressing all the derivatives of f at any point inside γ in terms of the values of the function on γ. The result below is the '$n = 0$ case' of the more general result to follow.

Theorem 3.13 (The Cauchy Integral Formula for circular paths).
Let D be a domain, $f \in \mathcal{O}(D)$, $r > 0$, $z_0 \in D$, $\Delta := \{z \in \mathbb{C} : |z - z_0| \leqslant r\} \subset D$, and $C(t) = z_0 + re^{2\pi i t}$, $t \in [0,1]$. Then for all $z \in D(z_0, r)$, $f(z) = \frac{1}{2\pi i} \int_C \frac{f(\zeta)}{\zeta - z} d\zeta$.

In order to prove this result, we will first prove the following technical fact, which will also prove to be useful later on. In this result we essentially prove the Cauchy Integral Formula for the centre z_0 of the disc Δ.

Proposition 3.14. *Let D be a domain, $z_0 \in D$, $f : D \to \mathbb{C}$ be holomorphic in $D \backslash \{z_0\}$, and continuous on D, $r > 0$, $\Delta := \{z \in \mathbb{C} : |z - z_0| \leqslant r\} \subset D$, and $C(t) = z_0 + re^{2\pi i t}$, $t \in [0,1]$. Then $f(z_0) = \frac{1}{2\pi i} \int_C \frac{f(\zeta)}{\zeta - z_0} d\zeta$.*

Proof. Let $\epsilon > 0$. Since f is continuous at z_0, there exists a $\delta \in (0, r)$ such that whenever $|\zeta - z_0| \leqslant \delta$, $|f(\zeta) - f(z_0)| < \epsilon$. Consider the circular path C_δ, with centre z_0 and radius δ traversed in the anticlockwise direction. Then C_δ and C are $D \backslash \{z_0\}$-homotopic (e.g., take the homotopy map H as a convex combination of C, C_δ: $H(\cdot, \tau) := (1 - \tau)C(\cdot) + \tau C_\delta(\cdot)$, $\tau \in [0,1]$).

By the Cauchy Integral Theorem, $\int_C \frac{f(\zeta)}{\zeta - z_0} d\zeta = \int_{C_\delta} \frac{f(\zeta)}{\zeta - z_0} d\zeta$. Hence,

$$\left| \frac{1}{2\pi i} \int_C \frac{f(\zeta)}{\zeta - z_0} d\zeta - f(z_0) \right| = \left| \frac{1}{2\pi i} \int_{C_\delta} \frac{f(\zeta)}{\zeta - z_0} d\zeta - f(z_0) \frac{1}{2\pi i} \int_{C_\delta} \frac{1}{\zeta - z_0} d\zeta \right|$$

$$= \left| \frac{1}{2\pi i} \int_{C_\delta} \frac{f(\zeta) - f(z_0)}{\zeta - z_0} d\zeta \right|$$

$$\leqslant \frac{1}{2\pi} \left(\max_{\zeta \in C_\delta} \frac{|f(\zeta) - f(z_0)|}{|\zeta - z_0|} \right) 2\pi\delta < \frac{1}{2\pi} \frac{\epsilon}{\delta} 2\pi\delta = \epsilon.$$

Since $\epsilon > 0$ was arbitrary, the claim follows. $\qquad \square$

Proof. (of Theorem 3.13.) Let z be such that $|z - z_0| < r$. Choose a $\delta > 0$ small enough so that the circular path C_δ with centre z and radius δ is contained in the interior of C. We will now show that C and C_δ are $D\backslash\{z\}$-homotopic.

We have $C(t) = z_0 + re^{2\pi it}$, and $C_\delta(t) := z + \delta e^{2\pi it}$, $t \in [0,1]$. Define the map $H : [0,1] \times [0,1] \to \mathbb{C}$ by $H(t,\tau) = (1 - \tau)C(t) + \tau C_\delta(t)$ for all $t, \tau \in [0,1]$. Then H is continuous. For all $t \in [0,1]$, $H(t,0) = C(t)$ and $H(t,1) = C_\delta(t)$. For all $\tau \in [0,1]$, $H(0,\tau) = (1 - \tau)C(0) + \tau C_\delta(0) = (1 - \tau)C(1) + \tau C_\delta(1) = H(1,\tau)$. Now we show that $H(t,\tau) \in D\backslash\{z\}$ for all $t, \tau \in [0,1]$. We have

$$\begin{aligned}
|H(t,\tau) - z_0| &= |(1 - \tau)C(t) + \tau C_\delta(t) - z_0| \\
&= |(1 - \tau)C(t) + \tau C_\delta(t) - ((1 - \tau)z_0 + \tau z_0)| \\
&\leqslant (1 - \tau)|C(t) - z_0| + \tau|C_\delta(t) - z_0| \\
&\leqslant (1 - \tau)r + \tau|C_\delta(t) - z + z - z_0| \\
&\leqslant (1 - \tau)r + \tau(\delta + |z - z_0|) \\
&\leqslant (1 - \tau)r + \tau r = r.
\end{aligned}$$

Thus $H(t,\tau) \in \Delta \subset D$. Also,

$$\begin{aligned}
|H(t,\tau) - z| &= |(1 - \tau)(z_0 + re^{2\pi it}) + \tau(z + \delta e^{2\pi it}) - z| \\
&= |re^{2\pi it} + (1 - \tau)(z_0 - z) + \tau(\delta - r)e^{2\pi it}| \\
&\geqslant r - |(1 - \tau)(z_0 - z) + \tau(\delta - r)e^{2\pi it}| \\
&\geqslant r - ((1 - \tau)|z_0 - z| + \tau(r - \delta)) \\
&= (1 - \tau)(r - |z_0 - z|) + \tau\delta \\
&\geqslant (1 - \tau)\delta + \tau\delta = \delta > 0.
\end{aligned}$$

So C and C_δ are $D\backslash\{z\}$-homotopic. Since $\frac{f(\cdot)}{\cdot - z} \in \mathcal{O}(D\backslash\{z\})$, by the Cauchy Integral Theorem, the second equality holds: $f(z) = \frac{1}{2\pi i} \int_{C_\delta} \frac{f(\zeta)}{\zeta - z} d\zeta = \frac{1}{2\pi i} \int_C \frac{f(\zeta)}{\zeta - z} d\zeta.$ $\square$

Exercise 3.32. Let $0 < a < 1$, and C be the unit circle with centre 0 traversed anticlockwise. Show that $\int_C \frac{i}{(z - a)(az - 1)} dz = \int_0^{2\pi} \frac{1}{1 + a^2 - 2a\cos t} dt$.

Use the Cauchy Integral Formula to deduce $\int_0^{2\pi} \frac{1}{1 + a^2 - 2a\cos t} dt = \frac{2\pi}{1 - a^2}$.

Exercise 3.33. Does $z \mapsto \frac{1}{z(1 - z^2)}$ have a primitive in $\{z \in \mathbb{C} : 0 < |z| < 1\}$?

Exercise 3.34. Let f be entire. Define the path C by $C(\theta) := e^{i\theta}$, for $\theta \in [0, 2\pi]$. Let $z \in \mathbb{C}$ such that $|z| < 1$. Prove that $(1 - |z|^2)f(z) = \frac{1}{2\pi i} \int_C f(\zeta) \frac{1 - \bar{\zeta}z}{\zeta - z} d\zeta$.

Deduce from the above that for all $z \in \mathbb{D}$, $(1 - |z|^2)|f(z)| \leqslant \frac{1}{2\pi} \int_0^{2\pi} |f(e^{i\theta})| d\theta$.

Exercise 3.35. Fill in the blanks.

(1) $\int_\gamma \frac{\exp z}{z-1} dz = $____, where γ is the circle $|z| = 2$ traversed in the anticlockwise direction.

(2) $\int_\gamma \frac{z^2+1}{z^2-1} dz = $____, where γ is the circle $|z-1| = 1$ traversed in the anticlockwise direction.

(3) $\int_\gamma \frac{z^2+1}{z^2-1} dz = $____, where γ is the circle $|z-i| = 1$ traversed in the anticlockwise direction.

(4) $\int_\gamma \frac{z^2+1}{z^2-1} dz = $____, where γ is the circle $|z+1| = 1$ traversed in the anticlockwise direction.

(5) $\int_\gamma \frac{z^2+1}{z^2-1} dz = $____, where γ is the circle $|z| = 3$ traversed in the anticlockwise direction.

Corollary 3.15 (Cauchy's Integral Formula for general paths).
Let D be a domain, $f \in \mathcal{O}(D)$, $z_0 \in D$, γ be a closed path in D which is $D\backslash\{z_0\}$-homotopic to a circular path C centred at z_0, such that C and its interior is contained in D. Then $f(z_0) = \frac{1}{2\pi i} \int_\gamma \frac{f(\zeta)}{\zeta - z_0} d\zeta$.

Proof. By the Cauchy Integral Formula for circular paths, $f(z_0) = \frac{1}{2\pi i} \int_C \frac{f(\zeta)}{\zeta - z_0} d\zeta$. As γ, C are $D\backslash\{z_0\}$-homotopic, it follows from the Cauchy Integral Theorem that $\frac{1}{2\pi i} \int_C \frac{f(\zeta)}{\zeta - z_0} d\zeta = \frac{1}{2\pi i} \int_\gamma \frac{f(\zeta)}{\zeta - z_0} d\zeta$. $\qquad \square$

This result highlights the 'rigidity' associated with holomorphic functions mentioned earlier. By this we mean that their highly structured nature (everywhere locally infinitesimally a rotation followed by a magnification) enables one to pin down their precise behaviour from very limited information. That is, even if we know the effect of a holomorphic function in a small portion of the plane (for example the values along a closed path), its values can be inferred at other far away points in a unique manner. The picture below illustrates this in the case of the Cauchy Integral Formula, where knowing the values of f on the curve γ enables one to determine the values at all points in the shaded region.

Exercise 3.36. Define F by $F(z) = \frac{\exp(iz)}{z^2+1}$ for all $z \in \mathbb{C}\backslash\{i, -i\}$. Let $R > 1$.

(1) Let σ be the closed path comprising the line segment $[-R, R]$ of the real axis from $-R$ to R, followed by the semicircle S of radius R in the upper half plane from R to $-R$. Show that $\int_\sigma F(z) dz = \frac{\pi}{e}$.

(2) Prove that $|\exp(iz)| \leq 1$ for z in the upper half plane, and conclude that for all large enough $|z|$, $|F(z)| \leq \frac{2}{|z|^2}$.

(3) Show that $\lim_{R \to \infty} \int_S F(z) dz = 0$, and so $\lim_{R \to \infty} \int_{[-R,R]} F(z) dz = \frac{\pi}{e}$.

(4) Parametrising $[-R, R]$ in terms of x, show that $\int_{-\infty}^{\infty} \frac{\cos x}{1+x^2} dx := \lim_{R \to \infty} \int_{-R}^{R} \frac{\cos x}{1+x^2} dx = \frac{\pi}{e}$.

Exercise 3.37. Evaluate $\int_0^{2\pi} e^{\cos\theta} \cos(\sin\theta) d\theta$. *Hint:* Consider $\exp(\exp(i\theta))$.

3.7. Holomorphic functions are infinitely complex differentiable

In this section it is shown that holomorphic functions are infinitely many times complex differentiable. Let us contrast this with the situation in Real Analysis. We had seen in Example 0.1 that the derivative may fail to be real differentiable at isolated points, and also that there are even more extreme examples of this phenomenon, namely functions $f : \mathbb{R} \to \mathbb{R}$ which are real differentiable everywhere, but f' is real differentiable nowhere.

Corollary 3.16. *If D be a domain, and $f \in \mathcal{O}(D)$, then $f' \in \mathcal{O}(D)$.*

We get the chain of implications: $f \in \mathcal{O}(D) \Rightarrow f' \in \mathcal{O}(D) \Rightarrow f'' \in \mathcal{O}(D) \Rightarrow \cdots$. So, if $f \in \mathcal{O}(D)$, then f is infinitely many times complex differentiable in D.

Here is a plan of how we will show this. From the Cauchy Integral Formula, $f(z) = \frac{1}{2\pi i} \int_C \frac{f(\zeta)}{\zeta - z} d\zeta$, where C is a circle centred at z with radius r such that the circle C and its interior are contained in D. If we formally differentiate under the integral sign, we get an expression for the derivative of f: $f'(z) = \frac{1}{2\pi i} \int_C \frac{f(\zeta)}{(\zeta - z)^2} d\zeta$. Having shown this formula, we will show that $\lim\limits_{w \to z} \frac{f'(w) - f'(z)}{z - w}$ exists by using the above expression for the derivative at z and w.

Proof. Let $z \in D$, and $r > 0$ be such that $\Delta := \{\zeta \in \mathbb{C} : |\zeta - z| \leqslant r\} \subset D$. Let $C(t) = z + re^{2\pi i t}$ for all $t \in [0, 1]$. We will show that $f'(z) = \frac{1}{2\pi i} \int_C \frac{f(\zeta)}{(\zeta - z)^2} d\zeta$. Define

$$g(\zeta) = \begin{cases} \frac{f(\zeta) - f(z)}{\zeta - z} & \text{if } z \in D \backslash \{z\}, \\ f'(z) & \text{if } \zeta = z. \end{cases}$$

Then g is holomorphic in $D \backslash \{z\}$ and continuous in D. By Proposition 3.14,

$$\begin{aligned} f'(z) = g(z) &= \frac{1}{2\pi i} \int_C \frac{g(\zeta)}{\zeta - z} d\zeta = \frac{1}{2\pi i} \int_C \frac{f(\zeta) - f(z)}{(\zeta - z)^2} d\zeta \\ &= \frac{1}{2\pi i} \int_C \frac{f(\zeta)}{(\zeta - z)^2} d\zeta - \frac{f(z)}{2\pi i} \int_C \frac{1}{(\zeta - z)^2} d\zeta \\ &= \frac{1}{2\pi i} \int_C \frac{f(\zeta)}{(\zeta - z)^2} d\zeta - 0. \qquad (\star) \end{aligned}$$

So we have shown the formula for $f'(z)$ at centres z of the closed discs that are contained in D. For $w \in D(z, r)$, let $\delta \in (0, r)$ be such that $D(w, \delta) \subset D(z, r)$. From the above, we get $f'(w) = \frac{1}{2\pi i} \int_{C_\delta} \frac{f(\zeta)}{(\zeta - w)^2} d\zeta = \frac{1}{2\pi i} \int_C \frac{f(\zeta)}{(\zeta - w)^2} d\zeta$ (the last equality follows from the Cauchy Integral Theorem, since $\frac{f(\cdot)}{(\cdot - w)^2} \in \mathcal{O}(D \backslash \{w\})$, and the paths C, C_δ are $D \backslash \{w\}$-homotopic).

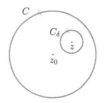

So for $w \neq z$ inside C,

$$\frac{f'(w)-f'(z)}{w-z} = \frac{1}{w-z}\left(\frac{1}{2\pi i}\int_C \frac{f(\zeta)}{(\zeta-w)^2}\,d\zeta - \frac{1}{2\pi i}\int_C \frac{f(\zeta)}{(\zeta-z)^2}\,d\zeta\right) = \frac{1}{2\pi i}\int_C \frac{f(\zeta)(2\zeta-z-w)}{(\zeta-w)^2(\zeta-z)^2}\,d\zeta.$$

When $w \approx z$, the integrand numerator $f(\zeta)(2\zeta - z - w) \approx 2f(\zeta)(\zeta - z)$, while the denominator $\approx (\zeta - z)^4$. So we expect that $\lim\limits_{w\to z} \frac{f'(w)-f'(z)}{w-z} = \frac{2}{2\pi i}\int_C \frac{f(\zeta)}{(\zeta-z)^3}\,d\zeta$. We prove this claim next. For $w \neq z$,

$$\frac{f'(w)-f'(z)}{w-z} - \frac{2}{2\pi i}\int_C \frac{f(\zeta)}{(\zeta-z)^3}\,d\zeta = (w-z)\frac{1}{2\pi i}\int_C \frac{(3\zeta-z-w)f(\zeta)}{(\zeta-w)^2(\zeta-z)^3}\,d\zeta. \qquad (*)$$

If we show that the integral is bounded by some constant for all w close enough to z, then since the integral is being multiplied by $w - z$, we see that as $w \to z$, we can make the overall expression as small as we please. To this end, let us consider a disc with centre z and having radius smaller than r, say $r/2$, and we will confine w (which is anyway supposed to be near z) to lie within this disc. As $f \in \mathcal{O}(D)$, in particular, its restriction to the C is continuous, and so is $|f| : C \to \mathbb{R}$. But by the Extreme Value Theorem, $M := \max_{\zeta \in \mathrm{ran}\, C} |f(\zeta)| = \max_{t\in[0,1]} |f(C(t))|$ exists. Moreover, for $\zeta \in \mathrm{ran}\, C$,

$$|\zeta - w| \geqslant \tfrac{r}{2}, \quad |3\zeta - z - 2w| = |3\zeta - 3z + 2z - 2w| \leqslant 3r + 2\tfrac{r}{2} = 4r, \quad |\zeta - z|^3 = r^3.$$

By the ML-inequality, $\left|\int_C \frac{(3\zeta-z-2w)f(\zeta)}{(\zeta-w)^2(\zeta-z)^3}\,d\zeta\right| \leqslant 2\pi r \frac{4rM}{(\frac{r}{2})^2 r^3} = \frac{32\pi M}{r^3}$. Using $(*)$,

$$\left|\frac{f'(w)-f'(z)}{w-z} - \frac{2}{2\pi i}\int_C \frac{f(\zeta)}{(\zeta-z)^3}\,d\zeta\right| \leqslant |w-z|\frac{2}{2\pi}\frac{32\pi M}{r^3} = \frac{32M}{r^3}|w-z| \xrightarrow{w\to z} 0.$$

Thus f' is complex differentiable at z. As $z \in D$ was arbitrary, $f' \in \mathcal{O}(D)$. $\qquad\square$

Exercise 3.38. If D is a domain and $f \in \mathcal{O}(D)$, show that for all $n \in \mathbb{N}$, $f^{(n)}$ has a continuous complex derivative.

Exercise 3.39. Let $U \subset \mathbb{C}$ be open, $f \in \mathcal{O}(U)$, $u := \mathrm{Re}\, f$ and $v := \mathrm{Im}\, f$. Show that $u, v \in C^\infty(U)$, where $C^\infty(U)$ denotes the set of all real-valued functions on U possessing continuous partial derivatives of all orders on U.

3.8. Liouville's Theorem, Fundamental Theorem of Algebra

Theorem 3.17 (Liouville's Theorem). *Every bounded entire function is constant.*

Let us contrast this with the situation in real analysis. For example $\mathbb{R} \ni x \mapsto \sin x$ is real differentiable on $\mathbb{R}$, and is bounded ($|\sin x| \leqslant 1$ for all $x \in \mathbb{R}$), but is not constant. On the other hand, in light of the above Liouville's Theorem, the entire (complex) function sin, being nonconstant, must necessarily be unbounded on $\mathbb{C}$! We had checked this earlier on page 24: $|\sin(iy)| \to \infty$ as $y \to \pm\infty$.

Proof. Let $M \geqslant 0$ be such that for all $z \in \mathbb{C}$, $|f(z)| \leqslant M$. Suppose $z \in \mathbb{C}$, $R > 0$, and let $C_R(t) := z + Re^{2\pi i t}$, $t \in [0,1]$. From $(\star)$ in the proof of Corollary 3.16, $f'(z) = \frac{1}{2\pi i}\int_{C_R} \frac{f(\zeta)}{(\zeta-z)^2}\,d\zeta$. By the ML-inequality,

$$|f'(z)| = \left|\frac{1}{2\pi i}\int_{C_R} \frac{f(\zeta)}{(\zeta-z)^2}\,d\zeta\right| \leqslant \frac{1}{2\pi}\frac{M}{R^2}2\pi R = \frac{M}{R}.$$

As $R > 0$ was arbitrary, $f'(z) = 0$. So $f'(z) = 0$ for all $z \in \mathbb{C}$. For $z \in \mathbb{C}$, let $[0, z]$ be the straight line path joining 0 to z. Then we get $f(z) - f(0) = \int_{[0,z]} f'(\zeta) \, d\zeta = 0$. So for all $z \in \mathbb{C}$, $f(z) = f(0)$, that is, f is constant. $\square$

The Fundamental Theorem of Algebra[1] can be proved using Liouville's Theorem.

Corollary 3.18 (Fundamental Theorem of Algebra).
Every polynomial of degree at least 1 has a zero in $\mathbb{C}$.

For a polynomial $p : \mathbb{C} \to \mathbb{C}$, given by $p(z) = c_0 + c_1 z + \cdots + c_d z^d$, $z \in \mathbb{C}$, where $c_0, c_1, \cdots, c_d \in \mathbb{C}$, $c_d \neq 0$, the number d is called the *degree of p*. A number $\zeta \in \mathbb{C}$ such that $p(\zeta) = 0$ is called a *zero of p*.

Proof. Let $p(z) = c_0 + c_1 z + \cdots + c_d z^d$ be a polynomial with degree $d \geqslant 1$, such that it has no zero in $\mathbb{C}$, i.e., for all $z \in \mathbb{C}$, $p(z) \neq 0$. Thus $p : \mathbb{C} \to \mathbb{C}\backslash\{0\}$. The function $g := \frac{1}{z} \in \mathcal{O}(\mathbb{C}\backslash\{0\})$, and so the composition map $f := g \circ p = \frac{1}{p}$, is entire. Recall Exercise 1.37, which gives an estimate on the growth of a polynomial: There exist $m, R > 0$ such that $|p(z)| \geqslant m|z|^d \geqslant mR^d$ for all $|z| > R$. So $|f(z)| \leqslant \frac{1}{mR^d}$ for $|z| > R$. By Weierstrass's Theorem, in the compact set $\{z \in \mathbb{C} : |z| \leqslant R\}$, the real-valued continuous function $|p|$ has a positive minimum k (as p is never 0). Thus $|f(z)| \leqslant \max\{\frac{1}{mR^d}, \frac{1}{k}\}$ for all $z \in \mathbb{C}$. By Liouville's Theorem, f must be constant, say C. So for all $|z| > R$, $\frac{1}{m|z|^d} \geqslant \frac{1}{|p(z)|} = |C|$, giving $|C| = 0$ and so $C = 0$. Hence for all $z \in \mathbb{C}$, $\frac{1}{p(z)} = C = 0$. Multiplying by $p(z)$, we get $1 = 0$, a contradiction. $\square$

Exercise 3.40. Let f be an entire function such that f is bounded away from 0, that is, there is a $\delta > 0$ such that for all $z \in \mathbb{C}$, $|f(z)| \geqslant \delta$. Show that f is a constant.

Exercise 3.41. Let $w_0 \in \mathbb{C}$ and $r > 0$. Show that an entire function, whose range is disjoint with the disc $\{w \in \mathbb{C} : |w - w_0| < r\}$, must be a constant.

Exercise 3.42. Let $f \in \mathcal{O}(\mathbb{C})$, and periodic in the real and in the imaginary directions, i.e., there exist positive T_1, T_2 in $\mathbb{R}$ such that $f(z) = f(z + T_1) = f(z + iT_2)$ for all $z \in \mathbb{C}$. Prove that f is constant. Give examples to show that periodicity in just the real direction or just the imaginary direction is not enough.

Exercise 3.43. A classical theme in the theory of entire functions is to try to characterise the entire function f based on the way $|f|$ grows for large $|z|$. Here is one instance of this.
(1) Show that if f is entire, and $|f(z)| \leqslant |\exp z|$ for all $z \in \mathbb{C}$, then in fact f is equal to $c \exp z$ for some complex constant c with $|c| \leqslant 1$. (Thus if a nonconstant entire function 'grows' no faster than the exponential function, it *is* an exponential function.)
(2) One may be tempted to argue that this cannot be right on the grounds that 'polynomials grow slower than the exponential function', but surely $p \neq \exp z$. Show that if p is a polynomial satisfying $|p(z)| \leqslant |\exp z|$ for all $z \in \mathbb{C}$, then $p \equiv 0$. *Hint:* Take $z = x < 0$.

---

[1]Despite its name, there is no purely algebraic proof of the theorem, since it turns out that any proof must use the completeness of the reals.

Exercise 3.44. Suppose f is an entire function, and satisfies $|f(z)| > |f'(z)|$ for all $z \in \mathbb{C}$. Prove that there exists a constant $C \in \mathbb{C}$ such that for all $z \in \mathbb{C}$, $f'(z) = Cf(z)$. Calculate $(e^{-Cz} f)'$, and hence determine all such f.

Exercise 3.45. Let $M \in \mathbb{R}$, and f be an entire function such $\operatorname{Re} f(z) \leqslant M$ for all $z \in \mathbb{C}$. Prove that f is a constant function.

Exercise 3.46. Suppose that f is an entire function. Let $u := \operatorname{Re} f$ and $v := \operatorname{Im} f$. Suppose further that $u(x, y) \leqslant v(x, y)$ for all $(x, y) \in \mathbb{C}$. Prove that $f'(z) = 0$ for all $z \in \mathbb{C}$. *Hint:* Consider $\exp(f + if)$.

Exercise 3.47. Let $f : \mathbb{C} \to \mathbb{C}$ be an entire function. Suppose that a_1, a_2 are complex numbers such that $a_1 \neq a_2$, and such that a_1, a_2 are contained in the interior of the circular path C with radius $R > 0$ and centre 0, traversed once in the counterclockwise direction.

(1) Prove that $\left| \int_C \frac{f(z)}{(z-a_1)(z-a_2)} dz \right| \leqslant \frac{2\pi R}{(R-|a_1|)(R-|a_2|)} \max_{z \in C} |f(z)|$.

(2) Find $\alpha, \beta \in \mathbb{C}$ such that for all $z \in \mathbb{C}$, $\frac{1}{(z-a_1)(z-a_2)} = \frac{\alpha}{z-a_1} + \frac{\beta}{z-a_2}$.

(3) Express $\int_C \frac{f(z)}{(z-a_1)(z-a_2)} dz$ in terms of $\int_C \frac{f(z)}{z-a_1} dz$ and $\int_C \frac{f(z)}{z-a_2} dz$.

Use the Cauchy Integral Formula to simplify these latter expressions.

(4) Deduce Liouville's Theorem.

Exercise 3.48. Throughout this exercise, the polynomial p has degree $d \geqslant 1$. Using the Division Algorithm and the Fundamental Theorem of Algebra, p can be factored as $p(z) = C(z - \alpha_1)^{m_1} \cdots (z - \alpha_k)^{m_k}$, where $m_1, \cdots, m_k \in \mathbb{N}$, $m_1 + \cdots + m_k = d$, $C \neq 0$, and $\alpha_1, \cdots, \alpha_k$ are the distinct zeroes of p. A zero α of p is called *simple* if $p(z) = (z - \alpha)q(z)$, where q is a polynomial such that $q(\alpha) \neq 0$. Show that α is a simple zero of p if and only if $p(\alpha) = 0$ and $p'(\alpha) \neq 0$. Show that there exists a $w \in \mathbb{C}$ such that the polynomial $\tilde{p}$, $\tilde{p}(z) = p(z) + w$ $(z \in \mathbb{C})$, has only simple zeroes.

Exercise 3.49. Let f be an entire function such that $f(0) = 0$ and $|f(z) - e^z \sin z| < 3$ for all $z \in \mathbb{C}$. Find f.

3.9. Morera's Theorem

The Cauchy Integral Theorem implies the following:

> Let D be a domain. If $f \in \mathcal{O}(D)$, then for any disc $\Delta \subset D$, and any closed piecewise smooth path γ in Δ, $\int_\gamma f(z) dz = 0$.

Morera's theorem is a converse to the above: If the contour integral is zero for some special paths, then f is holomorphic.

Theorem 3.19 (Morera's Theorem).
Let D be a domain, and $f : D \to \mathbb{C}$ be continuous. If for every disc $\Delta \subset D$, and for every closed rectangular path γ in D, we have that $\int_\gamma f(z) dz = 0$, then $f \in \mathcal{O}(D)$.

Proof. Let $z_0 \in D$, and let Δ be a disc with centre z_0 such that $\Delta \subset D$, and $\gamma_{z_0, z}$ is the path joining z_0 to z by first moving horizontally and then moving vertically. Define $F : \Delta \to \mathbb{C}$ by $F(z) = \int_{\gamma_{z_0, z}} f(\zeta) d\zeta$ for all $z \in \Delta$.

We will show that F is holomorphic in Δ, and its derivative is f. This shows that f is holomorphic in Δ! Why? Because f (being the derivative of a holomorphic function) is itself then holomorphic in Δ by Corollary 3.16.

Let $z \in \Delta$, and $\epsilon > 0$. As f is continuous, there exists $\delta > 0$ such that if $w \in \Delta$ satisfies $|w - z| < \delta$, then $|f(w) - f(z)| < \epsilon$. We have
$$F(w) - F(z) = \int_{\gamma_{z_0,w}} f(\zeta)\,d\zeta - \int_{\gamma_{z_0,z}} f(\zeta)\,d\zeta.$$
As the integral of f on closed rectangular paths is zero, $F(w) - F(z) = \int_{\gamma_{z,w}} f(\zeta)\,d\zeta$, where $\gamma_{z,w}$ is the path joining z to w by again first moving horizontally and then moving vertically. See the picture below which shows one particular case:
$$F(w) - F(z) = \int_{\gamma_{z_0,w}} f(\zeta)\,d\zeta - \int_{\gamma_{z_0,z}} f(\zeta)\,d\zeta = \int_A + \int_B + \int_C - (\int_A + \int_D) f(\zeta)\,d\zeta$$
$$= \left(\int_B + \int_C + \int_{-\gamma_{z,w}} + \int_{-D} \right) + \int_{\gamma_{z,w}} f(\zeta)\,d\zeta = (0) + \int_{\gamma_{z,w}} f(\zeta)\,d\zeta.$$

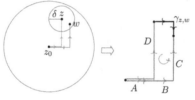

Thus for $0 < |w - z| < \delta$,
$$\frac{F(w)-F(z)}{w-z} - f(z) = \frac{1}{w-z} \int_{\gamma_{z,w}} f(\zeta)\,d\zeta - f(z)\frac{1}{w-z} \int_{\gamma_{z,w}} 1\,d\zeta = \frac{1}{w-z} \int_{\gamma_{z,w}} (f(\zeta) - f(z))\,d\zeta.$$
Here we have used the Fundamental Theorem of Contour Integration for the holomorphic function 1 to obtain $\int_{\gamma_{z,w}} 1\,d\zeta = w - z$. From the above, and using the fact that the length of $\gamma_{z,w}$ is $|\mathrm{Re}(w - z)| + |\mathrm{Im}(w - z)| \leqslant 2|w - z|$, we have
$$\left| \frac{F(w)-F(z)}{w-z} - f(z) \right| = \left| \frac{1}{w-z} \int_{\gamma_{z,w}} (f(\zeta) - f(z))\,d\zeta \right| \leqslant \frac{\epsilon}{|w-z|} (|\mathrm{Re}(w - z)| + |\mathrm{Im}(w - z)|) < 2\epsilon.$$
This completes the proof.   □

Exercise 3.50. Let $f : \mathbb{C} \to \mathbb{C}$ be continuous, and $f \in \mathcal{O}(\mathbb{C} \backslash (\mathbb{R} \times \{0\}))$. Show that $f \in \mathcal{O}(\mathbb{C})$.

3.10. Appendix

In this appendix, we will prove the Cauchy Integral Theorem without the simplifying assumptions used earlier. We first prove the following.

Theorem A (Cauchy-Goursat). Let U be an open subset of $\mathbb{C}$, and $f \in \mathcal{O}(U)$. Then for any triangle $\Delta \subset U$, $\int_{\partial\Delta} f(z)\,dz = 0$.

(Here a triangle Δ is considered open, i.e., the open region contained within the three sides constituting its boundary $\partial\Delta$. Also, we denote the counterclockwise path starting at any vertex traversed once along the boundary also by $\partial\Delta$.)

Proof. f is continuous in U. Let $\Delta_0 \subset U$ be a triangle. Set $I = \int_{\partial\Delta_0} f(z)\,dz$. Divide Δ_0 in 4 triangles $\Delta_0^1, \Delta_0^2, \Delta_0^3, \Delta_0^4$, using the midpoints of the sides of Δ_0. Then

$$I = \int_{\partial\Delta_0} f(z)\,dz = \sum_{j=1}^{4} \int_{\partial\Delta_0^j} f(z)\,dz. \text{ (Why?)}$$

So there is a j such that $|\int_{\partial\Delta_0^j} f(z)\,dz| \geq \frac{|I|}{4}$ (otherwise $|I| < 4\frac{|I|}{4}$, a contradiction). Let Δ_1 be one of the Δ_0^js with this property. The perimeter of $\partial\Delta_1$ is half that of $\partial\Delta_0$. Repeating this process of dividing a triangle into four smaller triangles, we get a decreasing sequence of triangles $\Delta_0 \supset \Delta_1 \supset \Delta_2 \supset \cdots$, such that

$$|\int_{\partial\Delta_k} f(z)\,dz| \geq \frac{|I|}{4^k}, \text{ and the perimeter } \ell(\partial\Delta_k) = \frac{\ell(\partial\Delta_0)}{2^k}.$$

Let $c \in \cap_{k\geq 0} \overline{\Delta_k} \neq^2 \varnothing$, where $\overline{\Delta_k} = \Delta_k \cup \partial\Delta_k$. Since f is complex differentiable at c, given $\epsilon > 0$, there exists a $\delta > 0$ such that whenever $|z - c| < \delta$, we have $|f(z) - f(c) - f'(c)(z-c)| \leq \epsilon|z-c|$. Take k large enough so that $\overline{\Delta_k} \subset D(c,\delta)$. By the Fundamental Theorem of Contour Integration, $\int_{\partial\Delta_k} (f(c) + f'(c)(z-c))dz = 0$. So

$$
\begin{aligned}
|\int_{\partial\Delta_k} f(z)\,dz| &= |\int_{\partial\Delta_k} (f(z) - f(c) - f'(c)(z-c))dz| \\
&\leq \left(\sup_{z\in\partial\Delta_k} |f(z) - f(c) - f'(c)(z-c)| \right) \ell(\partial\Delta_k) \\
&\leq \left(\sup_{z\in\partial\Delta_k} \epsilon|z - c| \right) \ell(\partial\Delta_k) \\
&\leq \epsilon\ell(\partial\Delta_k)\ell(\partial\Delta_k) \quad \text{(Why?)} \\
&= \epsilon\frac{(\ell(\partial\Delta_0))^2}{4^k}.
\end{aligned}
$$

For large enough k, $\frac{|I|}{4^k} \leq |\int_{\partial\Delta_k} f(z)\,dz| \leq \epsilon\frac{(\ell(\partial\Delta_0))^2}{4^k}$, that is, $|I| \leq \epsilon(\ell(\partial\Delta_0))^2$. As $\epsilon > 0$ was arbitrary, $|I| = 0$, and so $I = 0$. $\square$

Theorem B. Let $z_0 \in \mathbb{C}$, $r > 0$, and $f \in \mathcal{O}(D(z_0, r))$. Then there exists an $F \in \mathcal{O}(D(z_0, r))$ such that $F' = f$ in $D(z_0, r)$.

---

[2]For example, the centres c_k of $\overline{\Delta_k}$ form a bounded sequence in $\mathbb{R}^2$, and by the Bolzano–Weierstrass Theorem, this sequence has a convergent subsequence, say $(c_{n_k})_{k\in\mathbb{N}}$, with limit, say, $c \in \mathbb{R}^2$. For each $n \in \mathbb{N} \cup \{0\}$, there exists a $K \in \mathbb{N}$ such that for all $k \geq K$, $n_k > n$, and so $(c_{n_k})_{k\geq K}$ is a sequence in the closed set $\overline{\Delta_n}$, showing that its limit c also belongs to $\overline{\Delta_n}$. As $n \in \mathbb{N}$ was arbitrary, $c \in \cap_{n\geq 0}\overline{\Delta_n}$.

Proof. For $z \in D(z_0, r)$, let $[z_0, z]$ be the straight line segment joining z_0 to z. Define F by $F(z) = \int_{[z_0,z]} f(\zeta)\, d\zeta$ for all $z \in D(z_0, r)$. For $w \in D(z_0, r)$ and $w \neq z$, we get by Theorem A that $\int_{[z_0,z]} f(\zeta)\, d\zeta + \int_{[z,w]} f(\zeta)\, d\zeta + \int_{[w,z_0]} f(\zeta)\, d\zeta = 0$, i.e., $F(w) - F(z) = \int_{[z,w]} f(\zeta)\, d\zeta$. So for such w,

$$\left| \frac{F(w)-F(z)}{w-z} - f(z) \right| = \left| \frac{1}{w-z} \int_{[z,w]} f(\zeta)\, d\zeta - f(z) \right| = \frac{\left| \int_{[z,w]} (f(\zeta)-f(z))\, d\zeta \right|}{|w-z|}$$

$$\leqslant \frac{|w-z| \max\limits_{\zeta \in [z,w]} |f(\zeta)-f(z)|}{|w-z|} = \max\limits_{\zeta \in [z,w]} |f(\zeta) - f(z)|.$$

Given $\epsilon > 0$, let $\delta \in (0, r)$ be such that whenever $|\zeta - z| < \delta$, $|f(\zeta) - f(z)| < \epsilon$. Then for $0 < |w - z| < \delta$, every $\zeta \in [z, w]$ satisfies $|\zeta - z| < \delta$, and so we obtain $\left| \frac{F(w)-F(z)}{w-z} - f(z) \right| \leqslant \max\limits_{\zeta \in [z,w]} |f(\zeta) - f(z)| < \epsilon$. Thus $F'(z) = f(z)$. $\qquad\square$

Cauchy Integral Theorem

Let D be a domain, $\gamma_0, \gamma_1 : [0, 1] \to D$ be smooth, closed, D-homotopic paths, and $f \in \mathcal{O}(D)$. Then $\int_{\gamma_0} f(z)\, dz = \int_{\gamma_1} f(z)\, dz$.

Proof. Let $I = [0, 1]$ and $H : I^2 \to D$ be a homotopy taking γ_0 to γ_1. Since H is continuous and I^2 is compact, H is uniformly continuous and $H(I^2)$ is a compact subset of D. Define $\rho(w) = \inf_{z \in \mathbb{C} \backslash D} |w - z|$ for $w \in \mathbb{C}$. For $w, w' \in \mathbb{C}$, and $z \in \mathbb{C} \backslash D$, $\rho(w) \leqslant |w - z| \leqslant |w - w'| + |w' - z|$, giving $\rho(w) \leqslant |w - w'| + \rho(w')$. Interchanging w, w', we arrive at $|\rho(w) - \rho(w')| \leqslant |w - w'|$, showing that ρ is continuous. We claim that ρ is pointwise positive on the compact set $H(I^2)$. Let $w \in H(I^2)$. Clearly $\rho(w) \geqslant 0$. If $\rho(w) = 0$, then there exists a sequence $(z_n)_{n \in \mathbb{N}}$ in $\mathbb{C} \backslash D$ converging to w, and as $\mathbb{C} \backslash D$ is closed, $w \in \mathbb{C} \backslash D \subset \mathbb{C} \backslash H(I^2)$, a contradiction. So the minimum value of the continuous function ρ on the compact set $H(I^2)$ is positive. Thus $r := \inf\limits_{w \in H(I^2), z \in \mathbb{C} \backslash D} |w - z| > 0$.

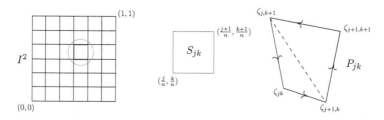

There is an $n \in \mathbb{N}$ such that if $(s - s')^2 + (t - t')^2 < \frac{4}{n^2}$, then $|H(s, t) - H(s', t')| < r$. Let $\zeta_{jk} := H(\frac{j}{n}, \frac{k}{n})$, $0 \leqslant j, k \leqslant n$, and put $S_{jk} = [\frac{j}{n}, \frac{j+1}{n}] \times [\frac{k}{n}, \frac{k+1}{n}]$, $0 \leqslant j, k \leqslant n-1$. The diameter of the square S_{jk} is $\frac{\sqrt{2}}{n}$, and so $H(S_{jk}) \subset D(\zeta_{jk}, r)$. So if we let P_{jk} be the (possibly degenerate) quadrilateral $[\zeta_{jk}, \zeta_{j+1,k}, \zeta_{j+1,k+1}, \zeta_{j,k+1}, \zeta_{jk}]$, then because discs are convex, $P_{jk} \subset D(\zeta_{jk}, r)$. But by Theorem A,

$$\int_{P_{jk}} f(z)\, dz = 0. \text{ (Why?)} \qquad\qquad (\star)$$

It can now be shown that $\int_{\gamma_0} f(z)\,dz = \int_{\gamma_1} f(z)\,dz$ by going up the ladder we have constructed, one rung at a time, as explained below.

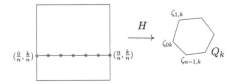

Let Q_k be the closed polygon $[\zeta_{0k}, \zeta_{1k}, \cdots, \zeta_{nk}]$, $0 \leqslant k \leqslant n$. We will show that
$$\int_{\gamma_0} f(z)\,dz = \int_{Q_0} f(z)\,dz = \int_{Q_1} f(z)\,dz = \cdots = \int_{Q_n} f(z)\,dz = \int_{\gamma_1} f(z)\,dz.$$
To see that $\int_{\gamma_0} f(z)\,dz = \int_{Q_0} f(z)\,dz$, observe that if $\sigma_j(t) = \gamma_0(t)$ for $\frac{j}{n} \leqslant t \leqslant \frac{j+1}{n}$, then $\sigma_j + [\zeta_{j+1,0}, \zeta_{j0}]$ is a closed smooth path in the disc $D(\zeta_{j0}, r) \subset D$.

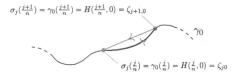

Hence by Theorem B and the Fundamental Theorem of Contour Integration,
$$\int_{\sigma_j} f(z)\,dz = -\int_{[\zeta_{j+1,0},\zeta_{j0}]} f(z)\,dz = \int_{[\zeta_{j0},\zeta_{j+1,0}]} f(z)\,dz.$$
Adding both sides of this equation for $0 \leqslant j \leqslant n$ yields $\int_{\gamma_0} f(z)\,dz = \int_{Q_0} f(z)\,dz$.
Similarly, $\int_{\gamma_1} f(z)\,dz = \int_{Q_1} f(z)\,dz$.

To see that $\int_{Q_k} f(z)\,dz = \int_{Q_{k+1}} f(z)\,dz$, $0 \leqslant k \leqslant n-1$, we use $(\star)$. Note that
$$0 = \sum_{j=0}^{n-1} \int_{P_{jk}} f(z)\,dz. \qquad (\star\star)$$

$\int_{P_{jk}} f(z)\,dz$ includes the integral over $[\zeta_{j+1,k}, \zeta_{j+1,k+1}]$, which is the negative of the integral over $[\zeta_{j+1,k+1}, \zeta_{j+1,k}]$, which in turn is a part of the integral $\int_{P_{j+1,k}} f(z)\,dz$.

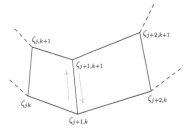

Moreover, $\zeta_{0,k} = H(0, \frac{k}{n}) = H(1, \frac{k}{n}) = \zeta_{n,k}$, so that $[\zeta_{0,k+1}, \zeta_{0,k}] = -[\zeta_{n,k}, \zeta_{n,k+1}]$. Taking these cancellations into account, $(\star\star)$ gives $0 = \int_{Q_k} f(z)\,dz - \int_{Q_{k+1}} f(z)\,dz$. $\square$

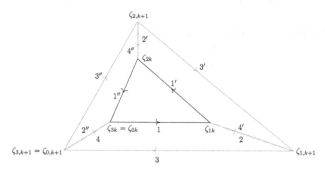

Illustrative example with $n = 3$, showing the 'cancellations' arising in $(\star\star)$:

$$0 = \int_{P_{0k}} + \int_{P_{1k}} + \int_{P_{2k}} f(z)\,dz$$

$$= \int_1 + \int_2 + \int_3 + \int_4 + \int_{1'} + \int_{2'} + \int_{3'} + \int_{4'} + \int_{1''} + \int_{2''} + \int_{3''} + \int_{4''} f(z)\,dz$$

$$= \int_1 + \int_{1'} + \int_{1''} + \int_3 + \int_{3'} + \int_{3''} f(z)\,dz = \int_{Q_k} - \int_{Q_{k+1}} f(z)\,dz.$$

Chapter 4

Taylor and Laurent series

In this chapter, firstly, we will learn that a complex power series

$$\sum_{n=0}^{\infty} c_n(z - z_0)^n$$

converges in some disc $D(z_0, R)$ and diverges for $|z - z_0| > R$. Also, in the disc $D(z_0, R)$, it defines a holomorphic function. Vice versa, if $f \in \mathcal{O}(D(z_0, R))$, then it has a power series expansion as above, called its 'Taylor series' (centred at z_0). We will use this to show the general Cauchy Integral Formula, and to understand the nature of zeroes of holomorphic functions. Moreover, we will prove two additional fundamental theorems about holomorphic functions: the Identity Theorem and the Maximum Modulus Theorem.

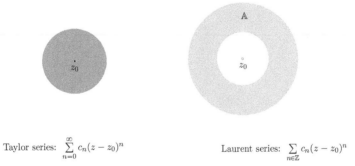

Taylor series: $\sum_{n=0}^{\infty} c_n(z - z_0)^n$ Laurent series: $\sum_{n \in \mathbb{Z}} c_n(z - z_0)^n$

In the second part of the chapter, we will study 'double-sided' series

$$\sum_{n \in \mathbb{Z}} c_n(z - z_0)^n,$$

where now negative integer powers of $z - z_0$ also occur. We will show that in general such a 'Laurent series' converges in an annulus $\mathbb{A} := \{z \in \mathbb{C} : r < |z - z_0| < R\}$, and defines a holomorphic function there. Vice versa, any $f \in \mathcal{O}(\mathbb{A})$ possesses a Laurent series expansion as above. We will use this result to classify 'singularities' of functions that are holomorphic in a punctured disc $D(z_0, R) \backslash \{z_0\}$, and to learn 'Residue Calculus', which in turn can be useful to evaluate some real integrals and real series, as we will see at the end of this chapter.

4.1. Series

Just as with real series, given a sequence $(a_n)_{n \in \mathbb{N}}$ of complex numbers, one can form the sequence $(s_n)_{n \in \mathbb{N}}$ of its *partial sums*, where $s_n := a_1 + \cdots + a_n$ for all $n \in \mathbb{N}$.

- The series $\sum\limits_{n=1}^{\infty} a_n$ *converges* if $(s_n)_{n \in \mathbb{N}}$ converges.

- If $\sum\limits_{n=1}^{\infty} a_n$ converges, then the *sum of the series is* $\sum\limits_{n=1}^{\infty} a_n := \lim\limits_{n \to \infty} s_n$.

- The series $\sum\limits_{n=1}^{\infty} a_n$ *diverges* if $(s_n)_{n \in \mathbb{N}}$ diverges.

- The series $\sum\limits_{n=1}^{\infty} a_n$ *converges absolutely* if the real series $\sum\limits_{n=1}^{\infty} |a_n|$ converges.

From the result in Exercise 1.38, which says that a complex sequence converges if and only if the sequences of its real and imaginary parts converge, it follows that

$$\sum_{n=1}^{\infty} a_n \text{ converges } \Leftrightarrow \text{ the real series } \sum_{n=1}^{\infty} \operatorname{Re} a_n \text{ and } \sum_{n=1}^{\infty} \operatorname{Im} a_n \text{ converge.}$$

Thus the results from real analysis are useful in testing the convergence of complex series. For example, it is easy to prove the following two facts, which we leave as exercises.

Exercise 4.1. If $\sum\limits_{n=1}^{\infty} a_n$ converges, then show that $\lim\limits_{n \to \infty} a_n = 0$.

Exercise 4.2. If $\sum\limits_{n=1}^{\infty} a_n$ converges absolutely, then show that $\sum\limits_{n=1}^{\infty} a_n$ converges.

Exercise 4.3. If $\sum\limits_{n=1}^{\infty} a_n$ converges, and $\epsilon > 0$, then show that there exists an $N \in \mathbb{N}$ such that $\sum\limits_{n=N+1}^{\infty} a_n$ converges and $\left| \sum\limits_{n=N+1}^{\infty} a_n \right| < \epsilon$.

Exercise 4.4. Show that if $|z| < 1$, then $\sum\limits_{n=0}^{\infty} z^n$ converges and that $\sum\limits_{n=0}^{\infty} z^n = \frac{1}{1-z}$.

Exercise 4.5. Show that if $|z| < 1$, then $\sum\limits_{n=1}^{\infty} n z^{n-1} = \frac{1}{(1-z)^2}$.

Exercise 4.6. Show that $\frac{1}{1^s} + \frac{1}{2^s} + \frac{1}{3^s} + \cdots$ converges for all complex s with $\operatorname{Re} s > 1$. Thus $s \mapsto \zeta(s) := \sum\limits_{n=1}^{\infty} \frac{1}{n^s}$ is a well-defined map in the half-plane $\{s \in \mathbb{C} : \operatorname{Re} s > 1\}$, and is called the *Riemann zeta function*.

The Riemann zeta function is linked with number theory through the *Euler Product Formula*: If $p_1, p_2, p_3, \cdots$ is the infinite list of primes in increasing order, then

$$\zeta(s) = \lim_{K \to \infty} \prod_{k=1}^{K} \frac{1}{1 - p_k^{-s}} \qquad (\operatorname{Re} s > 1).$$

Riemann (1826-66) showed that the function ζ can be extended holomorphically to $\mathbb{C} \backslash \{1\}$. It can be shown that the function ζ has zeroes at $-2, -4, -6, \cdots$, called 'trivial zeroes', but it also has other zeroes. All the non-trivial zeroes Riemann computed turned out to lie on the line $\operatorname{Re} s = \frac{1}{2}$. This led him to formulate the following conjecture (now called the *Riemann Hypothesis*), which is a famous unsolved problem in Mathematics: All non-trivial zeroes of the Riemann zeta function lie on the line $\operatorname{Re} s = \frac{1}{2}$.

4.2. Power series

Region of convergence.

Let $(c_n)_{n\in\mathbb{N}}$ be a complex sequence (thought of as a sequence of 'coefficients').

Then $\sum_{n=0}^{\infty} c_n z^n$ is called a *power series* in the complex variable z.

We imagine putting in specific values of z in the above series. Then for some $z \in \mathbb{C}$, the power series will converge, while for other values of z it may diverge.

A fundamental question is: For which $z \in \mathbb{C}$ does the power series $\sum_{n=0}^{\infty} c_n z^n$ converge?

Example 4.1. All polynomial expressions are power series, with only finitely many nonzero coefficients. Polynomials converge for *all* $z \in \mathbb{C}$.   ◇

Example 4.2. From Exercise 4.4, the power series $\sum_{n=0}^{\infty} z^n$ converges if $|z| < 1$.

As $|z^n - 0| = |z|^n \geqslant 1$ for all $n \in \mathbb{N}$, $\neg(\lim_{n\to\infty} z^n = 0)$. So $\sum_{n=0}^{\infty} z^n$ diverges if $|z| \geqslant 1$. ◇

The following result gives a general answer to the fundamental question above.

Theorem 4.1.
Let $(c_n)_{n\in\mathbb{N}}$ be a sequence in $\mathbb{C}$. Then exactly one of the following hold:

- *Either $\sum_{n=0}^{\infty} c_n z^n$ is absolutely convergent for all $z \in \mathbb{C}$,*

- *or there is a unique nonnegative real number R such that:*

  (1) *$\sum_{n=0}^{\infty} c_n z^n$ is absolutely convergent for all $z \in \mathbb{C}$ with $|z| < R$, and*

  (2) *$\sum_{n=0}^{\infty} c_n z^n$ is divergent for all $z \in \mathbb{C}$ with $|z| > R$.*

The unique $R > 0$ in the above theorem is called the *radius of convergence* of the power series, and if the power series converges for all $z \in \mathbb{C}$, then we say that the power series has an infinite radius of convergence, and write '$R = \infty$'.

What happens on the circle $|z| = R$? Complex power series may diverge at every point on the boundary (given by $|z| = R$), or diverge on some points of the boundary and converge at other points of the boundary, or converge at all the points on the boundary. There is no general result answering what happens at each point on the circle, and the specific power series at hand has to be investigated to find out the behaviour.

Proof of Theorem 4.1.

Let $S := \{y \in [0, \infty) : \exists z \in \mathbb{C}$ such that $y = |z|$ and $\sum_{n=0}^{\infty} c_n z^n$ converges$\}$. Then $0 \in S$.
Only two cases are possible:

$1°$ S is not bounded above. We will show that the radius of convergence is infinite.
Given $z \in \mathbb{C}$, there exists a $y \in S$ such that $|z| < y$. But as $y \in S$, there exists
a $z_0 \in \mathbb{C}$ such that $y = |z_0|$ and $\sum_{n=0}^{\infty} c_n z_0^n$ converges. So its nth term tends to 0
as $n \to \infty$, and in particular, the sequence of terms is bounded: $|c_n z_0^n| \leqslant M$
for all $n \in \mathbb{N}$. Then with $r := \frac{|z|}{|z_0|}$ (< 1), we have $|c_n z^n| = |c_n z_0^n|(\frac{|z|}{|z_0|})^n \leqslant Mr^n$
for all $n \in \mathbb{N}$. But $\sum_{n=0}^{\infty} Mr^n$ converges $(r < 1)$, and so by the Comparison Test,
$\sum_{n=0}^{\infty} c_n z^n$ is absolutely convergent. Since $z \in \mathbb{C}$ was arbitrary, the claim follows.

$2°$ S is bounded above. We will show that the radius of convergence is $\sup S$, i.e.,

    (a) if $|z| < \sup S$, then $\sum_{n=0}^{\infty} c_n z^n$ converges absolutely, and

    (b) if $|z| > \sup S$, then $\sum_{n=0}^{\infty} c_n z^n$ diverges.

If $z \in \mathbb{C}$ and $|z| < \sup S$, then by the definition of supremum, there exists a
$y \in S$ such that $|z| < y$. Then we repeat the proof in $1°$. Since $y \in S$, there
exists a $z_0 \in \mathbb{C}$ such that $y = |z_0|$ and $\sum_{n=0}^{\infty} c_n z_0^n$ converges. So its nth term tends
to 0 as $n \to \infty$, and in particular, the sequence of terms is bounded: $|c_n z_0^n| \leqslant M$.
Then with $r := \frac{|z|}{|z_0|}$ (< 1), we have $|c_n z^n| = |c_n z_0^n|(\frac{|z|}{|z_0|})^n \leqslant Mr^n$ for all $n \in \mathbb{N}$.
But $\sum_{n=0}^{\infty} Mr^n$ converges (as $r < 1$), and so $\sum_{n=0}^{\infty} c_n z^n$ is absolutely convergent.

If $z \in \mathbb{C}$ and $|z| > \sup S$, then setting $y := |z|$, $y \notin S$, and by the definition of S,
$\sum_{n=0}^{\infty} c_n z^n$ diverges (otherwise $y \in S$).

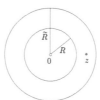

Uniqueness of R: Let $R, \widetilde{R}$ have the property described in the theorem and $R < \widetilde{R}$.
We have $R < r := \frac{R + \widetilde{R}}{2} < \widetilde{R}$. As $r < \widetilde{R}$, $\sum_{n=1}^{\infty} c_n r^n$ converges. As $R < r$, $\sum_{n=0}^{\infty} c_n r^n$
diverges, a contradiction. $\qquad\qquad\square$

Instead of the power series $\sum_{n=0}^{\infty} c_n z^n$ 'centred at 0', consider $\sum_{n=0}^{\infty} c_n (z - z_0)^n$ centred
at $z_0 \in \mathbb{C}$. Applying Theorem 4.1 to $\sum_{n=0}^{\infty} c_n w^n$, and taking $w = z - z_0$, we obtain:

Corollary 4.2.
Let $(c_n)_{n\in\mathbb{N}}$ be a sequence in $\mathbb{C}$, and $z_0 \in \mathbb{C}$. Then exactly one of the following hold:

- *Either $\sum_{n=0}^{\infty} c_n(z - z_0)^n$ is absolutely convergent for all $z \in \mathbb{C}$,*
- *or there is a unique nonnegative real number R such that:*
  - (1) *$\sum_{n=0}^{\infty} c_n(z - z_0)^n$ is absolutely convergent for all $z \in \mathbb{C}$ with $|z - z_0| < R$.*
  - (2) *$\sum_{n=0}^{\infty} c_n(z - z_0)^n$ is divergent for all $z \in \mathbb{C}$ with $|z - z_0| > R$.*

Exercise 4.7. Let R denote the radius of convergence of $\sum_{n=0}^{\infty} c_n(z - z_0)^n$. Show that:

- If $\sum_{n=0}^{\infty} c_n(z - z_0)^n$ converges for a $z_* \in \mathbb{C}$, then $R \geq |z_* - z_0|$.
- If $\sum_{n=0}^{\infty} c_n(z - z_0)^n$ diverges for a $z_* \in \mathbb{C}$, then $R \leq |z_* - z_0|$.

Exercise 4.8. Let $N \in \mathbb{N}$ and $z \in \mathbb{C}$. Show that $\sum_{n=1}^{N} |z^{n!}| \leq \sum_{n=1}^{N!} |z^n|$. Prove that $\sum_{n=1}^{\infty} z^{n!}$ converges in $\mathbb{D} := \{z \in \mathbb{C} : |z| < 1\}$. Determine the radius of convergence of $\sum_{n=1}^{\infty} z^{n!}$.

Theorem 4.3. *Consider the power series $\sum_{n=0}^{\infty} c_n z^n$. If $L := \lim_{n\to\infty} \left|\frac{c_{n+1}}{c_n}\right|$ exists, then:*

- *If $L \neq 0$, then the radius of convergence is $\frac{1}{L}$.*
- *If $L = 0$, then the radius of convergence is infinite.*

Proof.
$1°$ Let $L \neq 0$. For all nonzero z such that $|z| < \frac{1}{L}$ there exists a $q < 1$ and an N large enough such that $\frac{|c_{n+1}z^{n+1}|}{|c_n z^n|} = \left|\frac{c_{n+1}}{c_n}\right||z| \leq q < 1$ for all $n > N$. (This is because $\left|\frac{c_{n+1}}{c_n}z\right| \xrightarrow{n\to\infty} L|z| < 1$. So we may take $q = \frac{L|z|+1}{2} < 1$.) Thus by the Ratio Test, $\sum_{n=0}^{\infty} c_n z^n$ is absolutely convergent for such z.

If $|z| > \frac{1}{L}$, then there is an N large enough such that $\frac{|c_{n+1}z^{n+1}|}{|c_n z^n|} = \left|\frac{c_{n+1}}{c_n}\right||z| > 1$ for all $n > N$, as $\left|\frac{c_{n+1}}{c_n}z\right| \xrightarrow{n\to\infty} L|z| > 1$. By the Ratio Test, $\sum_{n=0}^{\infty} c_n z^n$ diverges.

$2°$ Let $L = 0$. For $z \in \mathbb{C}\backslash\{0\}$, there is a $q < 1$ such that $\frac{|c_{n+1}z^{n+1}|}{|c_n z^n|} = \left|\frac{c_{n+1}}{c_n}\right||z| \leq q < 1$ for all $n > N$. (This is because $\left|\frac{c_{n+1}}{c_n}z\right| \xrightarrow{n\to\infty} 0|z| = 0 < 1$. So we may take $q = \frac{1}{2} < 1$.) By the Ratio Test, $\sum_{n=0}^{\infty} c_n z^n$ is absolutely convergent for such z. $\square$

Example 4.3. As $\lim_{n\to\infty} \frac{\frac{1}{(n+1)^2}}{\frac{1}{n^2}} = 1$, $\sum_{n=1}^{\infty} \frac{z^n}{n^2}$ converges if $|z| < 1$ and diverges if $|z| > 1$.

If $|z| = 1$, then $\left|\frac{z^n}{n^2}\right| = \frac{1}{n^2}$, and as $\sum_{n=1}^{\infty} \frac{1}{n^2}$ converges, it follows that $\sum_{n=1}^{\infty} \frac{z^n}{n^2}$ is absolutely convergent. Thus at every point of the circle $|z| = 1$, the power series converges.

The previous example is in contrast to the geometric series $\sum_{n=0}^{\infty} z^n$, where we had convergence at no point of the circle $|z| = 1$.   ◇

Exercise 4.9. Consider the power series $\sum_{n=0}^{\infty} c_n x^n$. If $L := \lim_{n \to \infty} \sqrt[n]{|c_n|}$ exists, then:

- If $L \neq 0$, then the radius of convergence is $\frac{1}{L}$.
- If $L = 0$, then the radius of convergence is infinite.

Exercise 4.10. Show that $\sum_{n=1}^{\infty} \frac{z^n}{n^n}$ converges for all $z \in \mathbb{C}$.

Exercise 4.11. Show that $\sum_{n=1}^{\infty} n^n z^n$ converges only when $z = 0$.

Exercise 4.12. Find the radius of convergence of: $\sum_{n=1}^{\infty} \frac{(-1)^n}{n} z^n$, $\quad \sum_{n=0}^{\infty} n^{2022} z^n$, $\quad \sum_{n=0}^{\infty} \frac{1}{n!} z^n$.

Exercise 4.13. Let R be the radius of convergence of $\sum_{n=0}^{\infty} c_n z^n$. Show that:

(1) If $(\sqrt[n]{|c_n|})_{n \in \mathbb{N}}$ is not bounded, then $R = 0$.

(2) If $(\sqrt[n]{|c_n|})_{n \in \mathbb{N}}$ is bounded, and $M_n := \sup\{\sqrt[m]{|c_m|} : m \geqslant n\}$, $n \in \mathbb{N}$, then $(M_n)_{n \in \mathbb{N}}$ is convergent. Set $L = \lim_{n \to \infty} M_n = \limsup_{n \to \infty} \sqrt[n]{|c_n|}$. If $L = 0$, then $R = \infty$. If $L \neq 0$, then $R = \frac{1}{L}$.

Let $\sum_{n=0}^{\infty} a_n z^n$ and $\sum_{n=0}^{\infty} b_n z^n$ be two power series, with radii of convergence R_a, R_b. Their termwise sum is (absolutely) convergent in the common region of convergence, namely the disc $D(0, \min\{R_a, R_b\})$. Their *Cauchy product* is $\sum_{n=0}^{\infty} c_n z^n$, where

$$c_n := \sum_{k=0}^{n} a_{n-k} b_k, \quad \text{for } n \in \{0, 1, 2, 3, \cdots\}.$$

Using Exercise 4.14 below, it can be shown that the Cauchy product of two power series converges in the common region of convergence.

Exercise 4.14. Suppose $\sum_{n=0}^{\infty} a_n$ and $\sum_{n=0}^{\infty} b_n$ are absolutely convergent. Define $c_n = \sum_{k=0}^{n} a_{n-k} b_k$ for nonnegative integers n. Show that $\sum_{n=0}^{\infty} c_n$ converges, and $\sum_{n=0}^{\infty} c_n = (\sum_{n=0}^{\infty} a_n)(\sum_{n=0}^{\infty} b_n)$.

Hint: If A_n, B_n, C_n denote the nth partial sums, show that $|C_{2n} - A_n B_n| \to 0$ as $n \to \infty$.

Exercise 4.15. Let A and E be opposite vertices of an octagon $ABCDEFGH$. A frog starts at vertex A. From any vertex except E it jumps to one of the two adjacent vertices. When it reaches E it stops. Let a_n be the number of distinct paths of exactly $n \in \mathbb{N}$ jumps ending at E. To obtain a formula for a_n in terms of n, proceed as follows. Show that $a_{2n-1} = 0$ for all $n \in \mathbb{N}$. Also show that $a_2 = 0$ and $a_4 = 2$. Let b_n be the number of distinct paths of length n from C to E. Prove that for all $n > 1$, $a_{2n} = 2a_{2n-2} + 2b_{2n-2}$ and $b_{2n} = 2b_{2n-2} + a_{2n-2}$. Deduce that $a_{2n+2} = 4a_{2n} - 2a_{2n-2}$ for all $n > 1$. Define $A(z) := \sum_{n=0}^{\infty} a_{2n} z^n$, taking $a_0 := 0$. Show that $(1 - 4z + 2z^2) A(z) = 2z^2$ in the disc of convergence of the power series. Conclude that $a_{2n} = \frac{(2+\sqrt{2})^{n-1} - (2-\sqrt{2})^{n-1}}{\sqrt{2}}$ for all $n \in \mathbb{N}$. Explain using the Binomial Theorem why the right-hand side is an integer.

4.2.1. Power series are holomorphic. We have seen that polynomials are power series with an infinite radius of convergence, that is, they converge in the whole of $\mathbb{C}$. They are of course also holomorphic there. More generally, a power series $f(z) := \sum_{n=0}^{\infty} c_n z^n$, which is convergent for $|z| < R$, is holomorphic there, and

$$f'(z) = \frac{d}{dz}(c_0 + c_1 z + c_2 z^2 + \cdots) = c_1 + 2c_2 z + 3c_3 z^2 + \cdots = \sum_{n=1}^{\infty} c_n n z^{n-1}$$

as expected (differentiating termwise, as with finite sums, namely polynomials).

Theorem 4.4. *Let $R > 0$ and $f(z) := \sum_{n=0}^{\infty} c_n z^n$ converge for $|z| < R$.*
Then $f \in \mathcal{O}(D(0, R))$ and $f'(z) = \sum_{n=1}^{\infty} n c_n z^{n-1}$ for $|z| < R$.

Proof. First we will show that $g(z) := \sum_{n=1}^{\infty} n c_n z^{n-1} = c_1 + 2c_2 z + \cdots + n c_n z^{n-1} + \cdots$ is absolutely convergent for $|z| < R$. Fix z and let r satisfy $|z| < r < R$. By hypothesis $\sum_{n=0}^{\infty} c_n r^n$ converges, and so there is some positive number M such that $|c_n r^n| < M$ for all n. Let $\rho := \frac{|z|}{r}$. Then $0 \leqslant \rho < 1$, and

$$|n c_n z^{n-1}| = |c_n r^n| \frac{1}{r} n |\tfrac{z}{r}|^{n-1} \leqslant \frac{M n \rho^{n-1}}{r}.$$

As $\sum_{n=1}^{\infty} n \rho^{n-1}$ converges (to $\frac{1}{(1-\rho)^2}$; Exercise 4.5), by the Comparison Test, $\sum_{n=1}^{\infty} n c_n z^{n-1}$ is absolutely convergent.

Next, we will show that $f'(z_0) = g(z_0)$ for $|z_0| < R$, i.e., $\lim_{z \to z_0} \left(\frac{f(z) - f(z_0)}{z - z_0} - g(z_0) \right) = 0$. As before, let r be such that $|z_0| < r < R$. Below, we consider z satisfying $|z| < r$.

Let $\epsilon > 0$. As $\sum_{n=1}^{\infty} n c_n r^{n-1}$ converges absolutely, there exists an $N \in \mathbb{N}$ such that $\sum_{n=N+1}^{\infty} |n c_n r^{n-1}| < \frac{\epsilon}{4}$. We have $f(z) - f(z_0) = \sum_{n=1}^{\infty} c_n (z^n - z_0^n)$, and so for $z \neq z_0$,

$$\frac{f(z) - f(z_0)}{z - z_0} = \sum_{n=1}^{\infty} c_n \frac{z^n - z_0^n}{z - z_0} = \sum_{n=1}^{\infty} c_n (z^{n-1} + z^{n-2} z_0 + \cdots + z z_0^{n-2} + z_0^{n-1}).$$

Thus $\frac{f(z) - f(z_0)}{z - z_0} - g(z_0) = \sum_{n=1}^{\infty} c_n (z^{n-1} + z^{n-2} z_0 + \cdots + z_0^{n-1} - n z_0^{n-1})$. We let S_1 be the sum of the first N terms of this series (that is, from $n = 1$ to $n = N$) and S_2 be the sum of the remaining terms. Then since $|z|, |z_0| < r$, it follows that

$$|S_2| \leqslant \sum_{n=N+1}^{\infty} |c_n| (\underbrace{r^{n-1} + r^{n-1} + \cdots + r^{n-1}}_{n \text{ terms}} + n r^{n-1}) = \sum_{n=N+1}^{\infty} 2n |c_n| r^{n-1} < \frac{\epsilon}{2}.$$

$S_1 = \sum\limits_{n=1}^{N} c_n(z^{n-1} + z^{n-2}z_0 + \cdots + zz_0^{n-2} + z_0^{n-1} - nz_0^{n-1})$ is a polynomial in z and so

$\lim\limits_{z \to z_0} S_1 = \sum\limits_{n=1}^{N} c_n(z_0^{n-1} + z_0^{n-2}z_0 + \cdots + z_0 z_0^{n-2} + z_0^{n-1} - nz_0^{n-1}) = \sum\limits_{n=1}^{N} c_n(nz_0^{n-1} - nz_0^{n-1}) = 0.$

So there is a $\delta > 0$ such that whenever $|z - z_0| < \delta$, we have $|S_1| < \frac{\epsilon}{2}$. Thus for $|z| < r$ and $0 < |z - z_0| < \delta$, we have $|\frac{f(z)-f(z_0)}{z-z_0} - g(z_0)| \leqslant |S_1| + |S_2| < \frac{\epsilon}{2} + \frac{\epsilon}{2} = \epsilon$. This means that $f'(z_0) = g(z_0)$, as claimed. $\square$

Remark 4.1. If $(c_n)_{n \in \mathbb{N}}$ is a sequence of real numbers, then consider the *real* power series $\sum\limits_{n=0}^{\infty} c_n x^n$. From Real Analysis, we know that such a power series converges in an interval of the form $(-R, R)$ and diverges in $\mathbb{R} \backslash [-R, R]$ for some $R \geqslant 0$.

Theorems 4.1 and 4.4 imply that if we replace the real variable x by a complex variable z, then we can extend the real power series to a holomorphic function in the disc $D(0, R)$ in the complex plane. So we can view *real analytic* functions (namely functions of a real variable having a local power series expansion) as restrictions of holomorphic functions. This again highlights the interplay between the worlds of real analysis and complex analysis. We have seen a previous instance of this interaction when we studied the Cauchy-Riemann equations.   ✳

Corollary 4.5. *Let $R > 0$ and let $f(z) := \sum\limits_{n=0}^{\infty} c_n z^n$ converge for $|z| < R$. Then for $k \in \mathbb{N}$, and $|z| < R$, $f^{(k)}(z) = \sum\limits_{n=k}^{\infty} n(n-1)(n-2) \cdots (n-k+1)c_n z^{n-k}$.*   (∗) *In particular, $c_n = \frac{f^{(n)}(0)}{n!}$ for all $n \in \mathbb{N} \cup \{0\}$.*

Proof. A repeated application of Theorem 4.4 gives this: For $n, k \in \mathbb{N} \cup \{0\}$,

$$(z^n)^{(k)} = \begin{cases} n(n-1) \cdots (n-(k-1))z^{n-k} & \text{for } 0 \leqslant k \leqslant n, \\ 0 & \text{for } k > n. \end{cases}$$

For the last claim, we have $f(0) = c_0$, and for the $n \in \mathbb{N}$ cases, set $z = 0$ in (∗):

$$f^{(k)}(0) = k(k-1) \cdots 1 c_k + z \sum\limits_{n=k+1}^{\infty} n(n-1) \cdots (n-k+1)c_n z^{n-k-1}|_{z=0} = k! c_k. \quad \square$$

There is nothing special about taking power series centred at 0; we have:

Corollary 4.6. *Let $z_0 \in \mathbb{C}$, $R > 0$. If $f(z) := \sum\limits_{n=0}^{\infty} c_n(z-z_0)^n$ converges for $|z-z_0| < R$, then for $k \in \mathbb{N}$ and $|z - z_0| < R$, $f^{(k)}(z) = \sum\limits_{n=k}^{\infty} n(n-1) \cdots (n-k+1)c_n(z-z_0)^{n-k}$. In particular, $c_n = \frac{f^{(n)}(z_0)}{n!}$, for all $n \in \mathbb{N} \cup \{0\}$.*

Proof. $g(w) := \sum\limits_{n=0}^{\infty} c_n w^n$ converges for $w \in D(0, R)$. So $g \in \mathcal{O}(D(0,R))$, and $g'(w) = \sum\limits_{n=1}^{\infty} nc_n w^n$ for all $w \in D(0, R)$. Composing g with $(\cdot - z_0)|_{D(z_0, R)}$, we have that $f(\cdot) = g(\cdot - z_0) \in \mathcal{O}(D(z_0, R))$, and also it follows by the Chain Rule that $f'(z) = g'(z - z_0) \cdot 1 = \sum\limits_{n=1}^{\infty} nc_n(z-z_0)^{n-1}$ for all $z \in D(z_0, R)$. The general case for $n \geqslant 2$ now follows by a repeated application of this last result. $\square$

Remark 4.2 (Uniqueness of coefficients). Suppose $z_0 \in \mathbb{C}$, $R > 0$, and the power series $\sum_{n=0}^{\infty} c_n(z-z_0)^n$, $\sum_{n=0}^{\infty} \tilde{c}_n(z-z_0)^n$ both converge to the same function f in $D(z_0, R)$. Then from the above, for $n \geq 0$, we have $c_n = \frac{f^{(n)}(z_0)}{n!} = \tilde{c}_n$.   *

Exercise 4.16. For $|z| < 1$, what is $1^2 + 2^2 z + 3^2 z^2 + 4^2 z^3 + \cdots$?

Exercise 4.17. True or false? All statements refer to power series $\sum_{n=0}^{\infty} c_n z^n$.

(1) The set of points z for which the power series converges equals either the singleton set $\{0\}$ or some open disc of finite positive radius or the entire complex plane, but no other type of set.

(2) If the power series converges for $z = 1$, then it converges for all z with $|z| < 1$.

(3) If the power series converges for $z = 1$, then it converges for all z with $|z| = 1$.

(4) If the power series converges for $z = 1$, then it converges for $z = -1$.

(5) Some power series converge at all points of an open disc with centre 0 of some positive radius, and also at certain points on the boundary of the disc (that is the circle bounding the disc), and at no other points.

(6) There are power series that converge on a set of points which is exactly equal to the closed disc given by $|z| \leq 1$.

(7) If the power series diverges at $z = i$, then it diverges at $z = 1 + i$ as well.

Exercise 4.18 (Generalised Binomial Theorem). Let $a \in (0, 1)$. Show that the radius of convergence of the complex power series $1 + az + \frac{a(a-1)}{2!}z^2 + \frac{a(a-1)(a-2)}{3!}z^3 + \cdots$ is 1. Let $f \in \mathcal{O}(D(0,1))$ be the sum of the above power series. Prove that $(1 + z)f'(z) = af(z)$ in $D(0, R)$. Show that $(1 + \cdot)^a \in \mathcal{O}(D(0,1))$. Calculate $((1 + \cdot)^{-a} f)'$ in $D(0,1)$, and hence show that $f(z) = (1 + z)^a$ in $D(0,1)$.

Exercise 4.19. Show that the complex power series $J_0(z) = \sum_{n=0}^{\infty} \frac{(-1)^n}{2^{2n}(n!)^2} z^{2n}$ converges for all $z \in \mathbb{C}$. (*Hint:* Consider first $\sum_{n=0}^{\infty} \frac{(-1)^n}{2^{2n}(n!)^2} w^n$.) J_0 is called the *Bessel function of order* 0. Prove that J_0 solves the differential equation $f''(z) + \frac{1}{z}f'(z) + f(z) = 0$ in $\mathbb{C} \backslash \{0\}$.

Exercise 4.20. Let $(c_n)_{n \in \mathbb{N}}$ be a sequence in $\mathbb{C}$ such that $\sum_{n=0}^{\infty} |c_n|$ converges, but $\sum_{n=0}^{\infty} n|c_n|$ does not. Show that the radius of convergence of $\sum_{n=0}^{\infty} c_n z^n$ is equal to 1.

4.3. Taylor series

We have seen that complex power series $f := \sum_{n=0}^{\infty} c_n(\cdot - z_0)^n$ belong to $\mathcal{O}(D(z_0, R))$, where R is the radius of convergence of the power series. In this section, we will show that conversely, if $f \in \mathcal{O}(D(z_0, R))$, for a $z_0 \in \mathbb{C}$ and $R > 0$, then f has a power series expansion, $f(z) = \sum_{n=0}^{\infty} c_n(z - z_0)^n$ for $z \in D(z_0, R)$, where the coefficients c_n, $n \geq 0$, can be determined from the f. This power series is called the *Taylor series of f centred at z_0*. Thus every holomorphic f defined in a domain D possesses a power series expansion locally, in a disc around any point $z_0 \in D$. En route we will also prove the general Cauchy Integral Formula, expressing the derivatives $f^{(n)}$ for all $n \geqslant 0$ of a holomorphic function f inside a disc from the values of f along its bounding circle. The cases $n = 0, 1$ were seen earlier in Sections 3.6 and 3.7.

Theorem 4.7. *Let* $z_0 \in \mathbb{C}$, $R > 0$, *and* $f \in \mathcal{O}(D(z_0, R))$. *For any* $r \in (0, R)$, *define the circular path* $C(t) := z_0 + re^{2\pi i t}$ *for all* $t \in [0, 1]$, *and let* $c_n := \frac{1}{2\pi i} \int_C \frac{f(\zeta)}{(\zeta - z_0)^{n+1}} d\zeta$ *for all integers* $n \geqslant 0$. *Then* $f(z) = \sum_{n=0}^{\infty} c_n (z - z_0)^n$ *for all* $z \in D(z_0, R)$.

Proof. Let $z \in D(z_0, R)$. Initially, let r be such that $|z - z_0| < r < R$. Then by the Cauchy Integral Formula,

$$f(z) = \frac{1}{2\pi i} \int_C \frac{f(\zeta)}{\zeta - z} d\zeta = \frac{1}{2\pi i} \int_C \frac{f(\zeta)}{\zeta - z_0 + z_0 - z} d\zeta = \frac{1}{2\pi i} \int_C \frac{f(\zeta)}{(\zeta - z_0)(1 - \frac{z - z_0}{\zeta - z_0})} d\zeta.$$

If $w := \frac{z - z_0}{\zeta - z_0}$, then $|w| = \frac{|z - z_0|}{r} < 1$. Thus

$$\frac{1}{1 - \frac{z - z_0}{\zeta - z_0}} = \frac{1}{1 - w} = 1 + w + w^2 + w^3 + \cdots + w^{n-1} + \frac{w^n}{1 - w}$$
$$= 1 + \frac{z - z_0}{\zeta - z_0} + \cdots + \frac{(z - z_0)^{n-1}}{(\zeta - z_0)^{n-1}} + \frac{(z - z_0)^n}{(\zeta - z_0)^{n-1}(\zeta - z)},$$

and so plugging this in the above, we have, with $R_n(z) := \frac{1}{2\pi i} \int_C \frac{f(\zeta)(z - z_0)^n}{(\zeta - z_0)^n(\zeta - z)} d\zeta$,

$$f(z) = \frac{1}{2\pi i} \int_C f(\zeta) \left(\frac{1}{\zeta - z_0} + \cdots + \frac{(z - z_0)^{n-1}}{(\zeta - z_0)^n} + \frac{(z - z_0)^n}{(\zeta - z_0)^n(\zeta - z)} \right) d\zeta$$
$$= c_0 + c_1(z - z_0) + \cdots + c_{n-1}(z - z_0)^{n-1} + R_n(z).$$

So we will be done if we show that $R_n(z)$ goes to 0 as $n \to \infty$. As $|f|$ is a continuous real-valued function on the compact set $\mathrm{ran}\,C$, i.e., there is an $M > 0$ such that for all $\zeta \in \mathrm{ran}\,C$, $|f(\zeta)| < M$. Also, for $\zeta \in \mathrm{ran}\,C$, $|\frac{(z - z_0)^n}{(\zeta - z_0)^n}| = (\frac{|z - z_0|}{r})^n \xrightarrow{n \to \infty} 0$. The term $\frac{1}{|\zeta - z|}$ for $\zeta \in \mathrm{ran}\,C$, is bounded by the reciprocal of the 'distance between the circle $\mathrm{ran}\,C$ and z'. We have $|\zeta - z| = |\zeta - z_0 - (z - z_0)| \geqslant |\zeta - z_0| - |z - z_0| = r - |z - z_0|$. Thus $|R_n(z)| \leqslant (\frac{|z - z_0|}{r})^n \frac{M}{r - |z - z_0|} \xrightarrow{n \to \infty} 0$. Hence $\sum_{n=0}^{\infty} c_n(z - z_0)^n$ converges to $f(z)$.

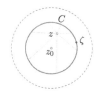

We have only shown the series expansion with $c_n = \frac{1}{2\pi i} \int_C \frac{f(\zeta)}{(\zeta - z_0)^{n+1}} d\zeta$ for r satisfying $|z - z_0| < r < R$, while the theorem statement said that we could have taken any $r \in (0, R)$. By the Cauchy Integral Theorem, $\int_C \frac{f(\zeta)}{(\zeta - z_0)^{n+1}} d\zeta$ is independent of r, and any value of $r \in (0, R)$ can be chosen here: Indeed, $\frac{f(\cdot)}{(\cdot - z_0)^{n+1}} \in \mathcal{O}(D(z_0, R) \backslash \{z_0\})$, and if, besides C, $\tilde{C}$ is another circular path with centre z_0 and some other radius $\tilde{r} \in (0, R)$ then $C, \tilde{C}$ are $D(z_0, R) \backslash \{z_0\}$-homotopic. $\quad\square$

From Corollary 4.6, whenever a power series $f(z) = \sum\limits_{n=0}^{\infty} c_n (z - z_0)^n$ is convergent for all $z \in D(z_0, R)$, we have $c_n = \frac{f^{(n)}(z_0)}{n!}$ for all integers $n \geqslant 0$.

Corollary 4.8 (Taylor Series). *If D is a domain, $f \in \mathcal{O}(D)$, $z_0 \in D$, and $R > 0$ is such that $D(z_0, R) \subset D$, then $f(z) = f(z_0) + \frac{f'(z_0)}{1!}(z - z_0) + \frac{f''(z_0)}{2!}(z - z_0)^2 + \cdots$ for all $z \in D(z_0, R)$. Also, for any $r \in (0, R)$, if $C(t) := z_0 + re^{2\pi i t}$ for all $t \in [0, 1]$, then $f^{(n)}(z_0) = \frac{n!}{2\pi i} \int_C \frac{f(z)}{(z - z_0)^{n+1}} dz$ for all integers $n \geqslant 0$.*

Proof. Theorem 4.7 implies that $f(z) = \sum\limits_{n=0}^{\infty} c_n (z - z_0)^n$ for all $z \in D(z_0, R)$, where $c_n = \frac{1}{2\pi i} \int_C \frac{f(z)}{(z - z_0)^{n+1}} dz$ for all $n \geqslant 0$, $C(t) := z_0 + re^{2\pi i t}$ for all $t \in [0, 1]$, and r is any number in $(0, R)$. By Corollary 4.6, for all $n \geqslant 0$, $\frac{f^{(n)}(z_0)}{n!} = c_n$. $\quad\square$

Corollary 4.9 (Cauchy Integral Formula). *Let D be a domain, $f \in \mathcal{O}(D)$, $z_0 \in D$, and $\Delta := \{z \in \mathbb{C} : |z - z_0| \leqslant R\} \subset D$. Let $C(t) = z_0 + Re^{2\pi i t}$ for all $t \in [0, 1]$. Then for all $z \in D(z_0, R)$, and all integers $n \geqslant 0$, $f^{(n)}(z) = \frac{n!}{2\pi i} \int_C \frac{f(\zeta)}{(\zeta - z)^{n+1}} d\zeta$.*

Proof. Let $z \in \Delta$. Then there exists a $r > 0$ small enough so that $D(z, r) \subset \Delta$. By Theorem 4.7, f has a power series expansion $f(w) = \sum\limits_{n=0}^{\infty} c_n (w - z)^n$ $(w \in D(z, r))$, where $c_n = \frac{1}{2\pi i} \int_{C_\rho} \frac{f(\zeta)}{(\zeta - z)^{n+1}} d\zeta = \frac{f^{(n)}(z)}{n!}$, $C_\rho(t) = z + \rho e^{2\pi i t}$ $(t \in [0, 1])$, and ρ is any number in $(0, r)$. But $\frac{f(\cdot)}{(\cdot - z)^{n+1}} \in \mathcal{O}(D \backslash \{z\})$ and C_ρ is $D \backslash \{z\}$-homotopic to C, and so by the Cauchy Integral Theorem, $\int_{C_\rho} \frac{f(\zeta)}{(\zeta - z)^{n+1}} d\zeta = \int_C \frac{f(\zeta)}{(\zeta - z)^{n+1}} d\zeta$. $\quad\square$

Example 4.4. $\exp \in \mathcal{O}(\mathbb{C})$. So $\exp z = \sum\limits_{n=0}^{\infty} \frac{\exp^{(n)}(0)}{n!} z^n$ for all $z \in \mathbb{C}$. As $\exp' z = \exp z$, $\exp^{(n)} = \exp$ $(n \geqslant 0)$ and $\exp^{(n)}(0) = \exp 0 = 1$. Thus $\exp z = \sum\limits_{n=0}^{\infty} \frac{z^n}{n!}$ $(z \in \mathbb{C})$. $\quad\diamond$

Example 4.5. $\text{Log} \in \mathcal{O}(D)$, where D denotes the 'plane with a cut' given by $\mathbb{C} \backslash ((-\infty, 0] \times \{0\})$. The largest open disc with centre $z_0 = 1$, which is contained in D, is $D(1, 1)$. Since $\text{Log}' z = \frac{1}{z}$ for $z \in D$, it follows that $\text{Log}^{(n)} z = \frac{(-1)^n (n-1)!}{z^n}$. In particular, $\text{Log}^{(n)} 1 = \frac{(-1)^n (n-1)!}{1^n}$. Also $\text{Log}\, 1 = 0$. Thus

$$\text{Log}\, z = \frac{z-1}{1} - \frac{(z-1)^2}{2} + \cdots + \frac{(-1)^n (z-1)^n}{n} + \cdots \text{ for all } z \in D(1, 1),$$

or equivalently, $\text{Log}(1 + w) = \frac{w}{1} - \frac{w^2}{2} + \cdots + \frac{(-1)^n w^n}{n} + \cdots$ for all $|w| < 1$. $\quad\diamond$

Exercise 4.21. Show that $\sin z = z - \frac{z^3}{3!} + \frac{z^5}{5!} - + \cdots$, $\cos z = 1 - \frac{z^2}{2!} + \frac{z^4}{4!} - + \cdots$ for all $z \in \mathbb{C}$.

Exercise 4.22. Find the Taylor series of the polynomial $z^6 - z^4 + z^2 - 1$ centred at 1.

Exercise 4.23. For the f below, find c_n, $n \geqslant 0$, in the Taylor series $\sum_{n=0}^{\infty} c_n z^n$ of f.

- $f(z) = \int_{[0,z]} e^{\zeta^2} d\zeta$ ($z \in \mathbb{C}$), where $[0, z]$ is the straight path from 0 to z. *Hint:* $f'(z) = e^{z^2}$.
- $f(z) = \frac{z^2}{(z+1)^2}$ ($z \in \mathbb{C} \backslash \{-1\}$). *Hint:* For $|z| < 1$, $\frac{1}{z+1} = 1 - z + z^2 - + \cdots$.

Exercise 4.24. Let $f \in \mathcal{O}(D(0,1))$ be such that $|f(z)| < 1$ for all $z \in D(0,1)$. Let $\sum_{n=0}^{\infty} c_n z^n$ be the Taylor series of f centred at 0. Prove that $|c_n| \leqslant 1$ for all $n \geqslant 0$.

Exercise 4.25. True or false? There exists an $r > 0$ and a function $f \in \mathcal{O}(D(0,r))$ such that $f^{(n)}(0) = (n!)^2$ for all $n \geqslant 0$.

Exercise 4.26. Determine all entire functions f which satisfy $f(z^2) = f(z)$ for all $z \in \mathbb{C}$.

Exercise 4.27. Let $a_0, a_1, \cdots$ be a sequence of real numbers tending to infinity, and let $f : \mathbb{C} \to \mathbb{C}$ be an entire function satisfying $|f^{(n)}(a_k)| \leqslant e^{-a_k}$ for all nonnegative integers k and n. Prove that $f(z) = c e^{-z}$ for some constant $c \in \mathbb{C}$ with $|c| \leqslant 1$.

Corollary 4.10 (Cauchy's inequality). *Let $z_0 \in \mathbb{C}$, $R > 0$, $f \in \mathcal{O}(D(z_0, R))$. Suppose there exists an $M > 0$ such that $|f(z)| \leqslant M$ for all $z \in D(z_0, R)$. Then for all $n \geqslant 0$, $|f^{(n)}(z_0)| \leqslant \frac{n! M}{R^n}$.*

Proof. Let $C(t) = z_0 + r e^{2\pi i t}$, $t \in [0,1]$, where $r \in (0, R)$. By Corollary 4.9,

$$|f^{(n)}(z_0)| = |\frac{n!}{2\pi i} \int_C \frac{f(z)}{(z-z_0)^{n+1}} dz| \leqslant \frac{n!}{2\pi} \max_{z \in C} |\frac{f(z)}{(z-z_0)^{n+1}}| 2\pi r = \frac{n!}{2\pi} \frac{M}{r^{n+1}} 2\pi r = \frac{n! M}{r^n}.$$

The claim now follows by passing to the limit as $r \nearrow R$. $\qquad \square$

Exercise 4.28. Let $f \in \mathcal{O}(\mathbb{C})$ such that there exists an $M > 0$ and an integer $n \geqslant 0$ such that for all $z \in \mathbb{C}$, $|f(z)| \leqslant M|z|^n$. Use Cauchy's inequality to prove $f^{(n+1)}(z) = 0$ for all $z \in \mathbb{C}$, and show that f is a polynomial of degree at most n. What happens when $n = 0$?

Exercise 4.29. Evaluate $\int_C \frac{\sin z}{z^{2023}} dz$, where C is the circular path with centre 0 and radius 1 traversed once in the anticlockwise direction.

4.4. Classification of zeroes

Let $f \in \mathcal{O}(D)$, where D is a domain. A point $z_0 \in D$ is a *zero* of f if $f(z_0) = 0$. Theorem 4.11 says that the zeroes of a nonzero f are 'isolated'. This need not happen with continuous functions (which are not as 'rigid' as holomorphic functions): Zeroes of nonzero continuous functions need not be isolated.

Example 4.6. Define $\varphi : \mathbb{R} \to \mathbb{R}$ by $\varphi(x) = x \sin \frac{1}{x}$ if $x \neq 0$, and $\varphi(0) = 0$. Then φ is continuous at any $x \neq 0$. Moreover, $|\varphi(x) - \varphi(0)| = |x \sin \frac{1}{x} - 0| \leqslant |x| \cdot 1 = |x - 0|$ for all $x \neq 0$, and so φ is continuous at 0. For all $n \in \mathbb{N}$, $\varphi(\frac{1}{n\pi}) = \frac{1}{n\pi} \sin(n\pi) = 0$, while $\varphi(\frac{2}{(4n+1)\pi}) = \frac{2}{(4n+1)\pi} \sin(2n\pi + \frac{\pi}{2}) = \frac{2}{(4n+1)\pi} 1 = \frac{2}{(4n+1)\pi}$. Define $f : \mathbb{C} \to \mathbb{C}$ by $f(z) = \varphi(\text{Re } z) + i\varphi(\text{Im } z)$ for all $z \in \mathbb{C}$. Then f is continuous. Since we have $f(0) = \varphi(0) + i\varphi(0) = 0$, 0 is a zero of f. Also, 0 is not an isolated zero of f, since $f(\frac{1}{n\pi}) = \varphi(\frac{1}{n\pi}) + i\varphi(0) = 0$ for all $n \in \mathbb{N}$. Finally, f is not identically zero in any disc centred at 0: For all $n \in \mathbb{N}$, $f(\frac{2}{(4n+1)\pi}) = \varphi(\frac{2}{(4n+1)\pi}) + i\varphi(0) = \frac{2}{(4n+1)\pi} \neq 0$. $\diamond$

Example 4.7. exp has no zeroes in $\mathbb{C}$. Indeed, $|\exp z| = e^{\operatorname{Re} z} > 0$ for all $z \in \mathbb{C}$.
$\cos z - 3$ has infinitely many zeroes in $\mathbb{C}$ at $2\pi n \pm i \log(3 + 2\sqrt{2})$, $n \in \mathbb{Z}$, all of which lie on a horizontal line, and they are isolated, with a distance of 2π between any two isolated adjacent zeroes. We had seen this in Exercise 1.58.

The polynomial p, $p(z) = (z+1)^3 z^9 (z-1)^9$, has zeroes at $-1, 0, 1$. $\diamond$

If p is a nonzero polynomial such that $p(z_0) = 0$, then by the Division Algorithm there exists a polynomial q (the quotient) such that $p(z) = (z - z_0)q(z)$ (that is, the remainder is 0). Now we have two possible cases:

1° $q(z_0) \neq 0$. Then z_0 is an isolated zero of p.

2° $q(z_0) = 0$. Then we repeat the above procedure with p replaced by q.

Eventually, we obtain $p(z) = (z - z_0)^m q(z)$ for some $m \geqslant 1$ and $q(z_0) \neq 0$. Then m is called the *order of z_0* (as a zero of p). Note that $0 = p'(z_0) = \cdots = p^{(m-1)}(z_0)$ and $p^{(m)}(z_0) \neq 0$. Theorem 4.11 below says that the same thing holds for holomorphic functions f (replacing the polynomial p), except that we end up with another holomorphic function g (instead of the polynomial q). This is not completely surprising, since power series are analogues of polynomials, and every holomorphic function has a local power series expansion.

Let D be a domain and $f \in \mathcal{O}(D)$. We say $z_0 \in D$ is a *zero of f of order $m \in \mathbb{N}$* if $0 = f(z_0) = f'(z_0) = \cdots = f^{(m-1)}(z_0)$, but $f^{(m)}(z_0) \neq 0$.

Theorem 4.11 (Classification of zeroes). *Let D be a domain, $f \in \mathcal{O}(D)$, and $z_0 \in D$ be a zero of f. Then exactly one of the following hold:*

1° $f \equiv 0$.

2° *There exists an $m \in \mathbb{N}$ such that z_0 is a zero of f of order m, and there exists a $g \in \mathcal{O}(D)$ such that $g(z_0) \neq 0$ and $f(z) = (z - z_0)^m g(z)$ for all $z \in D$.*

In 2°, as g is continuous and as $g(z_0) \neq 0$, there is a small $r > 0$ such that g is never zero in $D(z_0, r)$, and so $f(z) = (z - z_0)^m g(z) \neq 0$ for all $z \in D(z_0, r) \backslash \{z_0\}$. Thus z_0 is the *only* zero of f in $D(z_0, r)$, that is, z_0 is isolated.

Proof. First we will just show the theorem replacing 1° by the statement that there is an $R > 0$ such that $f(z) = 0$ for all $z \in D(z_0, R)$. Later, after showing the Identity Theorem, we will prove the more general statement.

Let $R > 0$ be such that $D(z_0, R) \subset D$. Then f has a Taylor series expansion, $f(z) = c_0 + c_1(z - z_0) + c_2(z - z_0)^2 + \cdots$ for $z \in D(z_0, R)$. Since $f(z_0) = 0$, $c_0 = 0$. Then exactly one of the following hold:

1° All the c_n are zero. Then $f(z) = 0$ for all $z \in D(z_0, R)$.

2° There is a first integer $m \geqslant 1$ such that $c_m \neq 0$. Then $c_0 = c_1 = \cdots = c_{m-1} = 0$. As $c_n = \frac{f^{(n)}(z_0)}{n!}$ for all integers $n \geqslant 0$, z_0 is a zero of order m. For $z \in D(z_0, R)$,

$$f(z) = c_m(z - z_0)^m + c_{m+1}(z - z_0)^{m+1} + \cdots = (z - z_0)^m \sum_{k=0}^{\infty} c_{m+k}(z - z_0)^k. \quad (*)$$

Define $g : D \to \mathbb{C}$ by $g(z) = \begin{cases} \frac{f(z)}{(z-z_0)^m} & \text{for } z \in D\backslash\{z_0\}, \\ \sum\limits_{k=0}^{\infty} c_{m+k}(z-z_0)^k & \text{for } z \in D(z_0, R). \end{cases}$

From $(*)$, the two definitions give the same value whenever both are applicable, and so g is well-defined. Moreover, we have:

- $g \in \mathcal{O}(D)$. For $z \in D\backslash\{z_0\}$, $g = \frac{f}{(\cdot - z_0)^m}$, and since $f, \frac{1}{(\cdot - z_0)^m} \in \mathcal{O}(D\backslash\{z_0\})$, we have $g \in \mathcal{O}(D\backslash\{z_0\})$. For $z \in D(z_0, R)$, g is given by a power series, and so g is holomorphic in $D(z_0, R)$.

- $g(z_0) = c_m \neq 0$, by the definition of m.

- $f(z) = (z - z_0)^m g(z)$ for $z \in D\backslash\{z_0\}$ follows from the definition of g. If $z = z_0$, then both sides are 0. So for all $z \in D$, $f(z) = (z - z_0)^m g(z)$. $\quad\square$

Clearly, if $f \in \mathcal{O}(D)$, where D is a domain, and $f \equiv 0$ in a neighbourhood of its zero z_0, then $f^{(n)}(z_0) = 0$ for all $n \geqslant 0$. So z_0 does not have a finite order.

Exercise 4.30. Let $z_0 \in \mathbb{C}$, $R > 0$, and $f, g \in \mathcal{O}(D(z_0, R))$ be such that $g(z_0) \neq 0$, and there exists an $m \in \mathbb{N}$ such that for all $z \in D(z_0, R)$, $f(z) = (z - z_0)^m g(z)$. Prove that z_0 is a zero of f of order m.

Example 4.8. For each $n \in \mathbb{Z}$, $n\pi$ is a zero of $\sin \in \mathcal{O}(\mathbb{C})$. Also, $\sin$ is not identically zero in any neighbourhood of $n\pi$ (consider $\sin|_{\mathbb{R}}$). As $\sin' z = \cos z$ and $\cos z|_{z=n\pi} = (-1)^n \neq 0$, $n\pi$ is a zero of $\sin z$ of order 1. So for each $n \in \mathbb{Z}$, there exists a $g_n \in \mathcal{O}(\mathbb{C})$ such that $g(n\pi) \neq 0$ and for all $z \in \mathbb{C}$, $\sin z = (z - n\pi)g_n(z)$. [1]

$\exp(z^3) - 1$ has a zero at 0 since $\exp(0^3) - 1 = 1 - 1 = 0$. As $\exp(z^3) = 1 + \frac{z^3}{1!} + \frac{z^6}{2!} + \cdots$ for all $z \in \mathbb{C}$, $\exp(z^3) - 1 = z^3 g(z)$, where g is defined by $g(z) := \frac{1}{1!} + \frac{z^3}{2!} + \cdots$. As g is a power series that converges in $\mathbb{C}$, g is entire. Also, $g(0) = 1 \neq 0$. Thus the order of 0 as a zero of $\exp(z^3) - 1$ is 3 by Exercise 4.30. $\quad\diamond$

Exercise 4.31. Find the order of the zero z_0 for the function f in each case:
- $z_0 = i$ and $f(z) = (1 + z^2)^4$.
- $z_0 = 2n\pi i$, where n is an integer, and $f(z) = \exp z - 1$.
- $z_0 = 0$ and $f(z) = \cos z - 1 + \frac{1}{2}(\sin z)^2$.

Exercise 4.32. Let f be holomorphic in a disc that contains a circle C in its interior. Suppose that in the disc, there is exactly one zero z_0 of order 1 of f, which lies in the interior of C. Prove that $z_0 = \frac{1}{2\pi i}\int_C \frac{zf'(z)}{f(z)}dz$.

Exercise 4.33. Let D be a domain and $f \in \mathcal{O}(D)$ have a zero of order m at $z_0 \in D$. Prove that z_0 is a zero of $(f(\cdot))^2$ of order $2m$, and if $m > 1$, then z_0 is a zero of f' of order $m-1$.

Exercise 4.34. Let f be an entire function having z_0 as a zero of order 1. Define g by $g(z) = \cos(f(z)) - 1$ for all $z \in \mathbb{C}$. Show that z_0 is a zero of g with order 2.

Exercise 4.35. Let $f \in \mathcal{O}(D(0,1))$ satisfy $|f(\frac{1}{\log n})| \leqslant \frac{1}{n}$ for all $n = 2, 3, \cdots$. Show that f must be identically 0. *Hint:* First show $f(0) = 0$.

---

[1] In fact, $\sin z = z \prod\limits_{n=1}^{\infty}(1 - \frac{z^2}{n^2\pi^2})$ for all $z \in \mathbb{C}$. This can be shown using the 'Weierstrass Factorisation Theorem; see e.g. [**3**, Chap. VII, §5 and §6].

4.5. The Identity Theorem

In this section, we will learn the Identity Theorem, which once again highlights the rigidity of holomorphic functions. It says roughly that if a holomorphic function has an accumulation of zeroes in its domain, then it is identically 0.

Theorem 4.12 (Identity Theorem).
Let D be a domain, $f \in \mathcal{O}(D)$, $(z_n)_{n \in \mathbb{N}}$ be a sequence of distinct zeroes of f which converges to $z_* \in D$. Then f is identically zero in D.

Proof. First note that z_* is itself a zero of f because by the continuity of f, we have $f(z_*) = f(\lim_{n \to \infty} z_n) = \lim_{n \to \infty} f(z_n) = \lim_{n \to \infty} 0 = 0$. Moreover, as $(z_n)_{n \in \mathbb{N}}$ is a sequence of distinct zeroes of f converging to z_*, it is clear that z_* is not an isolated zero of f. So $f \equiv 0$ in some disc $D(z_*, r)$, where $r > 0$. We want to show that $f \equiv 0$ in D. Let $w \in D$. Then there is a path $\gamma : [0, 1] \to D$ that joins z_* to w: $\gamma(0) = z_*$ and $\gamma(1) = w$. Let $S := \{t \in [0, 1] : f(\gamma(\tau)) = 0 \text{ for } 0 \leqslant \tau \leqslant t\}$, and $T := \sup S$ (which[2] exists: $0 \in S \neq \varnothing$, and S is bounded above by 1). Also, $T > 0$ since $f \circ \gamma : [0, 1] \to D$ is continuous, and 0 belongs to the open subset $(f \circ \gamma)^{-1}(D(z_*, R))$ of $[0, 1]$. Let $(t_n)_{n \in \mathbb{N}}$ be[3] a sequence in S such that $\lim_{n \to \infty} t_n = T$. There are two possible cases:

1° $T = 1$. By continuity, $f(w) = f(\gamma(T)) = \lim_{n \to \infty} f(\gamma(t_n)) = \lim_{n \to \infty} 0 = 0$.

2° $T \in (0, 1)$. Then $f(\gamma(T)) = \lim_{n \to \infty} f(\gamma(t_n)) = 0$. So $\gamma(T)$ is a zero of f.

But $\gamma(T)$ is not an isolated zero[4]. So there is a $\rho > 0$ so that $f \equiv 0$ on $D(\gamma(T), \rho)$. Then as T belongs to the open set $\gamma^{-1}(D(\gamma(T), \rho)) \cap (0, 1)$, there is an $\epsilon > 0$ so that $T + \Delta t \in \gamma^{-1}(D(\gamma(T), \rho)) \cap (0, 1)$ for all $\Delta t < \epsilon$. But this means $T + \frac{\epsilon}{2} \in S$, contradicting $T = \sup S$. So this case is impossible.

Hence 1° holds, and as w was arbitrary, $f \equiv 0$ in D. $\qquad\square$

Example 4.9. Using the definitions of $\cos, \sin : \mathbb{C} \to \mathbb{C}$, we had shown that for all $z \in \mathbb{C}$, $(\cos z)^2 + (\sin z)^2 = 1$. Here is a different proof. Define $f \in \mathcal{O}(\mathbb{C})$ by $f(z) = (\cos z)^2 + (\sin z)^2 - 1$ for all $z \in \mathbb{C}$. By Pythagoras' Theorem, for all $x \in \mathbb{R}$, $f(x) = (\cos x)^2 + (\sin x)^2 - 1 = 0$. The Identity Theorem now implies that $f \equiv 0$ in $\mathbb{C}$. $\qquad\diamond$

---

[2]If we think of t as time while we travel from z_* (time $t = 0$) to w (time $t = 1$) along γ, then T is the largest time such that f has been 0 along the path covered so far.

[3]This can be constructed by considering $T - \frac{T}{n} < T$, $n \in \mathbb{N}$, which while failing to be upper bounds of S, guarantee the existence of $t_n \in S$ so that $T - \frac{T}{n} < t_n \leqslant T$, and by using the Sandwich Theorem.

[4]Otherwise, there is a $\delta' > 0$ so that f is never zero in $D(\gamma(T), \delta') \backslash \{\gamma(T)\}$. This is a contradiction, since for all n large enough, $|\gamma(t_n) - \gamma(T)| < \delta'$, i.e., $\gamma(t_n) \in D(\gamma(T), \delta')$ and $f(\gamma(t_n)) = 0$.

The following is equivalent to the Identity Theorem.

Theorem 4.13 (Identity Theorem). *Let D be a domain, $f, g \in \mathcal{O}(D)$, $(z_n)_{n \in \mathbb{N}}$ be a sequence of distinct points in D which converges to $z_* \in D$, and such that for all $n \in \mathbb{N}$, $f(z_n) = g(z_n)$. Then $f(z) = g(z)$ for all $z \in D$.*

Proof. Define $h : D \to \mathbb{C}$ by $h(z) = f(z) - g(z)$, $z \in D$, and note that the z_n, $n \in \mathbb{N}$, are zeroes of the holomorphic function h. By Theorem 4.12, h must be identically zero in D, and so the claim follows.   □

Vice versa, Theorem 4.12 follows from the above by taking $g \equiv 0$.

Example 4.10. Recall that we defined $\exp : \mathbb{C} \to \mathbb{C}$ by $\exp z := e^x(\cos y + i \sin y)$ for $z = x + iy \in \mathbb{C}$, where $x, y \in \mathbb{R}$. Then $\exp$ is an entire function such that $\exp x = e^x$ for $x \in \mathbb{R}$. In other words, $\exp$ is an entire extension of the usual real exponential function. Is there any other entire extension possible? We show that the answer is no! Suppose $g : \mathbb{C} \to \mathbb{C}$ is entire and $g(x) = e^x$ for all real x. But then $\exp x = g(x)$ for all $x \in \mathbb{R}$. In particular, $\exp \frac{1}{n} = g(\frac{1}{n})$, for all $n \in \mathbb{N}$, and $\frac{1}{n} \to 0 \in \mathbb{C}$ as $n \to \infty$. By the Identity Theorem, $\exp z = g(z)$ for all $z \in \mathbb{C}$. So there is *only one* entire function whose restriction to $\mathbb{R}$ is e^x. This explains the naturalness of the definition of the complex exponential in Section 1.4.1.   ◇

We can now also complete the proof of the result on the classification of zeroes (Theorem 4.11). We had shown the theorem replacing 1° by the statement that there is an $R > 0$ such that $f(z) = 0$ for all $z \in D(z_0, R)$. But from the Identity Theorem, $f \equiv 0$ in D e.g. by taking $z_n = z_0 + \frac{R}{2n}$ for all $n \in \mathbb{N}$.

Exercise 4.36. Use the Identity Theorem to show $\cos(z+w) = (\cos z)(\cos w) - (\sin z)(\sin w)$ for all $z, w \in \mathbb{C}$, by appealing to the corresponding identity when z, w are real numbers.

Exercise 4.37. Let D be a domain. Then it is easy to check that $\mathcal{O}(D)$ is a commutative ring[5] with the pointwise operations: $(f + g)(z) = f(z) + g(z)$ and $(f \cdot g)(z) = f(z)g(z)$ for all $z \in D$ and all $f, g \in \mathcal{O}(D)$. Check that $\mathcal{O}(D)$ is an *integral domain*, that is, a nonzero ring having no zero divisors (if $f \cdot g = 0$ for $f, g \in \mathcal{O}(D)$, then either $f = 0$ or $g = 0$).
If instead of $\mathcal{O}(D)$, we consider the set $C(D)$ of all complex-valued *continuous* functions on D, then $C(D)$ is again a commutative ring with pointwise operations. Is $C(D)$ an integral domain? This shows that continuous functions are not as 'rigid' as holomorphic functions.

Exercise 4.38. Does there exist a nonzero entire function with an uncountable set of zeroes? *Hint:* $\mathbb{C} = \bigcup_{n \in \mathbb{N}} D(0, n)$. A countable union of finite sets is finite or countable.

Exercise 4.39. Let $\mathbb{D} := \{z \in \mathbb{C} : |z| < 1\}$, and $f \in \mathcal{O}(\mathbb{D})$ be such that f has at least one zero in $\mathbb{D} \backslash \{0\}$, and $f(z)f(-z) = f(z^2)$ for all $z \in \mathbb{D}$. Prove that if $z_0 \in \mathbb{D} \backslash \{0\}$ is a zero of f, then z_0^2 is a *new* zero of f in $\mathbb{D} \backslash \{0\}$. Determine all possible such f. *Hint:* Construct a sequence of zeros and invoke the identity theorem.

---

[5]A *commutative ring* R, is a set R with two laws of composition $+$ and $\cdot$ such that $(R, +)$ is an Abelian group, $\cdot$ is associative, commutative and has an identity, and the distributive law holds: for $a, b, c \in R$, $(a + b) \cdot c = a \cdot c + b \cdot c$.

Exercise 4.40. Let D be a domain and $f, g \in \mathcal{O}(D)$. Which of the following conditions imply $f = g$ identically in D?
(1) There is a sequence $(z_n)_{n \in \mathbb{N}}$ of distinct terms in D such that $f(z_n) = g(z_n)$ for all $n \in \mathbb{N}$.
(2) There is a convergent sequence $(z_n)_{n \in \mathbb{N}}$ of distinct points in D with its limit in D such that $f(z_n) = g(z_n)$ for all $n \in \mathbb{N}$.
(3) γ is a smooth path in D joining distinct points $z, w \in D$ and $f = g$ on γ.
(4) $w \in D$ is such that $f^{(n)}(w) = g^{(n)}(w)$ for all $n \geqslant 0$.

Exercise 4.41. Let $f : (0, \infty) \to \mathbb{R}$ be given by $f(x) = \sin(\log x)$ for all $x > 0$. Find all $x > 0$ such that $f(x) = 0$. Does there exist an entire function F which extends f?

Exercise 4.42. Suppose $r > 0$, and that $g \in \mathcal{O}(D(0, r))$ is such that $\sum\limits_{n=0}^{\infty} g^{(n)}(0)$ converges. Prove that g has a unique entire extension.

Exercise 4.43. Let the entire f be such that in every power series (i.e., for every $z_0 \in \mathbb{C}$) $f(z) = \sum\limits_{n=0}^{\infty} c_n (z - z_0)^n$, at least one coefficient is 0. Prove that f is a polynomial.

Exercise 4.44. Let $\mathbb{D} := D(0, 1)$, and $f, g \in \mathcal{O}(\mathbb{D})$ be nonzero everywhere in $\mathbb{D}$, and such that $\frac{f'(\frac{1}{n})}{f(\frac{1}{n})} = \frac{g'(\frac{1}{n})}{g(\frac{1}{n})}$ for all integers $n \geqslant 2$. Prove that $f = \lambda g$ for some $\lambda \in \mathbb{C} \backslash \{0\}$.

Exercise 4.45. Find all entire f satisfying $f(\frac{1}{n}) = \frac{1}{n^2}$ for all $n \in \mathbb{N}$. *Hint:* $g(z) := f(z) - z^2$. Show that there is no entire function f satisfying $f(\frac{1}{n^2}) = \frac{1}{n}$ for all $n \in \mathbb{N}$.

Exercise 4.46 (Argument Principle). Let D be a domain, $f \in \mathcal{O}(D)$, $z_0 \in D$, $R > 0$ be such that $\{z \in \mathbb{C} : |z - z_0| \leqslant R\} \subset D$, $C(t) := z_0 + Re^{2\pi i t}$, $t \in [0, 1]$, and $f(C(t)) \neq 0$ for all $t \in [0, 1]$. Let $Z(f) := \{z \in D(z_0, R) : f(z) = 0\}$ and $m(z)$ be the order of the zero $z \in Z(f)$ of f. Then $\frac{1}{2\pi i} \int_C \frac{f'(z)}{f(z)} dz = \sum\limits_{z \in Z(f)} m(z)$.

4.6. The Maximum Modulus Theorem

We now prove the Maximum Modulus Theorem, saying that for a nonconstant $f \in \mathcal{O}(D)$, $|f|$ cannot have a maximiser in the domain D.

Theorem 4.14 (Maximum Modulus Theorem). *Let D be a domain, $f \in \mathcal{O}(D)$, $z_0 \in D$ be such that for all $z \in D$, $|f(z_0)| \geqslant |f(z)|$. Then f is constant on D.*

Proof. Let $r > 0$ be such that $D(z_0, 2r) \subset D$. Define C_r by $C_r(t) = z_0 + re^{2\pi i t}$ for all $t \in [0, 1]$. By the Cauchy Integral Formula,

$$f(z_0) = \frac{1}{2\pi i} \int_{C_r} \frac{f(z)}{z - z_0} dz = \frac{1}{2\pi i} \int_0^1 \frac{f(z_0 + re^{2\pi i t})}{re^{2\pi i t}} 2\pi i re^{2\pi i t} dt = \int_0^1 f(z_0 + re^{2\pi i t}) dt.$$

(The last expression can be viewed as the 'average' of the values of f on C_r.) Thus $\int_0^1 |f(z_0)| dt = |f(z_0)| = |\int_0^1 f(z_0 + re^{2\pi i t}) dt| \leqslant \int_0^1 |f(z_0 + re^{2\pi i t})| dt$. Rearranging, $\int_0^1 (|f(z_0 + re^{2\pi i t})| - |f(z_0)|) dt \geqslant 0$. Since $|f(z_0 + re^{2\pi i t})| \leqslant |f(z_0)|$ for all $t \in [0, 1]$, the integrand is pointwise $\leqslant 0$. So $-\int_0^1 (|f(z_0)| - |f(z_0 + re^{2\pi i t})|) dt = 0$. As the integrand is continuous and pointwise nonnegative, the above implies that $|f(z_0 + re^{2\pi i t})| = |f(z_0)|$ for all $t \in [0, 1]$. By replacing r by any smaller number, the same conclusion holds. Thus f maps the disc $\Delta := \{z \in \mathbb{C} : |z - z_0| \leqslant r\}$ into the circle $\{w \in \mathbb{C} : |w| = |f(z_0)|\}$. By Example 2.10, f is constant on Δ. The Identity Theorem shows that f is constant on D. $\qquad\square$

Example 4.11. Let $\mathbb{H} = \{z \in \mathbb{C} : \operatorname{Re} z \geqslant 0\}$. Define $f : \mathbb{H} \to \mathbb{C}$ by $f(z) = \frac{e^{-z}}{z+1}$, $z \in \mathbb{H}$. It can be shown $\|f\|_\infty := \max\{|f(z)| : z \in \mathbb{H}\}$ exists. Without worrying about the existence of this maximum, let us see how, assuming its existence, the Maximum Modulus Theorem enables us to calculate its value. Suppose $z_0 \in \mathbb{H}$ is a maximiser of $|f|$ on $\mathbb{H}$. If $\operatorname{Re} z_0 > 0$, then z_0 is a maximiser of $|f|$ on the open right-half plane $\mathbb{H}^\circ := \{z \in \mathbb{C} : \operatorname{Re} z > 0\}$, and since f is not constant (e.g. $f(1) = \frac{1}{2e} \neq \frac{1}{3e^2} = f(2)$), we get a contradiction to the Maximum Modulus Theorem. So $z_0 \in i\mathbb{R}$, that is, $z_0 = iy_0$ for some $y_0 \in \mathbb{R}$. But $|f(iy)| = |\frac{e^{-iy}}{iy+1}| = \frac{1}{\sqrt{y^2+1}} \leqslant \frac{1}{\sqrt{0^2+1}}$ for all $y \in \mathbb{R}$. Thus

$$\|f\|_\infty = \max_{z \in \mathbb{H}} |f(z)| = \max_{y \in \mathbb{R}} |f(iy)| = \max_{y \in \mathbb{R}} \frac{1}{\sqrt{y^2+1}} = \frac{1}{\sqrt{0^2+1}} = 1.$$ ◇

Exercise 4.47. Let D be a domain and let $f \in \mathcal{O}(D)$ be a nonconstant map. Prove that there is no maximiser of $|f|$ in D.

Exercise 4.48 (Minimum Modulus Theorem). Let D be a domain and let $f \in \mathcal{O}(D)$ be such that there is a $z_0 \in D$ such that for all $z \in D$, $|f(z)| \geqslant |f(z_0)|$. Then prove that $f(z_0) = 0$ or f is constant on D.

Exercise 4.49. Consider the function f defined by $f(z) = z^2 - 2$. Find the maximum and minimum value of $|f(z)|$ on $\{z \in \mathbb{C} : |z| \leqslant 1\}$.

Exercise 4.50. Prove or disprove: There exists a $f \in \mathcal{O}(D(0,1))$ such that $|f(z)| = e^{|z|}$ for all $z \in D(0,1)$.

4.7. Laurent series

Laurent series generalise Taylor series. Indeed, while a Taylor series

$$\sum_{n=0}^{\infty} c_n (z - z_0)^n = c_0 + c_1 (z - z_0)^1 + c_2 (z - z_0)^2 + \cdots$$

has nonnegative powers of the term $z - z_0$, and converges in a *disc*, a *Laurent series* is an expression of the following type, which has negative powers of $z - z_0$ too:

$$\sum_{n \in \mathbb{Z}} c_n (z - z_0)^n = \cdots + c_{-2}(z - z_0)^{-2} + c_{-1}(z - z_0)^{-1} + c_0 + c_1(z - z_0)^1 + c_2(z - z_0)^2 + \cdots,$$

We will see that a Laurent series 'converges' in an annulus $\{z \in \mathbb{C} : r < |z - z_0| < R\}$ with centre z_0 and gives a holomorphic function there, and conversely, if we have a holomorphic function in an annulus with centre z_0, then the function has a Laurent series expansion in the annulus. E.g., for all $z \in \mathbb{C}$, $\exp z = 1 + \frac{z}{1!} + \frac{z^2}{2!} + \frac{z^3}{3!} + \cdots$, and so for $z \neq 0$, we have the 'Laurent series expansion' $\exp \frac{1}{z} = 1 + \frac{1}{z} + \frac{1}{2!} \frac{1}{z^2} + \frac{1}{3!} \frac{1}{z^3} + \cdots$. Note that $\exp \frac{1}{z} \in \mathcal{O}(\mathbb{C} \backslash \{0\})$, which is a degenerate annulus centred at 0 with inner radius $r = 0$ and outer radius $R = +\infty$. Let us first define what we mean by the convergence of $\sum_{n \in \mathbb{Z}} c_n (z - z_0)^n$.

Definition 4.1. Let $\mathbb{Z} \ni n \mapsto c_n \in \mathbb{C}$ be a function. We say that the *Laurent series* $\sum_{n \in \mathbb{Z}} c_n(z - z_0)^n$ *converges* (for z) if both $\sum_{n=1}^{\infty} c_{-n}(z - z_0)^{-n}$, $\sum_{n=0}^{\infty} c_n(z - z_0)^n$ converge.
If $\sum_{n \in \mathbb{Z}} c_n(z-z_0)^n$ converges, we write $\sum_{n \in \mathbb{Z}} c_n(z-z_0)^n = \sum_{n=1}^{\infty} c_{-n}(z - z_0)^{-n} + \sum_{n=0}^{\infty} c_n(z - z_0)^n$,
and call this the *sum* of the Laurent series.

Example 4.12. Let us find out the set of complex numbers z for which the Laurent series $\cdots + \frac{1}{8z^3} + \frac{1}{4z^2} + \frac{1}{2z} + 1 + z + z^2 + z^3 + \cdots$ converges. We have:

- $1 + z + z^2 + z^3 + \cdots$ converges for $|z| < 1$, and it diverges for $|z| \geqslant 1$.
- $\frac{1}{2z} + \frac{1}{4z^3} + \frac{1}{8z^5} + \cdots$ converges for $|\frac{1}{2z}| < 1$ and diverges for $|\frac{1}{2z}| \geqslant 1$, that is, it converges for $|z| > \frac{1}{2}$ and diverges for $|z| \leqslant \frac{1}{2}$.

So the Laurent series converges when $|z| < 1$ and $|z| > \frac{1}{2}$, i.e., inside the annulus $\mathbb{A} := \{z \in \mathbb{C} : \frac{1}{2} < |z| < 1\}$, and diverges when $|z| \geqslant 1$ or $|z| \leqslant \frac{1}{2}$, i.e., outside $\mathbb{A}$. $\Diamond$

For which z does $\sum_{n \in \mathbb{Z}} c_n(z - z_0)^n$ converge?

- From Theorem 4.1, there is some nonnegative R (possibly $R = +\infty$) such that $\sum_{n=0}^{\infty} c_n(z - z_0)^n$ converges for $|z - z_0| < R$ and diverges for $|z - z_0| > R$.

- What about the series $\sum_{n=1}^{\infty} c_{-n}(z - z_0)^{-n}$? Again by Theorem 4.1, there is some $\widetilde{R} \geqslant 0$ such that the power series $\sum_{n=1}^{\infty} c_{-n}w^n$ converges for $|w| < \widetilde{R}$ and diverges for $|w| > \widetilde{R}$. For $z \neq z_0$, set $w := (z - z_0)^{-1}$. Then $\sum_{n=1}^{\infty} c_{-n}(z - z_0)^{-n}$ converges for $\frac{1}{|z-z_0|} < \widetilde{R}$, i.e. for $|z - z_0| > \frac{1}{\widetilde{R}} =: r$, and diverges for $|z - z_0| < r$.

Hence the Laurent series converges in the annulus $\mathbb{A} := \{z \in \mathbb{C} : r < |z - z_0| < R\}$ and diverges if either $|z - z_0| < r$ or $|z - z_0| > R$.

Is $\sum_{n \in \mathbb{Z}} c_n(z - z_0)^n$ holomorphic in the annulus $\mathbb{A}$ where it converges?

- $\sum_{n=0}^{\infty} c_n(\cdot - z_0)^n \in \mathcal{O}(D(z_0, R))$ and so also $\sum_{n=0}^{\infty} c_n(\cdot - z_0)^n \in \mathcal{O}(\mathbb{A})$.

- $w \xmapsto{g} \sum_{n=1}^{\infty} c_{-n}w^n$ is holomorphic in $\{w \in \mathbb{C} : |w| < \widetilde{R}\}$. Also $f = \frac{1}{\cdot - z_0} \in \mathcal{O}(\mathbb{C} \backslash \{z_0\})$. So $g \circ f$ is holomorphic in $\{z \in \mathbb{C} : |z - z_0| > r\}$, i.e., $\sum_{n=1}^{\infty} c_{-n}(\cdot - z_0)^{-n}$ is holomorphic in $\{z \in \mathbb{C} : r < |z - z_0|\}$, and so also in the annulus $\mathbb{A}$.

Hence the sum $z \mapsto \sum_{n=0}^{\infty} c_n(z - z_0)^n + \sum_{n=1}^{\infty} c_{-n}(z - z_0)^{-n} = \sum_{n \in \mathbb{Z}} c_n(z-z_0)^n$ is holomorphic in the annulus $\{z \in \mathbb{C} : r < |z - z_0| < R\}$.

Summarising, a Laurent series $\sum_{n \in \mathbb{Z}} c_n(z - z_0)^n$ converges in a possibly empty annulus $\mathbb{A} = \{z \in \mathbb{C} : r < |z - z_0| < R\}$ for some r, R, and $\sum_{n \in \mathbb{Z}} c_n(\cdot - z_0)^n \in \mathcal{O}(\mathbb{A})$.

Conversely, a function holomorphic in an annulus has a Laurent series expansion.

Theorem 4.15. *Let* $z_0 \in \mathbb{C}$, $0 \leqslant r < R \leqslant \infty$, $\mathbb{A} = \{z \in \mathbb{C} : r < |z - z_0| < R\}$, $f \in \mathcal{O}(\mathbb{A})$. *For any* $\rho \in (r, R)$, *define the circular path* $C(t) := z_0 + \rho\, e^{2\pi i t}$ *for all* $t \in [0, 1]$. *Let* $c_n := \frac{1}{2\pi i} \int_C \frac{f(\zeta)}{(\zeta - z_0)^{n+1}} d\zeta$ $(n \in \mathbb{Z})$. *Then* $f(z) = \sum_{n \in \mathbb{Z}} c_n (z - z_0)^n$ *for all* $z \in \mathbb{A}$. *Moreover, the coefficients are unique in such a Laurent series expansion of f in $\mathbb{A}$.*

Example 4.13. Let $f(z) := z^3 \exp \frac{1}{z}$ for all $z \neq 0$. Then f is holomorphic in the annulus $\mathbb{A} := \mathbb{C} \backslash \{0\} = \{z \in \mathbb{C} : 0 < |z| < +\infty\}$. By Theorem 4.15, f has a Laurent series expansion $\sum_{n \in \mathbb{Z}} c_n z^n$ in $\mathbb{C} \backslash \{0\}$. For $z \neq 0$, since we have $\exp \frac{1}{z} = 1 + \frac{1}{z} + \frac{1}{2!z^2} + \cdots$, we obtain $f(z) = z^3 \exp \frac{1}{z} = z^3 + z^2 + \frac{z}{2!} + \frac{1}{3!} + \frac{1}{4!z} + \cdots$. $\diamond$

Proof of Theorem 4.15. (Existence.) Fix $z \in \mathbb{A}$. Suppose that $\tilde{r}$ and $\tilde{R}$ are such that $r < \tilde{r} < |z - z_0| < \tilde{R} < R$. Define circular paths $C_{\tilde{r}}$ and $C_{\tilde{R}}$ by $C_{\tilde{r}}(t) = z_0 + \tilde{r}e^{it}$ and $C_{\tilde{R}}(t) = z_0 + \tilde{R}e^{it}$ for all $t \in [\theta, 2\pi + \theta]$, and $\theta := (\text{Arg}\, z) + \frac{\pi}{2}$. Let $\sigma : [\tilde{r}, \tilde{R}] \to \mathbb{A}$ be the path $\sigma(t) = ti\frac{z - z_0}{|z - z_0|}$. (This is a straight line path joining $C_{\tilde{r}}$ and $C_{\tilde{R}}$, and the multiplication by i produces a rotation by $90°$, ensuring that σ avoids z.)

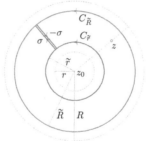

Then $C_{\tilde{R}} - \sigma - C_{\tilde{r}} + \sigma$ is equivalent to a reparameterised path $\gamma : [0, 1] \to \mathbb{A} \backslash \{z\}$, which is $\mathbb{A} \backslash \{z\}$-homotopic to a small circle C_δ centred at z. Also, $\frac{f(\cdot)}{\cdot - z} \in \mathcal{O}(\mathbb{A} \backslash \{z\})$, and so $\int_\gamma \frac{f(\zeta)}{\zeta - z} d\zeta = \int_{C_\delta} \frac{f(\zeta)}{\zeta - z} d\zeta = f(z)2\pi i$, where the first equality follows from the Cauchy Integral Theorem, and the second equality follows from the Cauchy Integral Formula. Since the integral along σ cancels with that along $-\sigma$, we obtain that

$$f(z) = \frac{1}{2\pi i} \int_\gamma \frac{f(\zeta)}{\zeta - z} d\zeta = \frac{1}{2\pi i} \int_{C_{\tilde{R}}} - \int_\sigma - \int_{C_{\tilde{r}}} + \int_\sigma \frac{f(\zeta)}{\zeta - z} d\zeta$$

$$= \frac{1}{2\pi i} \int_{C_{\tilde{R}}} \frac{f(\zeta)}{\zeta - z} d\zeta - \frac{1}{2\pi i} \int_{C_{\tilde{r}}} \frac{f(\zeta)}{\zeta - z} d\zeta = (\text{I}) + (\text{II}),$$

where $(\text{I}) := \frac{1}{2\pi i} \int_{C_{\tilde{R}}} \frac{f(\zeta)}{\zeta - z} d\zeta$ and $(\text{II}) = -\frac{1}{2\pi i} \int_{C_{\tilde{r}}} \frac{f(\zeta)}{\zeta - z} d\zeta$. We will now show that (I) equals $\sum_{n=0}^\infty c_n (z - z_0)^n$, while (II) equals $\sum_{n=1}^\infty c_{-n} (z - z_0)^{-n}$, and these put together yield the desired Laurent series expansion of f.

Step 1. In this step we will show that $\frac{1}{2\pi i} \int_{C_{\tilde{R}}} \frac{f(\zeta)}{\zeta - z} d\zeta = \sum_{n=0}^\infty c_n (z - z_0)^n$. For $\zeta \in C_{\tilde{R}}$, we have $\frac{f(\zeta)}{\zeta - z} = \frac{f(\zeta)}{\zeta - z_0 + z_0 - z} = \frac{f(\zeta)}{(\zeta - z_0)(1 - \frac{z - z_0}{\zeta - z_0})} = \frac{f(\zeta)}{(\zeta - z_0)(1 - w)}$, where $w := \frac{z - z_0}{\zeta - z_0}$. So $|w| = \frac{|z - z_0|}{|\zeta - z_0|} = \frac{|z - z_0|}{\tilde{R}} < 1$, and so $\frac{1}{1 - w} = 1 + w + w^2 + w^3 + \cdots + w^{n-1} + \frac{w^n}{1 - w}$. Thus

$$\frac{f(\zeta)}{\zeta-z} = \frac{f(\zeta)}{\zeta-z_0}(1+w+\cdots+w^{n-1}+\frac{w^n}{1-w}) = \frac{f(\zeta)}{\zeta-z_0}+\cdots+\frac{f(\zeta)}{(\zeta-z_0)^n}(z-z_0)^{n-1}+\frac{f(\zeta)(z-z_0)^n}{(\zeta-z_0)^n(\zeta-z)}.$$

Hence

$$\frac{1}{2\pi i}\int_{C_{\tilde{R}}}\frac{f(\zeta)}{\zeta-z}d\zeta$$

$$= \frac{1}{2\pi i}\int_{C_{\tilde{R}}}\frac{f(\zeta)}{\zeta-z_0}d\zeta+\cdots+\frac{1}{2\pi i}\int_{C_{\tilde{R}}}\frac{f(\zeta)}{(\zeta-z_0)^n}d\zeta(z-z_0)^{n-1}+\frac{1}{2\pi i}\int_{C_{\tilde{R}}}\frac{f(\zeta)}{(\zeta-z_0)^n(\zeta-z)}d\zeta(z-z_0)^n$$

$$\overset{(*)}{=} c_0 + c_1(z-z_0)+\cdots+c_{n-1}(z-z_0)^{n-1}+R_n(z),$$

where $R_n(z) := \frac{1}{2\pi i}\int_{C_{\tilde{R}}}\frac{f(\zeta)(z-z_0)^n}{(\zeta-z_0)^n(\zeta-z)}d\zeta$. To see $(*)$, note that a reparametrisation of $C_{\tilde{R}}$ is A-homotopic to a circular path $C(t) = z_0+\rho e^{2\pi i t}$, $t\in[0,1]$, for any $\rho\in(r,R)$, and so by the Cauchy Integral Theorem, $\int_{C_{\tilde{R}}}\frac{f(\zeta)}{(\zeta-z_0)^k}d\zeta = \int_C \frac{f(\zeta)}{(\zeta-z_0)^k}d\zeta = c_k$ for $k=1,\cdots,n-1$. To show finish the proof of Step 1, we will show $\lim_{n\to\infty}R_n = 0$.

Let $M>0$ be such that for all $\zeta\in C_{\tilde{R}}$, $|f(\zeta)| < M$. We have $|\zeta-z_0| = \tilde{R}$ and also $|\zeta-z| = |\zeta-z_0-(z-z_0)| \geqslant \tilde{R}-|z-z_0|$. So $|R_n(z)| \leqslant (\frac{|z-z_0|}{\tilde{R}})^n\frac{M}{\tilde{R}-|z-z_0|}\overset{n\to\infty}{\longrightarrow} 0$. Thus $c_0 + c_1(z-z_0)+c_2(z-z_0)^2+\cdots = \frac{1}{2\pi i}\int_{\gamma_2}\frac{f(\zeta)}{\zeta-z}d\zeta$.

Step 2. We will show that $-\frac{1}{2\pi i}\int_{C_{\tilde{r}}}\frac{f(\zeta)}{\zeta-z}d\zeta = \sum_{n=1}^{\infty}c_{-n}(z-z_0)^{-n}$. We have

$$-\frac{1}{2\pi i}\int_{C_{\tilde{r}}}\frac{f(\zeta)}{\zeta-z}d\zeta = \frac{1}{2\pi i}\int_{C_{\tilde{r}}}\frac{f(\zeta)}{(z-z_0)-(\zeta-z_0)}d\zeta = \frac{1}{2\pi i}\int_{C_{\tilde{r}}}\frac{f(\zeta)}{(z-z_0)(1-\frac{\zeta-z_0}{z-z_0})}d\zeta.$$

If $w := \frac{\zeta-z_0}{z-z_0}$, then $|w| = \frac{\tilde{r}}{|z-z_0|} < 1$, and $\frac{1}{1-\frac{\zeta-z_0}{z-z_0}} = 1+w+\cdots+w^{n-1}+\frac{w^n}{1-w}$. So

$$-\frac{1}{2\pi i}\int_{C_{\tilde{r}}}\frac{f(\zeta)}{\zeta-z}d\zeta$$

$$= \frac{1}{2\pi i}\int_{C_{\tilde{r}}}f(\zeta)(\frac{1}{z-z_0}+\cdots+\frac{(\zeta-z_0)^{n-1}}{(z-z_0)^n}+\frac{(\zeta-z_0)^n}{(z-z_0)^n(z-\zeta)})d\zeta$$

$$= \frac{1}{2\pi i}\int_{C_{\tilde{r}}}f(\zeta)d\zeta\cdot\frac{1}{z-z_0}+\cdots+\frac{1}{2\pi i}\int_{C_{\tilde{r}}}\frac{f(\zeta)}{(\zeta-z_0)^{-n+1}}d\zeta\cdot\frac{1}{(z-z_0)^n}+\frac{1}{2\pi i}\int_{C_{\tilde{r}}}\frac{f(\zeta)(\zeta-z_0)^n}{(z-z_0)^n(z-\zeta)}d\zeta$$

$$\overset{(\star)}{=} c_{-1}(z-z_0)^{-1}+\cdots+c_{-n}(z-z_0)^{-n}+\tilde{R}_n(z),$$

where $\tilde{R}_n(z) := \frac{1}{2\pi i}\int_{C_{\tilde{r}}}\frac{f(\zeta)(\zeta-z_0)^n}{(z-z_0)^n(z-\zeta)}d\zeta$. To see $(\star)$, note that a reparameterisation of $C_{\tilde{r}}$ is A-homotopic to C ($C(t) = z_0+\rho e^{2\pi i t}$, $t\in[0,1]$, where $\rho\in(r,R)$). By the Cauchy Integral Theorem, $\int_{C_{\tilde{r}}}\frac{f(\zeta)}{(\zeta-z_0)^k}d\zeta = \int_C\frac{f(\zeta)}{(\zeta-z_0)^k}d\zeta = c_k$ for $k = -1,\cdots,-n$. Let $M>0$ be such that for all $\zeta\in C_{\tilde{r}}$, $|f(\zeta)| < M$. We have $|\zeta-z_0| = \tilde{r}$ and $|z-\zeta| = |(z-z_0)-(\zeta-z_0)| \geqslant |z-z_0|-\tilde{r}$, and so $|\tilde{R}_n(z)| \leqslant (\frac{\tilde{r}}{|z-z_0|})^n\frac{M\tilde{r}}{|z-z_0|-\tilde{r}}\overset{n\to\infty}{\longrightarrow} 0$. Thus $c_{-1}(z-z_0)^{-1}+\cdots+c_{-n}(z-z_0)^{-n}+\cdots = -\frac{1}{2\pi i}\int_{C_{\tilde{r}}}\frac{f(\zeta)}{\zeta-z}d\zeta$.

This completes the proof of the existence part of the Laurent expansion.

Uniqueness of coefficients.

Cauchy's Integral Formula allows us to show that the Laurent expansion is unique, i.e., if $f(z) = \sum_{n\in\mathbb{Z}}\tilde{c}_n(z-z_0)^n$ for $r < |z-z_0| < R$, then for all $n\in\mathbb{Z}$, $\tilde{c}_n = c_n$.

If $n \neq -1$, then $(z - z_0)^n = \frac{d}{dz}\left(\frac{(z-z_0)^{n+1}}{n+1}\right)$. So $\int_C (z - z_0)^n dz = 0$ $(n \neq -1)$, where C is given by $C(t) = z_0 + \rho e^{2\pi i t}$, $t \in [0, 1]$. But $\int_C \frac{1}{z-z_0} dz = 2\pi i$. By term-by-term integration,

$$\int_C \frac{f(z)}{(z-z_0)^{m+1}} dz = \int_C (z-z_0)^{-m-1} \sum_{n \in \mathbb{Z}} \tilde{c}_n (z-z_0)^n dz = \sum_{n \in \mathbb{Z}} \tilde{c}_n \int_C (z-z_0)^{n-m-1} dz = 2\pi i \tilde{c}_m,$$

proving the claim about the uniqueness of coefficients. The term-by-term integration is justified as follows. We have

$$\sum_{n \in \mathbb{Z}} \tilde{c}_n (z - z_0)^{n-m-1} = \left(\cdots + \frac{\tilde{c}_{m-1}}{(z-z_0)^2}\right) + \frac{\tilde{c}_m}{z-z_0} + \left(\tilde{c}_{m+1} + \tilde{c}_{m+2}(z - z_0) + \cdots\right)$$

$$= f_1(z) + \frac{\tilde{c}_m}{z-z_0} + f_2(z).$$

We need only show that f_1, f_2 have a primitive in the annulus, because then

$$\int_C \sum_{n \in \mathbb{Z}} \tilde{c}_n (z-z_0)^{n-m-1} dz = \int_C \left(f_1(z) + \frac{\tilde{c}_m}{z-z_0} + f_2(z)\right) dz = 0 + 2\pi i \tilde{c}_m + 0 = 2\pi i \tilde{c}_m.$$

We have $f_2(z) = \sum_{n=1}^{\infty} \tilde{c}_{m+n}(z - z_0)^{n-1}$ for $z \in D(z_0, R)$. If $F_2(z) := \sum_{n=1}^{\infty} \frac{\tilde{c}_{m+n}}{n}(z - z_0)^n$ for $z \in D(z_0, R)$, then $\frac{d}{dz} F_2(z) = f_2(z)$, and so F_2 is a primitive of f_2. Note that since for all $n \in \mathbb{N}$, $\left|\frac{\tilde{c}_{m+n}}{n}(z - z_0)^n\right| \leq R|\tilde{c}_{m+n}(z - z_0)^{n-1}|$, it follows by the comparison test that $\sum_{n=1}^{\infty} \frac{\tilde{c}_{m+n}}{n}(z - z_0)^n$ converges absolutely in $D(z_0, R)$.

For $R > |z - z_0| > r$, $f_1(z) = \sum_{n=1}^{\infty} \frac{\tilde{c}_{m-n}}{(z-z_0)^{n+1}} = \sum_{n=1}^{\infty} \tilde{c}_{m-n} w^{n+1}$, where $w = \frac{1}{z-z_0}$. So $\sum_{n=1}^{\infty} \tilde{c}_{m-n} w^{n+1}$ converges for $|w| < \frac{1}{r}$. Let $G(w) := -\sum_{n=1}^{\infty} \frac{\tilde{c}_{m-n}}{n} w^n$ for $|w| < \frac{1}{r}$. As $\sum_{n=1}^{\infty} \tilde{c}_{m-n} w^{n+1}$ is absolutely convergent for $\frac{1}{R} < |w| < \frac{1}{r}$, and since for all $n \in \mathbb{N}$, $\left|\frac{\tilde{c}_{m-n}}{n} w^n\right| \leq R|\tilde{c}_{m-n} w^{n+1}|$, by the comparison test, $\sum_{n=1}^{\infty} \frac{\tilde{c}_{m-n}}{n} w^n$ is absolutely convergent, and being a power series, this holds also for all $w \in D(0, \frac{1}{r})$. We have $\frac{d}{dw} G(w) = -\sum_{n=1}^{\infty} \tilde{c}_{m-n} w^{n-1}$. If $F_1(z) := G(\frac{1}{z-z_0}) = -\sum_{n=1}^{\infty} \frac{\tilde{c}_{m-n}}{n}(z - z_0)^{-n}$ for $z \in \mathbb{A}$, then $\frac{d}{dz} F_1(z) = \left(G'(\frac{1}{z-z_0})\right)\left(-\frac{1}{(z-z_0)^2}\right) = (z - z_0)^{-2} \sum_{n=1}^{\infty} \tilde{c}_{m-n}(z - z_0)^{-n+1} = f_1(z)$. $\square$

The uniqueness of coefficients is valid only if we consider a particular fixed annulus. It can happen that the same function has different Laurent expansions, but valid in *different* annuli, as shown in the following example.

Example 4.14. Consider f defined by $f(z) = \frac{1}{z(z-1)}$, $z \in \mathbb{C} \setminus \{0, 1\}$.

f is holomorphic in the annulus $\mathbb{A} := \{z \in \mathbb{C} : 0 < |z| < 1\}$. Since $|z| < 1$,
$$f(z) = \tfrac{1}{z(z-1)} = -\tfrac{1}{z}\left(1 + z + z^2 + z^3 + \cdots\right) = -\tfrac{1}{z} - 1 - z - z^2 - z^3 - \cdots.$$
The Laurent series coefficients are $c_{-2} = c_{-3} = \cdots = 0$, $c_{-1} = c_0 = c_1 = \cdots = -1$.

But f is also holomorphic in the annulus $\widetilde{\mathbb{A}} := \{z \in \mathbb{C} : 1 < |z|\}$. So f has a Laurent series expansion in $\widetilde{\mathbb{A}}$ too. Since $|z| > 1$, we have
$$f(z) = \tfrac{1}{z(z-1)} = \tfrac{1}{z^2(1-\frac{1}{z})} = \tfrac{1}{z^2}\left(1 + \tfrac{1}{z} + \tfrac{1}{z^2} + \tfrac{1}{z^3} + \cdots\right) = \tfrac{1}{z^2} + \tfrac{1}{z^3} + \cdots.$$
The Laurent series coefficients are $\widetilde{c}_{-2} = \widetilde{c}_{-3} = \cdots = 1$, $c_{-1} = c_0 = c_1 = \cdots = 0$. So the coefficients are different, but this is not surprising, since the annuli for the Laurent series expansions were different too. $\diamond$

Exercise 4.51. In Example 4.14 find Laurent series expansions for f also in the annuli $\mathbb{A}_1 := \{z \in \mathbb{C} : 0 < |z - 1| < 1\}$ and $\widetilde{\mathbb{A}}_1 := \{z \in \mathbb{C} : 1 < |z - 1|\}$.

Exercise 4.52. Let $w \in \mathbb{C}$. Define f_w by $f_w(z) = \exp(\tfrac{w}{2}(z - \tfrac{1}{z}))$ for all $z \in \mathbb{C}\backslash\{0\}$. Then $f_w \in \mathcal{O}(\mathbb{C}\backslash\{0\})$. So f_w has a Laurent series expansion $f_w(z) = \sum\limits_{n=-\infty}^{\infty} J_n(w)z^n$, $z \in \mathbb{C}\backslash\{0\}$, defining the *Bessel functions* J_n for all $n \in \mathbb{Z}$. Show the following:
$$f_{-w}(-z) = f_w(z), \quad J_n(-w) = (-1)^n J_n(w), \quad J_n(w) = \tfrac{1}{2\pi}\int_0^{2\pi} \exp(iw\sin\theta - in\theta)d\theta.$$

Exercise 4.53. Using the Laurent series theorem, evaluate $\int_C z^5 e^{-\frac{1}{z^2}} dz$, where C is the circular path given by $C(t) = e^{it}$ for all $t \in [0, 2\pi]$.

4.8. Classification of singularities

If we look at the three functions $\frac{\sin z}{z}$, $\frac{1}{z^3}$, $\exp\frac{1}{z}$, then we notice that each of them is not defined at 0. We refer to 0 as a 'singularity' of these functions, because they are not defined at that point. However, the functions are holomorphic in a region obtained by deleting that point from its neighbourhood. We will see that these functions behave very differently near their common singularity. In other words, the 'nature of the singularity' differs in each case. We will explain precisely how the behaviour is different in each case, and this is what we mean by the classification of singularities. Moreover, we will learn two results, which will allow us to find out the type of singularity at hand. Of these two characterisation results for singularities, one result will be in terms of limits, while the other will be in terms of what happens with Laurent coefficients. We first give the following definition.

Definition 4.2. Let f be a complex-valued function, not defined at $z_0 \in \mathbb{C}$, such that f is holomorphic in a punctured disc $\{z \in \mathbb{C} : 0 < |z - z_0| < R\}$ centred at z_0 with a radius $R > 0$. Then we call z_0 an *isolated singularity* of f.

Example 4.15. $\frac{\sin z}{z}$, $\frac{1}{z^3}$, $\exp\frac{1}{z}$, each have an isolated singularity at 0. $\frac{1}{\sin\frac{1}{z}}$ has a singularity at 0, but 0 is not an isolated singularity (since e.g. at $z = \frac{1}{n\pi}$, $n \in \mathbb{N}$, the function $\frac{1}{\sin\frac{1}{z}}$ is not defined). $\diamond$

Definition 4.3. Let $z_0 \in \mathbb{C}$ be an isolated singularity of $f \in \mathcal{O}(D(z_0, R)\backslash\{z_0\})$, $R > 0$. Then z_0 is called

- a *removable singularity of* f if $\exists F \in \mathcal{O}(D(z_0, R))$ such that $F|_{D(z_0,R)\backslash\{z_0\}} = f$.
- a *pole of* f if $\lim\limits_{z\to z_0} |f(z)| = \infty$
  (i.e., $\forall M > 0 \ \exists \delta > 0$ such that whenever $0 < |z - z_0| < \delta$, we have $|f(z)| > M$).
- an *essential singularity of* f if z_0 is neither removable nor a pole.

Example 4.16.

(1) f given by $f(z) = \frac{\sin z}{z}$ has a removable singularity at 0: For $z \neq 0$,
$$\frac{\sin z}{z} = \frac{1}{z}\Big(z - \frac{z^3}{3!} + \frac{z^5}{5!} - + \cdots\Big) = \sum_{n=0}^{\infty} \frac{(-1)^n}{(2n+1)!} z^{2n}.$$

The right-hand side, being a power series with an infinite radius of convergence (why?), defines an entire function F. Since this entire function F coincides with f in the punctured plane $\mathbb{C}\backslash\{0\}$, f has a removable singularity at 0. From here, we obtain $\lim\limits_{z\to 0} \frac{\sin z}{z} = \lim\limits_{z\to 0} F(z) = F(0) = 1$.

(2) The function $\frac{1}{z^3}$ has a pole at 0, since $\lim\limits_{z\to 0} \frac{1}{|z|^3} = +\infty$. (If $M > 0$, then taking $\delta := \frac{1}{M^{1/3}} > 0$, whenever $0 < |z - 0| < \delta$, we have $|f(z)| = \frac{1}{|z|^3} > M$.)

(3) The function $\exp \frac{1}{z}$ has an essential singularity at 0. Indeed,

- 0 is not a removable singularity, because for example $\lim\limits_{x\to 0+} e^{\frac{1}{x}} = +\infty$.
- 0 is also not a pole, as $\lim\limits_{x\to 0-} e^{\frac{1}{x}} = 0$, and so $\neg\big(\lim\limits_{z\to 0} |f(z)| = +\infty\big)$. ◇

Exercise 4.54. Let $f(z) := \cos(e^{\frac{1}{z}})$ for $z \neq 0$. Show that 0 is an essential singularity of f. *Hint*: Consider $z_n = \frac{1}{\log(n\pi)}$, $n \in \mathbb{N}$.

We will now learn the characterisation result for singularities, in terms of limiting behaviour. Below $\neg$ is the symbol for negation, read as 'it is not the case that'.

Theorem 4.16 (Classification of singularities via limits).
Suppose z_0 is an isolated singularity of f. Then

z_0 *is removable*	$\Leftrightarrow$	$\lim\limits_{z\to z_0} (z - z_0)f(z) = 0.$
z_0 *is a pole*	$\Leftrightarrow$	• $\neg\big(\lim\limits_{z\to z_0} (z - z_0)f(z) = 0\big)$ *and* • $\exists n \in \mathbb{N}$ *such that* $\lim\limits_{z\to z_0}(z - z_0)^{n+1}f(z) = 0.$ (*The smallest such n is called the order of the pole z_0 of f.*)
z_0 *is essential*	$\Leftrightarrow$	$\forall n \in \mathbb{N} \ \neg\big(\lim\limits_{z\to z_0} (z - z_0)^n f(z) = 0\big).$

Proof. (1) z_0 removable $\Rightarrow \lim\limits_{z\to z_0} (z - z_0)f(z) = 0$:

Let z_0 be removable. Let $F \in \mathcal{O}(D(z_0, R))$, $R > 0$, be such that $F|_{D(z_0,R)\backslash\{z_0\}} = f$. Then $(\cdot - z_0)F$ is continuous at z_0, and its value there is 0. Given $\epsilon > 0$, there exists a $\delta > 0$ such that if $|z - z_0| < \delta$, then $|(z - z_0)F(z) - 0| < \epsilon$, and so in particular, if $0 < |z - z_0| < \delta$, then $|(z - z_0)f(z) - 0| < \epsilon$. Thus $\lim\limits_{z\to z_0} (z - z_0)f(z) = 0$.

$\lim\limits_{z\to z_0}(z-z_0)f(z)=0 \Rightarrow z_0$ is removable:

Let $\lim\limits_{z\to z_0}(z-z_0)f(z)=0$, and $f\in\mathcal{O}(D(z_0,R)\setminus\{z_0\})$, where $R>0$. Then f has a Laurent series expansion $f(z)=\sum\limits_{n\in\mathbb{Z}}c_n(z-z_0)^n$, $0<|z-z_0|<R$, where for $n\in\mathbb{Z}$,

$c_n=\frac{1}{2\pi i}\int_{C_r}\frac{f(z)}{(z-z_0)^{n+1}}dz$, and $C_r(t)=z_0+re^{2\pi it}$, $t\in[0,1]$, for any $r\in(0,R)$.

We will show $c_{-n}=0$ for $n\in\mathbb{N}$. Given $\epsilon>0$, let $r>0$ be such that $|(z-z_0)f(z)|<\epsilon$ on C_r. Then, for $n\in\mathbb{N}$,

$|c_{-n}|=\left|\frac{1}{2\pi i}\int_{C_r}\frac{f(z)}{(z-z_0)^{-n+1}}dz\right|=\left|\frac{1}{2\pi i}\int_{C_r}\frac{(z-z_0)f(z)}{(z-z_0)^{-n+2}}dz\right|\leqslant\frac{1}{2\pi}\frac{\epsilon}{r^{-n+2}}2\pi r=\epsilon\,r^{n-1}\leqslant\epsilon R^{n-1}.$

As $\epsilon>0$ was arbitrary, $c_{-n}=0$ for all $n\in\mathbb{N}$. With $F(z):=\sum\limits_{n=0}^{\infty}c_n(z-z_0)^n$, $F\in\mathcal{O}(D(z_0,R))$, and $F=f$ in the punctured disc $\{z\in\mathbb{C}:0<|z-z_0|<R\}$.

(2) z_0 is a pole $\Rightarrow$ $\begin{cases} \neg\left(\lim\limits_{z\to z_0}(z-z_0)f(z)=0\right) \text{ and} \\ \exists n\in\mathbb{N}\text{ such that }\lim\limits_{z\to z_0}(z-z_0)^{n+1}f(z)=0. \end{cases}$

Let z_0 be a pole. Then z_0 is not removable (otherwise if $F\in\mathcal{O}(D(z_0,R))$, $R>0$, is such that $F|_{D(z_0,R)\setminus\{z_0\}}=f$, then $\lim\limits_{z\to z_0}|f(z)|=\lim\limits_{z\to z_0}|F(z)|=|F(z_0)|<+\infty$, a contradiction). By the part (1) above, $\neg\left(\lim\limits_{z\to z_0}(z-z_0)f(z)=0\right)$.

There is some $r>0$ such that $|f(z)|>1$ for all z in the punctured disc $D(z_0,r)\setminus\{z_0\}$. In particular, $f\neq 0$ in $D(z_0,r)\setminus\{z_0\}$, and so $g:=\frac{1}{f}\in\mathcal{O}(D(z_0,r)\setminus\{z_0\})$. Since z_0 is a pole of f, it follows that $\lim\limits_{z\to z_0}(z-z_0)g(z)=0\cdot 0=0$. By part (1) above, z_0 is a removable singularity of g. So there exists a $G\in\mathcal{O}(D(z_0,r))$ such that $G|_{D(z_0,r)\setminus\{z_0\}}=g$. We have $G(z_0)=\lim\limits_{z\to z_0}G(z)=\lim\limits_{z\to z_0}g(z)=0$. Hence z_0 is a zero of G. Since G is not identically zero in a neighbourhood of z_0, by the result on the classification of zeroes, z_0 has some order $m\in\mathbb{N}$, and there exists an $H\in\mathcal{O}(D(z_0,r))$ such that $H(z_0)\neq 0$ and $G(z)=(z-z_0)^mH(z)$ in $D(z_0,r)$. So for $z\in D(z_0,r)\setminus\{z_0\}$, we have $f(z)=\frac{1}{g(z)}=\frac{1}{G(z)}=\frac{1}{(z-z_0)^mH(z)}$. Consequently, $\lim\limits_{z\to z_0}(z-z_0)^{m+1}f(z)=\lim\limits_{z\to z_0}\frac{z-z_0}{H(z)}=\frac{0}{H(z_0)}=0.$

$\left.\begin{array}{l} \neg\left(\lim\limits_{z\to z_0}(z-z_0)f(z)=0\right)\text{ and} \\ \exists n\in\mathbb{N}\text{ such that }\lim\limits_{z\to z_0}(z-z_0)^{n+1}f(z)=0 \end{array}\right\}\Rightarrow z_0$ is a pole.

Let m be the smallest such n: $\lim\limits_{z\to z_0}(z-z_0)^{m+1}f(z)=0$, $\neg\left(\lim\limits_{z\to z_0}(z-z_0)^mf(z)=0\right)$. By part (1), $(\cdot-z_0)^mf$ has a removable singularity at $z=z_0$. Thus there exists an $F\in\mathcal{O}(D(z_0,R))$, $R>0$, such that $F|_{D(z_0,R)\setminus\{z_0\}}=(z-z_0)^mf$. Thus we have $F(z_0)=\lim\limits_{z\to z_0}F(z)=\lim\limits_{z\to z_0}(z-z_0)^mf(z)\neq 0$ (definition of m). From $f(z)=\frac{F(z)}{(z-z_0)^m}$ for $z\in D(z_0,R)\setminus\{z_0\}$, $\lim\limits_{z\to z_0}|f(z)|=\lim\limits_{z\to z_0}\frac{|F(z)|}{|z-z_0|^m}=|F(z_0)|\lim\limits_{z\to z_0}\frac{1}{|z-z_0|^m}=+\infty$. So z_0 is a pole of f.

(3) This follows from the first two parts. If z_0 is an essential singularity, then it is not removable, and so $\neg\big(\lim_{z\to z_0}(z-z_0)f(z)=0\big)$, proving the claim for $n=1$. As z_0 is not a pole, by (2), the claim is true for all other $n\in\mathbb{N}$ as well.

Conversely, if for all $n\in\mathbb{N}$, $\neg\big(\lim_{z\to z_0}(z-z_0)^nf(z)=0\big)$, then in particular, from the $n=1$ case, we get by (1) that z_0 cannot be a removable singularity. From the cases for all other natural numbers $n\geqslant 2$, using (2), we conclude that z_0 cannot be a pole either. So z_0 must be an essential singularity of f. $\qquad\square$

Corollary 4.17. *If z_0 is an isolated singularity of f and there exists an $m\in\mathbb{N}$ such that $\lim_{z\to z_0}(z-z_0)^mf(z)=L\neq 0$, then z_0 is a pole of order m of f.*

Proof. $\lim_{z\to z_0}(z-z_0)^{m+1}f(z)=\big(\lim_{z\to z_0}(z-z_0)\big)\lim_{z\to z_0}((z-z_0)^mf(z))=0\cdot L=0$. For $(\mathbb{N}\ni)\,\ell<m+1$ we will show $\neg\big(\lim_{z\to z_0}(z-z_0)^\ell f(z)=0\big)$. For $\ell=m$, this is given. If $\lim_{z\to z_0}(z-z_0)^\ell f(z)=0$ for some $\ell<m$, then we arrive at the contradiction that $0\neq L=\lim_{z\to z_0}(z-z_0)^mf(z)=\big(\lim_{z\to z_0}(z-z_0)^{m-\ell}\big)\lim_{z\to z_0}((z-z_0)^\ell f(z))=0\cdot 0=0$. $\quad\square$

Example 4.17. Let us reconsider Example 4.16.

We have $\lim_{z\to 0}z\frac{\sin z}{z}=\lim_{z\to 0}\sin z=\sin 0=0$. So 0 is a removable singularity of $\frac{\sin z}{z}$.

As $\lim_{z\to 0}z^3\frac{1}{z^3}=1\neq 0$, 0 is a pole of order 3 of $\frac{1}{z^3}$.

For $x>0$, $e^{\frac{1}{x}}=1+\frac{1}{1!x}+\frac{1}{2!x^2}+\frac{1}{3!x^3}+\cdots>\frac{1}{n!x^n}$ for all $n\in\mathbb{N}$. Thus for all $n\in\mathbb{N}$, $x^ne^{\frac{1}{x}}>\frac{1}{n!}$ and $\neg\big(\lim_{x\to 0+}x^ne^{\frac{1}{x}}=0\big)$. So for all $n\in\mathbb{N}$, $\neg(\lim_{z\to 0}z^n\exp\frac{1}{z}=0)$. Thus 0 is an essential singularity of $\exp\frac{1}{z}$. $\qquad\qquad\qquad\diamond$

Exercise 4.55. As $\mathrm{Log}(1+z)\in\mathcal{O}(D(0,1))$ and $\frac{1}{z^2}\in\mathcal{O}(\mathbb{C}\backslash\{0\})$, $\frac{\mathrm{Log}(1+z)}{z^2}\in\mathcal{O}(D(0,1)\backslash\{0\})$. Prove that $z=0$ is a pole of order 1 of $\frac{\mathrm{Log}(1+z)}{z^2}$.

Exercise 4.56 (Bernoulli numbers). Determine the set Z of complex zeroes of e^z-1. For each $z\in Z$, determine the order of z as a zero of e^z-1. Define f by $f(z)=\frac{z}{e^z-1}$ for $z\in\mathbb{C}\backslash Z$. Show that 0 is an isolated singularity of f, and that it is removable. Prove that there exist $r>0$ and complex B_n $(n\geqslant 0)$, such that

$$\frac{z}{e^z-1}=\sum_{n=0}^{\infty}\frac{B_n}{n!}z^n\text{ for }0<|z|<r.$$

Determine B_0. Determine all the coefficients with odd indices, namely $B_1,B_3,B_5,\cdots$.
Hint: Use $\frac{z}{e^z-1}-\frac{(-z)}{e^{-z}-1}=-z$.

Exercise 4.57. Let $f(z):=\frac{1}{\cos z-1}$ for all $0<|z|<2\pi$. Prove that 0 is a pole of f, and find its order.

We will now learn our second characterisation result for singularities, in terms of the Laurent series coefficients.

Theorem 4.18 (Classification via Laurent coefficients). *Let z_0 be an isolated singularity of f, and $f(z) = \sum_{n \in \mathbb{Z}} c_n(z - z_0)^n$ for $0 < |z - z_0| < R$, for some $R > 0$. Then*

z_0 is removable $\Leftrightarrow$	For all $n \in \mathbb{N}$, $c_{-n} = 0$
z_0 is a pole of order $m \in \mathbb{N}$ $\Leftrightarrow$	The index $m \in \mathbb{N}$ satisfies • $c_{-m} \neq 0$ and • for all $n > m$, $c_{-n} = 0$.
z_0 is essential $\Leftrightarrow$	There are infinitely many $n \in \mathbb{N}$ such that $c_{-n} \neq 0$.

Proof. (1) z_0 is removable $\Rightarrow$ (for all $n \in \mathbb{N}$, $c_{-n} = 0$).

Let z_0 be a removable singularity. Then f has an extension $F \in \mathcal{O}(D(z_0, R))$. Let the Taylor series expansion of F be given by $F(z) = \sum_{n=0}^{\infty} \tilde{c}_n(z - z_0)^n$ for $z \in D(z_0, R)$. For $0 < |z - z_0| < R$, $f(z) = \sum_{n=0}^{\infty} \tilde{c}_n(z - z_0)^n = \sum_{n \in \mathbb{Z}} c_n(z - z_0)^n$. By the uniqueness of the Laurent series expansion, $c_n = \tilde{c}_n$ for $n \geq 0$, and $c_n = 0$ for all $n < 0$.

(For all $n \in \mathbb{N}$, $c_{-n} = 0$) $\Rightarrow$ z_0 is removable.

Let $c_n = 0$ for all $n < 0$. For $0 < |z - z_0| < R$, $f(z) = \sum_{n \in \mathbb{Z}} c_n(z - z_0)^n = \sum_{n=0}^{\infty} c_n(z - z_0)^n$. Define $F(z) = \sum_{n=0}^{\infty} c_n(z - z_0)^n$ for $z \in D(z_0, R)$. As F is given by a power series, $F \in \mathcal{O}(D(z_0, R))$. Moreover $F|_{D(z_0, R) \setminus \{z_0\}} = f$. Hence z_0 is removable.

(2) z_0 is a pole of order $m \Rightarrow f(z) = \frac{c_{-m}}{(z-z_0)^m} + \cdots + \frac{c_{-1}}{z-z_0} + \sum_{n=0}^{\infty} c_n(z - z_0)^n$, $c_{-m} \neq 0$.

Let z_0 be a pole of order m. Using the classification of singularities via limits (Theorem 4.16), $\lim_{z \to z_0}(z - z_0)((z - z_0)^m f(z)) = \lim_{z \to z_0}(z - z_0)^{m+1} f(z) = 0$, and so $(\cdot - z_0)^m f$ has a removable singularity at z_0. For $z \in D(z_0, R) \setminus \{z_0\}$, we have $(z - z_0)^m f(z) = (z - z_0)^m \sum_{n \in \mathbb{Z}} c_n(z - z_0)^n = \sum_{n \in \mathbb{Z}} c_n(z - z_0)^{n+m}$, and by the previous part, this last series has all coefficients of negative powers of $z - z_0$ equal to 0, that is, $0 = c_{-(m+1)} = c_{-(m+2)} = \cdots$. Hence for $0 < |z - z_0| < R$,

$$(z - z_0)^m f(z) = c_{-m} + c_{-m+1}(z - z_0) + c_{-m+2}(z - z_0)^2 + \cdots,$$

and so $f(z) = \frac{c_{-m}}{(z-z_0)^m} + \cdots + \frac{c_{-1}}{(z-z_0)} + c_0 + c_1(z - z_0) + c_2(z - z_0)^2 + \cdots$. Moreover, the power series $c_{-m} + c_{-m+1}(z - z_0) + c_{-m+2}(z - z_0)^2 + \cdots$ is holomorphic in $D(z_0, R)$, and so it has the limit c_{-m} as $z \to z_0$. Thus

$$\lim_{z \to z_0}(z - z_0)^m f(z) = \lim_{z \to z_0}(c_{-m} + c_{-m+1}(z - z_0) + c_{-m+2}(z - z_0)^2 + \cdots) = c_{-m}.$$

As z_0 is a pole of f order m, $\neg \big(\lim_{z \to z_0}(z - z_0)^m f(z) = 0\big)$. So $c_{-m} \neq 0$.

$f(z) = \frac{c_{-m}}{(z-z_0)^m} + \cdots + \frac{c_{-1}}{z-z_0} + c_0 + \sum_{n=0}^{\infty} c_n(z-z_0)^n$, $c_{-m} \neq 0 \Rightarrow z_0$ pole of order m.

Suppose that there is some $m \in \mathbb{N}$ such that $c_{-m} \neq 0$ and $c_n = 0$ for all $n < -m$. Then $(z-z_0)^m f(z) = c_{-m} + c_{-m+1}(z-z_0) + c_{-m+2}(z-z_0)^2 + \cdots$, and since the right hand side defines a holomorphic function, say h, in $D(z_0, R)$, it follows that

$$\lim_{z \to z_0} (z-z_0)^m f(z) = \lim_{z \to z_0} h(z) = h(z_0) = c_{-m} \neq 0, \quad \text{and}$$
$$\lim_{z \to z_0} (z-z_0)^{m+1} f(z) = 0 \cdot c_{-m} = 0.$$

Thus z_0 is a pole of order m of f, by Corollary 4.17.

(3) This is immediate from the previous two parts and the fact that an essential singularity is neither a removable singularity nor a pole. $\square$

Example 4.18. Let us reconsider Example 4.16.

For $z \neq 0$, $\frac{\sin z}{z} = \frac{1}{z}(z - \frac{z^3}{3!} + \frac{z^5}{5!} - \frac{z^7}{7!} + - \cdots) = 1 - \frac{z^2}{3!} + \frac{z^4}{5!} - \frac{z^6}{7!} + - \cdots$. Since there are no negative powers of z appearing in the Laurent series expansion, it follows that 0 is a removable singularity of $\frac{\sin z}{z}$.

For $z \neq 0$, $\frac{1}{z^3} = \cdots + 0 + \frac{1}{z^3} + 0 + \cdots$. Thus $c_{-3} = 1 \neq 0$, while $0 = c_{-4} = c_{-5} = \cdots$. So 0 is a pole of $\frac{1}{z^3}$ of order 3.

For $z \neq 0$, $\exp\frac{1}{z} = 1 + \frac{1}{1!z} + \frac{1}{2!z^2} + \frac{1}{3!z^3} + \cdots$. For infinitely many $n \in \mathbb{N}$ (in fact for all of them), $c_{-n} = \frac{1}{n!} \neq 0$. Thus 0 is an essential singularity of $\exp\frac{1}{z}$.

Also, for $z \neq 0$, $\exp\frac{1}{z^2} = 1 + \frac{1}{1!z^2} + \frac{1}{2!z^4} + \frac{1}{3!z^6} + \cdots$. For infinitely many $n \in \mathbb{N}$, $c_{-n} \neq 0$. Thus 0 is an essential singularity of $\exp\frac{1}{z^2}$. $\diamond$

Exercise 4.58. Let $f \in \mathcal{O}(D)$, where D is a domain, and suppose f has a zero $z_0 \in D$ of order $m \in \mathbb{N}$. Show that $\frac{1}{f}$ is well-defined locally in a punctured disc $D(z_0, r) \backslash \{z_0\} \subset D$, for some $r > 0$, and that z_0 is a pole of $\frac{1}{f}$ of order m.

Exercise 4.59. Let $z_0 \in \mathbb{C}$, $r > 0$. Suppose $f \in \mathcal{O}(D(z_0, r) \backslash \{z_0\})$ is such that $f(z) \neq 0$ for all $z \in D(z_0, r) \backslash \{z_0\}$, and z_0 is a pole of f of order m. Show that $\frac{1}{f} \in \mathcal{O}(D(z_0, r) \backslash \{z_0\})$ has a holomorphic extension g to $D(z_0, r)$, and that g has a zero of order m at z_0.

Exercise 4.60. Let $r > 0$, and $z_0 \in \mathbb{C}$ be a pole of order m of $f \in \mathcal{O}(D(z_0, r) \backslash \{z_0\})$. Let f have the Laurent series expansion $f(z) = \sum_{n \in \mathbb{Z}} c_n(z-z_0)^n$ for $0 < |z-z_0| < r$. Show that $c_{-1} = \frac{1}{(m-1)!} \lim_{z \to z_0} \frac{d^{m-1}}{dz^{m-1}}((z-z_0)^m f(z))$.

Exercise 4.61. True or false?

(1) If f has a Laurent expansion $z^{-1} + c_0 + c_1 z + \cdots$, convergent in some punctured disc about the origin, then f has a pole at 0.

(2) A function may have different Laurent series centred at z_0, depending on the annulus of convergence selected.

(3) If f has an isolated singularity at z_0, then it has a Laurent series centred at z_0 and is convergent in some punctured disc $0 < |z-z_0| < R$.

(4) If a Laurent series for f is convergent in an annulus $R_1 < |z-z_0| < R_2$ is actually a Taylor series (no negative powers of $z-z_0$), then it converges (at least) in $D(z_0, R_2)$.

Exercise 4.62. Decide the nature of the singularity, if any, at 0 for the following functions. If the function is holomorphic or the singularity is isolated, expand the function in appropriate powers of z convergent in a punctured disc given by $0 < |z| < R$.

$$\sin z, \quad \sin \tfrac{1}{z}, \quad \tfrac{\sin z}{z}, \quad \tfrac{\sin z}{z^2}, \quad \tfrac{1}{\sin \tfrac{1}{z}}, \quad z \sin \tfrac{1}{z}$$

Exercise 4.63. If z_0 is a pole of order $m \in \mathbb{N}$ of f, then show that $\lim\limits_{z \to z_0} (z - z_0)^m f(z) \neq 0$.

Exercise 4.64. True or false?

(1) $\lim\limits_{z \to 0} |\exp \tfrac{1}{z}| = +\infty$.

(2) If f has a pole of order m at z_0, then there exists a polynomial p such that $f - \frac{p}{(z - z_0)^m}$ has a holomorphic extension to a disc around z_0.

(3) If f is not identically zero and holomorphic in a neighbourhood of 0, then there is an nonnegative integer m such that $\frac{f}{z^n}$ has a pole at 0 whenever $n > m$.

(4) If f, g have poles of order m_f, m_g respectively at z_0, then their pointwise product $f \cdot g$ has a pole of order $m_f + m_g$ at z_0.

Exercise 4.65. Give an example of a function holomorphic in all of $\mathbb{C}$ except for essential singularities at the two points 0 and 1.

Exercise 4.66. The function f given by $f(z) = \frac{1}{z-1}$ clearly does not have a singularity at 0. As it has the Laurent series $z^{-1} + z^{-2} + z^{-3} + \cdots$ for $|z| > 1$, one might then say that this series has infinitely many negative powers of z, and fallaciously conclude that the point 0 is an essential singularity of f. Point out the flaw in this argument.

Exercise 4.67. Prove or disprove: If f and g have a pole and an essential singularity respectively at the point z_0, then fg has an essential singularity at z_0.

Wild behaviour near essential singularities. We now show a result illustrating the 'wild' behaviour of a function f at its essential singularity z_0. It says that given any $w \in \mathbb{C}$, any $\epsilon > 0$, and any arbitrarily small punctured disc Δ with centre z_0, there is a point z in Δ such that $f(z)$ lies within a distance ϵ from w. So the image of any punctured disc centred at the essential singularity is dense in $\mathbb{C}$. Or in even more descriptive terms, f comes arbitrarily close to any complex value in every neighbourhood of z_0.

Theorem 4.19 (Casorati-Weierstrass). *If z_0 is an essential singularity of f, then $\forall w \in \mathbb{C}, \forall \delta > 0, \forall \epsilon > 0, \exists z \in \mathbb{C}$ such that $0 < |z - z_0| < \delta$ and $|f(z) - w| < \epsilon$.*

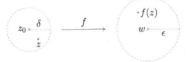

Proof. Suppose the statement is false. Then there exist $w \in \mathbb{C}$, $\delta > 0$ and $\epsilon > 0$ such that whenever $z \in D(z_0, \delta) \backslash \{z_0\}$, we have $|f(z) - w| \geq \epsilon$. Let $g(z) := \frac{1}{f(z) - w}$ for all $z \in D(z_0, \delta) \backslash \{z_0\}$. Then $g \in \mathcal{O}(D(z_0, \delta) \backslash \{z_0\})$. As $0 \leq |g(z)| = \frac{1}{|f(z) - w|} \leq \frac{1}{\epsilon}$ for all $z \in D(z_0, \delta) \backslash \{z_0\}$, we obtain $\lim\limits_{z \to z_0} (z - z_0) g(z) = 0$.

So g has a removable singularity at z_0. Denote its extension to $D(z_0, \delta)$ by g again. Let m be the order of the zero of g at z_0. (Set $m = 0$ if $g(z_0) \neq 0$.) Then $g(z) = (z - z_0)^m h(z)$, for some $h \in \mathcal{O}(D(z_0, \delta))$ and such that $h(z_0) \neq 0$. Then for $0 < |z - z_0| < \delta$,

$$(z - z_0)^{m+1} f(z) = (z - z_0)^{m+1}(f(z) - w + w) = (z - z_0)^{m+1}\frac{1}{g(z)} + (z - z_0)^{m+1}w$$
$$= \frac{z - z_0}{h(z)} + (z - z_0)^{m+1}w \xrightarrow{z \to z_0} \frac{0}{h(z_0)} + 0 \cdot w = 0.$$

Thus either f has a removable singularity at z_0 (when $m = 0$) or a pole at z_0 (when $m \in \mathbb{N}$). Hence z_0 cannot be an essential singularity of f, a contradiction. $\qquad\square$

Example 4.19. $\exp \frac{1}{z}$ has an essential singularity at 0. We will show that it assumes any given nonzero complex w ($= \rho e^{i\theta}$, $\theta \in \mathbb{R}$, $\rho > 0$), in any arbitrarily small neighbourhood of 0. Let $z = re^{it}$, $r > 0$, $t \in \mathbb{R}$, be such that $e^{\frac{1}{z}} = e^{\frac{\cos t}{r} - i\frac{\sin t}{r}} = \rho e^{i\theta}$. Equating absolute values, $\frac{\cos t}{r} = \log \rho$. Equating arguments, *one* solution is given when $-\frac{\sin t}{r} = \theta$. As $(\cos t)^2 + (\sin t)^2 = 1$, $r = \frac{1}{\sqrt{(\log \rho)^2 + \theta^2}}$. Also, $t = \cos^{-1}\frac{\log \rho}{\sqrt{(\log \rho)^2 + \theta^2}}$. As θ can be increased by integral multiplies of 2π, without changing w, r can be made as small as we please. $\qquad\diamond$

This example illustrates a much stronger result than the Casorati-Weierstrass Theorem, due to Picard, which says that the image of any punctured disc centred at an essential singularity misses at most one point of $\mathbb{C}$. In our example above, the exceptional value is $w = 0$. A proof of Picard's Theorem is beyond the scope of this book, but can be found in [3].

Exercise 4.68. Prove that if $f \in \mathcal{O}(D(z_0, r) \backslash \{z_0\})$, where $z_0 \in \mathbb{C}$ and $r > 0$, has an essential singularity at z_0 and $w \in \mathbb{C}$, then there exists a sequence $(z_n)_{n \in \mathbb{N}}$ in $D(z_0, r) \backslash \{z_0\}$ such that $\lim_{n \to \infty} z_n = z_0$ and $\lim_{n \to \infty} f(z_n) = w$.

Exercise 4.69. Prove that if f is entire and not a polynomial, and $w \in \mathbb{C}$, then there exists a complex sequence $(z_n)_{n \in \mathbb{N}}$ such that $\lim_{n \to \infty} z_n = \infty$ and $\lim_{n \to \infty} f(z_n) = w$.

4.9. Residue Theorem

Let D be a domain, $z_0 \in D$, and $f \in \mathcal{O}(D \backslash \{z_0\})$ have an isolated singularity at z_0. If $D(z_0, R) \subset D$ for an $R > 0$, and for $z \in D(z_0, R) \backslash \{z_0\}$, f has the Laurent series $f(z) = \sum_{n \in \mathbb{Z}} c_n (z - z_0)^n$, then the coefficient c_{-1} is called *the residue of f at z_0*, and is denoted by $\mathrm{Res}(f, z_0)$. In everyday language, 'residue' means something which is 'left over'. One can think of c_{-1} as something which is left over in the following manner: If $r \in (0, R)$, and $C(t) := z_0 + re^{2\pi i t}$, $t \in [0, 1]$, then as

$$\int_C c_n (z - z_0)^n dz = \begin{cases} 0 & \text{for } n \neq -1, \\ 2\pi i c_{-1} & \text{for } n = -1, \end{cases}$$

if termwise integration ($*$) is allowed below, then

$$\int_C f(z)dz = \int_C \sum_{n \in \mathbb{Z}} c_n (z - z_0)^n dz \overset{(*)}{=} 2\pi i c_{-1}.$$

In fact, from the Laurent series theorem, $2\pi i\, c_{-1} = \int_C \frac{f(z)}{(z-z_0)^{-1+1}}dz = \int_C f(z)dz$. Why make a fuss about the residue? $2\pi i c_{-1} = \int_C f(z)dz$ gives a way of computing contour integrals via calculating the residue of f at z_0 (which amounts to finding the value of the coefficient c_{-1} in the Laurent expansion of f). So if there is way of calculating c_{-1} easily, then we can compute $\int_\gamma f(z)dz = \int_C f(z)dz = 2\pi i c_{-1}$ for any closed path γ which is $D\backslash\{z_0\}$-homotopic to C. In Exercise 4.60, there *is* a way of calculating c_{-1} via the formula $c_{-1} = \frac{1}{(m-1)!} \lim_{z\to z_0} \frac{d^{m-1}}{dz^{m-1}}((z-z_0)^m f(z))$, when z_0 is a pole of order m of f. It turns out that some awkward real integrals can be computed by first relating them to a contour integral $\int_\gamma f(z)dz$ for an appropriate holomorphic f in some domain and some path γ, and then using this route via the residue to evaluate the contour integral, and eventually also the real integral.

Example 4.20. We will view the real integral $\int_0^{2\pi} \frac{1}{5+3\cos\theta}d\theta$ as a contour integral along a circular path γ as follows. If $z = e^{i\theta}$, then $\cos\theta = \frac{e^{i\theta}+e^{-i\theta}}{2} = \frac{z+z^{-1}}{2} = \frac{z^2+1}{2z}$. Thus if $\gamma(\theta) := e^{i\theta}$, $\theta \in [0,2\pi]$, then $\gamma'(\theta)d\theta = ie^{i\theta}d\theta$, and so

$$\int_0^{2\pi} \frac{1}{5+3\cos\theta}d\theta = \int_\gamma \frac{1}{5+3\frac{z^2+1}{2z}} \frac{1}{iz}dz = \int_\gamma -\frac{2i}{(3z+1)(z+3)}dz.$$

Let $f(z) := -\frac{2i}{(3z+1)(z+3)}$, $z \in \mathbb{C}\backslash\{-3, -\frac{1}{3}\}$. Then f has only two singularities, which are isolated, and they are poles, at $-\frac{1}{3}$ and at -3, both of order 1. Of these, only the one at $-\frac{1}{3}$ lies inside γ. A reparametrisation of γ is $\mathbb{C}\backslash\{-3,-\frac{1}{3}\}$-homotopic to the circle C, where $C(t) = -\frac{1}{3} + re^{2\pi it}$, $t \in [0,1]$, and e.g. $r = \frac{1}{3}$. So $\int_0^{2\pi} \frac{1}{5+3\cos\theta}d\theta = \int_\gamma f(z)dz = \int_C f(z)dz = 2\pi i \operatorname{Res}(f, -\frac{1}{3})$. In $D(-\frac{1}{3}, r)\backslash\{-\frac{1}{3}\}$, we have $f(z) = \frac{c_{-1}}{z+\frac{1}{3}} + c_0 + \cdots$. So $(z+\frac{1}{3})f(z) = c_{-1} + c_0(z+\frac{1}{3}) + \cdots \in \mathcal{O}(D(-\frac{1}{3}, r))$. Thus $\operatorname{Res}(f, -\frac{1}{3}) = c_{-1} = \lim_{z\to -\frac{1}{3}} (z+\frac{1}{3})f(z) = \lim_{z\to -\frac{1}{3}} -\frac{2i}{3(z+3)} = -\frac{i}{4}$. So $\int_0^{2\pi} \frac{1}{5+3\cos\theta}d\theta = \frac{\pi}{2}$. ◇

More generally, if f has a finite number of *poles* in D, then the following holds.

Theorem 4.20 (Residue Theorem). *Let D be a domain, $p_1, \cdots, p_K \in D$ for some $K \in \mathbb{N}$, $f \in \mathcal{O}(D\backslash\{p_1, \ldots, p_K\})$, $p_1, \ldots, p_K$ be poles of f of order $m_1, \ldots, m_K \in \mathbb{N}$, respectively. Let $R_1, \cdots, R_K > 0$ be such that the discs $D(p_k, R_k) \subset D$, and they are mutually disjoint. For $k = 1, \cdots K$, let C_k be a circular path centred at p_k traversed once in the counterclockwise direction, contained in $D(p_k, R_k)$. Let γ be a closed path in $D\backslash\{p_1, \ldots, p_K\}$, such that for each $k = 1, \ldots, K$, a reparametrisation of γ is $D\backslash\{p_k\}$-homotopic to C_k. Then $\int_\gamma f(z)dz = 2\pi i \sum_{k=1}^{K} \operatorname{Res}(f, p_k)$.*

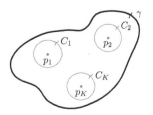

Proof. In $D(p_k, R_k)\backslash\{p_k\}$, $f(z) = \sum\limits_{n=1}^{m_k} c_{-n}^{(k)}(z - p_k)^{-n} + \sum\limits_{n=0}^{\infty} c_n^{(k)}(z - p_k)^n = r_k + h_k$,
where we denote the sum with negative powers of $z - p_k$ by r_k, and the sum with
nonnegative powers of $z - p_k$ by h_k, $k = 1, \cdots K$. Then $f - r_k = h_k \in \mathcal{O}(D(p_k, R_k))$,
and r_k is a rational function, defined in $\mathbb{C}\backslash\{p_k\}$, having only one singularity, namely
a pole at p_k. Set

$$g(z) := \begin{cases} f(z) - (r_1(z) + \cdots + r_K(z)) & \text{if } z \in D\backslash\{p_1, \cdots, p_K\}, \\ h_k - \sum\limits_{j\in\{1,\cdots,K\}\backslash\{k\}} r_j(z) & \text{if } z \in D(p_k, R_k), \ k = 1, \cdots, K. \end{cases}$$

Then g is well-defined as $f - r_k = h_k$, $k \in \{1, \cdots, K\}$. Also, $g \in \mathcal{O}(D)$. Firstly,
$g \in \mathcal{O}(D\backslash\{p_1, \cdots, p_K\})$. Moreover, for any $k \in \{1, \cdots, K\}$, as $h_k \in \mathcal{O}(D(p_k, R_k))$,
and $r_j \in \mathcal{O}(D(p_k, R_k))$ for all $j \neq k$, it follows that $g \in \mathcal{O}(D(p_k, R_k))$. As a
reparametrisation of γ is $D\backslash\{p_1\}$-homotopic to the circle C_1, by the Cauchy Integral
Theorem, $0 = \int_{C_1} g(z)dz = \int_{\gamma} g(z)dz = \int_{\gamma}(f - (r_1 + \cdots + r_K))dz$. Thus

$$\int_{\gamma} f\, dz = \sum_{k=1}^{K} \int_{\gamma} r_k\, dz \overset{(\star)}{=} \sum_{k=1}^{K} \int_{C_k} r_k\, dz \overset{(\star\star)}{=} \sum_{k=1}^{K} 2\pi i\, c_{-1,k} = 2\pi i \sum_{k=1}^{K} \text{Res}(f, p_k),$$

where $(\star)$ follows by the Cauchy Integral Theorem (since for each $k = 1, \cdots, K$,
$r_k \in \mathcal{O}(D\backslash\{p_k\})$, and a reparametrisation of γ is $D\backslash\{p_k\}$-homotopic to C_k), and
$(\star\star)$ follows since for all $k = 1, \cdots K$, $\int_{C_k} r_k\, dz = \sum\limits_{n=1}^{m_k} c_{-n}^{(k)} \int_{C_k} \frac{1}{(z-p_k)^n}\, dz = c_{-1}^{(k)}$. $\square$

Exercise 4.70. Evaluate $\int_{\gamma} \frac{\text{Log}\, z}{1+\exp z}dz$ along the path γ shown below.

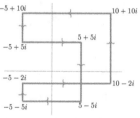

Exercise 4.71. Evaluate $\int_{-\pi}^{\pi} \frac{1}{2+\cos\theta}d\theta$.

Exercise 4.72. Evaluate $\int_0^{2\pi} \frac{\cos\theta}{5+4\cos\theta}d\theta$.

Exercise 4.73. Deduce $\int_0^{2\pi} e^{\cos\theta}\cos(n\theta - \sin\theta)d\theta = \frac{2\pi}{n!}$, $n \in \mathbb{N}$, by finding $\int_C \frac{\exp z}{z^{n+1}}dz$,
where $C(\theta) := e^{i\theta}$, $\theta \in [0, 2\pi]$.

Exercise 4.74. Let f have a zero of order 1 at z_0, so that $\frac{1}{f}$ has a pole of order 1 at z_0.
Prove that $\text{Res}(\frac{1}{f}, z_0) = \frac{1}{f'(z_0)}$.

Exercise 4.75. Prove that $\text{Res}(\frac{1}{\sin z}, k\pi) = (-1)^k$, $k \in \mathbb{Z}$.

Exercise 4.76. Using Residue Calculus, show that $\int_0^{2\pi}(\sin t)^{2n}dt = \frac{\pi}{2^{2n-1}}\binom{2n}{n}$ for all $n \in \mathbb{N}$.
Hint: Use the Binomial Theorem for finding the coefficient of z^0 in $(z - z^{-1})^{2n}$.

Exercise 4.77. Let g be an entire function.
(1) Show that the residue of $g(z)(1 + \frac{1}{z^2})$ at 0 is $g'(0)$.
(2) Show that $\int_0^{2\pi} g(e^{it})\cos t\, dt = \pi g'(0)$. *Hint:* $2\cos t = z + \frac{1}{z}$, where $z = e^{it}$.

Exercise 4.78. The nth Fibonacci number f_n, where $n \geqslant 0$, is defined by the following recurrence relation: $f_0 = 1$, $f_1 = 1$, $f_n = f_{n-1} + f_{n-2}$ for $n \geqslant 2$. Let $F(z) := \sum_{n=0}^{\infty} f_n z^n$.

(1) Prove by induction that $f_n \leqslant 2^n$ for all $n \geqslant 0$.

(2) Using the estimate $f_n \leqslant 2^n$, deduce that the radius of convergence of F is at least $\frac{1}{2}$.

(3) Using the recurrence relation, show that $F(z) = \frac{1}{1-z-z^2}$.

(4) Verify that $\text{Res}(\frac{1}{z^{n+1}(1-z-z^2)}, 0) = f_n$.

(5) Using the Residue Theorem, prove that $f_n = \frac{1}{\sqrt{5}}((\frac{1+\sqrt{5}}{2})^{n+1} - (\frac{1-\sqrt{5}}{2})^{n+1})$.
   Hint: If $I_R = \int_{C_R} \frac{1}{z^{n+1}(1-z-z^2)} dz$, $C_R(t) = Re^{it}$, $t \in [0, 2\pi]$, then show that $\lim_{R \to \infty} I_R = 0$.

Exercise 4.79. Consider the rational function $f(z) := \frac{1}{z^{2022}-1}$ of the complex variable z.

(1) Determine the poles of f and their order.

(2) What is the residue $\text{Res}(f, p)$ at each of the poles p?

(3) Show that $\int_C f(z)dz = 0$, where $C(t) := 2e^{it}$, $t \in [0, 2\pi]$.

Integral of a real rational function using the Residue Theorem. As we mentioned earlier, the Residue Theorem can be used to calculate contour integrals, and sometimes gives an easy way to calculate some real integrals. Let us see how it can be used to calculate the improper integrals of rational functions. Consider an integral of the type $\int_{-\infty}^{\infty} f(x)dx$, where $f : \mathbb{R} \to \mathbb{R}$ is a continuous function. Recall that such an integral, for which the interval of integration is not finite, is called an *improper integral*, defined as

$$\int_{-\infty}^{\infty} f(x)dx = \lim_{a \to -\infty} \int_a^0 f(x)dx + \lim_{b \to +\infty} \int_0^b f(x)dx,$$

when both the limits on the right hand side exist. In this case,

$$\int_{-\infty}^{\infty} f(x)dx = \lim_{r \to \infty} \int_{-r}^0 f(x)dx + \lim_{r \to +\infty} \int_0^r f(x)dx$$
$$= \lim_{r \to \infty} \left(\int_{-r}^0 f(x)dx + \int_0^r f(x)dx \right) = \lim_{r \to +\infty} \int_{-r}^r f(x)dx.$$

We call $\lim_{r \to +\infty} \int_{-r}^r f(x)dx$ the *Cauchy principal value* of the integral[6].

We now assume that the integrand f is a real rational function whose denominator is different from 0 for all real x, and the denominator has degree at least two more than the degree of the numerator. Then it can be seen that $\int_{-\infty}^{\infty} f(x)dx$ exists as follows. Let $f = \frac{p}{q}$, where $p, q \in \mathbb{R}[x]$ have degrees $\deg p, \deg q$, respectively. By the result in Exercise 1.37, there exist constants $m, M, R > 0$ such that $|p(x)| \leqslant M|x|^{\deg p}$ and $|q(x)| \geqslant m|x|^{\deg q}$ for all $|x| > R$, giving $|f(x)| \leqslant \frac{M}{m} \frac{1}{x^2}$. As $\int_R^{\infty} \frac{1}{x^2} dx < \infty$, it follows by comparison that $\int_{-\infty}^{\infty} |f(x)|dx < \infty$, and hence also $\int_{-\infty}^{\infty} |f(x)|dx$ exists. We will compute $\int_{-r}^r f(x)dx$ using the Residue Theorem, and determine $\int_{-\infty}^{\infty} f(x)dx$ by passing to the limit as $r \to \infty$.

---

[6]Even if the Cauchy principal value exists, $\int_{-\infty}^{\infty} f(x)dx$ need not: e.g., $\lim_{r \to +\infty} \int_{-r}^r x\,dx = 0$, $\int_0^b x\,dx = \frac{b^2}{2}$.

Consider $\int_\gamma f(z)dz$, around a path $\gamma = s + \sigma$ as shown in the following picture, where s denotes the straight line path $[-r, r] \ni x \mapsto s(x) := x$, and σ is the semicircular path starting at r and ending at $-r$ in the upper half-plane.

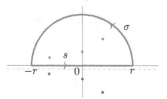

Since f is rational, it has finitely many poles in the upper half plane, and choosing r large enough, γ encloses all of these poles in its interior. We now apply the Residue Theorem to f, taking the domain D to be the half-plane above the dotted line shown in picture, separating the real axis and the poles of f in the lower half-plane. Then

$$\int_\gamma f(z)dz = \int_\sigma f(z)dz + \int_{-r}^r f(x)dx = 2\pi i \sum_{k:\,\mathrm{Im}(p_k)>0} \mathrm{Res}(f, p_k),$$

where the sum consists of terms for all the poles that lie in the upper half-plane. From this, we obtain $\int_{-r}^r f(x)dx = 2\pi i \sum_{k:\,\mathrm{Im}(p_k)>0} \mathrm{Res}(f, p_k) - \int_\sigma f(z)dz$.

We show that as r increases, the value of the integral over the corresponding semicircular arc σ approaches 0. Indeed, as the degree of the denominator of f is at least two more than the degree of the numerator, there exist M, r_0 large enough such that $|f(z)| \leqslant \frac{M}{|z|^2}$ for all $|z| > r_0$. Hence for $r > r_0$, $|\int_\sigma f(z)\,dz| \leqslant \frac{M}{r^2}\pi r = \frac{M\pi}{r}$. Consequently, $\lim_{r\to+\infty} \int_\sigma f(z)dz = 0$, and so $\int_{-\infty}^\infty f(x)dx = 2\pi i \sum_{k:\,\mathrm{Im}(p_k)>0} \mathrm{Res}(f, p_k)$.

Example 4.21. We will show that $\int_0^\infty \frac{1}{1+x^4}dx = \frac{\pi}{2\sqrt{2}}$. With $f(z) := \frac{1}{1+z^4}$, f has four poles of order 1: $p_1 = \exp\frac{\pi i}{4}$, $p_2 = \exp\frac{3\pi i}{4}$, $p_3 = \exp\frac{5\pi i}{4}$, $p_4 = \exp\frac{7\pi i}{4}$.

The first two of these poles lie in the upper half plane. We have

$$\mathrm{Res}(f, p_1) = \lim_{z\to p_1} \frac{z-p_1}{1+z^4} = \lim_{z\to p_1} \frac{1}{\frac{(1+z^4)-(1+p_1^4)}{z-p_1}} = \frac{1}{\frac{d}{dz}(1+z^4)|_{z=p_1}} = \frac{1}{4p_1^3} = -\frac{p_1}{4},$$

$$\mathrm{Res}(f, p_2) = \lim_{z\to p_2} \frac{z-p_2}{1+z^4} = \lim_{z\to p_2} \frac{1}{\frac{(1+z^4)-(1+p_2^4)}{z-p_2}} = \frac{1}{\frac{d}{dz}(1+z^4)|_{z=p_2}} = \frac{1}{4p_2^3} = -\frac{p_2}{4}.$$

Thus $\int_{-\infty}^\infty \frac{1}{1+x^4}dx = 2\pi i(-\frac{p_1}{4} - \frac{p_2}{4}) = 2\pi i(-\frac{\sqrt{2}i}{4}) = \frac{\pi}{\sqrt{2}}$. As f is an even function (i.e., $f(x) = f(-x)$ for all $x \in \mathbb{R}$), we obtain $\int_0^\infty \frac{1}{1+x^4}dx = \frac{1}{2}\int_{-\infty}^\infty \frac{1}{1+x^4}dx = \frac{\pi}{2\sqrt{2}}$. $\diamond$

Exercise 4.80. Evaluate the following integrals:

(1) $\int_0^\infty \frac{1}{1+x^2}\,dx$.

(2) $\int_0^\infty \frac{1}{(a^2+x^2)(b^2+x^2)}\,dx$, where $a > b > 0$.

(3) $\int_0^\infty \frac{1}{(1+x^2)^2}\,dx$.

(4) $\int_0^\infty \frac{1+x^2}{1+x^4}\,dx$.

Exercise 4.81. Use Residue Calculus to determine $\int_{-\infty}^\infty \frac{e^{-i\xi x}}{1+x^2}\,dx$, where $\xi \in \mathbb{R}$.
Hint: Consider the two cases $\xi \geq 0$ and $\xi < 0$ separately.

Exercise 4.82 (Basel Problem). The Residue Theorem can sometimes be used for summing series. In this exercise, we will show that $\sum_{n=1}^\infty \frac{1}{n^2} = \frac{\pi^2}{6}$. Let $f(z) := \frac{1}{z^2}\frac{\cos(\pi z)}{\sin(\pi z)}$. The zeroes of $\sin(\pi z)$ are located at the integers. It follows that f has poles at the integers.

(1) Show that 0 is a pole of f of order 3. Find $\mathrm{Res}(f, 0)$.

(2) Show that each $n \in \mathbb{Z}\backslash\{0\}$ is a pole of f of order 1. Find $\mathrm{Res}(f, n)$.

Consider the square path S_N with vertices at $(N + \frac{1}{2})(\pm 1 \pm i)$, traversed once in the counterclockwise direction. For a $z = x + iy$, $x, y \in \mathbb{R}$, belonging to the top or bottom sides of S_N, we have the crude estimate $|y| \geq \frac{1}{2}$. Thus

$$\left|\frac{\cos(\pi z)}{\sin(\pi z)}\right| = \left|\frac{e^{i\pi z}+e^{-i\pi z}}{e^{i\pi z}-e^{-i\pi z}}\right| \leq \frac{|e^{i\pi z}|+|e^{-i\pi z}|}{||e^{i\pi z}|-|e^{-i\pi z}||} = \frac{e^{-\pi y}+e^{\pi y}}{|e^{-\pi y}-e^{\pi y}|} = \frac{\cosh(\pi y)}{|\sinh(\pi y)|} = \frac{1}{|\tanh(\pi y)|} \leq \frac{1}{\tanh\frac{\pi}{2}},$$

since $\tanh$ is strictly increasing ($\tanh' t = \frac{1}{(\cosh t)^2} > 0$ for all $t \in \mathbb{R}$). Thus for z belonging to the top or bottom sides of S_N, we have $|f(z)| \leq \frac{A}{|z|^2}$, where $A = \frac{1}{\tanh\frac{\pi}{2}}$.

(3) If z belongs to the right or left side of S_N and $y := \mathrm{Im}\,z$, then using the trigonometric addition formulae, show that $|\cos(\pi z)| = |\sin(i\pi y)|$ and $|\sin(\pi z)| = |\cos(i\pi y)|$.

So for a z belonging to the right or left side of S_N, if $y := \mathrm{Im}\,z$, then

$$\left|\frac{\cos(\pi z)}{\sin(\pi z)}\right| = \left|\frac{\sin(i\pi y)}{\cos(i\pi y)}\right| = \left|\frac{e^{-\pi y}-e^{\pi y}}{e^{-\pi y}+e^{\pi y}}\right| = \left|\frac{e^{\pi|y|}-e^{-\pi|y|}}{e^{\pi|y|}+e^{-\pi|y|}}\right| = \frac{1-e^{-2\pi|y|}}{1+e^{-2\pi|y|}} \leq 1.$$

Thus for z belonging to the right or left sides of S_N, we have $|f(z)| \leq \frac{1}{|z|^2}$.

(4) Show that $\lim_{N\to\infty} \int_{S_N} f(z)\,dz = 0$ using the ML-inequality and the estimates above.

(5) Deduce that $\sum_{n=1}^\infty \frac{1}{n^2} = \frac{\pi^2}{6}$.

Chapter 5

Harmonic functions

In this last chapter, we will study harmonic functions, which are real-valued functions that solve a certain partial differential equation, called the Laplace equation.

We will show that the real and imaginary parts of holomorphic functions are harmonic, and that the converse holds locally, and even globally on simply connected domains.

Finally, some consequences of the above interplay between harmonic and holomorphic functions, in particular in the context of a certain 'boundary value problem' for the Laplace equation, called the Dirichlet problem.

5.1. What is a harmonic function?

Definition 5.1. Let U be an open subset of $\mathbb{R}^2$. A function $u : U \to \mathbb{R}$ is called *harmonic* if $u \in C^2(U)$ (i.e., it has continuous partial derivatives up to order 2), and u satisfies the *Laplace equation*, i.e., $(\Delta u)(x, y) := \frac{\partial^2 u}{\partial x^2}(x, y) + \frac{\partial^2 u}{\partial y^2}(x, y) = 0$ for all $(x, y) \in U$.

Example 5.1. Let $U = \mathbb{R}^2$. If $u(x, y) = x^2 - y^2$ for $(x, y) \in \mathbb{R}^2$, then $\frac{\partial u}{\partial x} = 2x$, $\frac{\partial^2 u}{\partial x^2} = 2$, $\frac{\partial u}{\partial y} = -2y$, $\frac{\partial^2 u}{\partial y^2} = -2$. Thus $\frac{\partial^2 u}{\partial x^2}(x, y) + \frac{\partial^2 u}{\partial y^2}(x, y) = 2 - 2 = 0$. Since $u \in C^2(\mathbb{R}^2)$ and $\Delta u = 0$ in $\mathbb{R}^2$, u is harmonic in $\mathbb{R}^2$. $\diamond$

Example 5.2. If $\tilde{u}(x, y) = x^2 + y^2$ for $(x, y) \in \mathbb{R}^2$, then $\frac{\partial^2 \tilde{u}}{\partial x^2}(x, y) + \frac{\partial^2 \tilde{u}}{\partial y^2}(x, y) = 2 + 2 \neq 0$. Since $\Delta \tilde{u}$ is never 0 in $\mathbb{R}^2$, $\tilde{u}$ is not harmonic in any open subset of $\mathbb{R}^2$. $\diamond$

Exercise 5.1. Show that the functions u below are harmonic in the given open set U.
(1) $u(x, y) = \log(x^2 + y^2)$, $U = \mathbb{R}^2 \backslash \{(0, 0)\}$.
(2) $u(x, y) = e^x \sin y$, $U = \mathbb{R}^2$.

Exercise 5.2. Show that the set H(U) of all harmonic functions on an open set U forms a real vector space with pointwise operations. Is the pointwise product of two harmonic functions also necessarily harmonic?

Why study harmonic functions? Harmonic functions are important because they satisfy the Laplace equation. The Laplace equation is the prototype of an important class of partial differential equations, namely 'elliptic equations', which is one of the three main classes of second order linear PDEs.

Class of PDE	Main example	
Elliptic	Laplace equation	$\frac{\partial^2 u}{\partial x^2} + \frac{\partial^2 u}{\partial y^2} = 0$
Parabolic	Diffusion equation	$\frac{\partial u}{\partial t} - \frac{\partial^2 u}{\partial x^2} = 0$
Hyperbolic	Wave equation	$\frac{\partial^2 u}{\partial t^2} - \frac{\partial^2 u}{\partial x^2} = 0$

The Laplace equation arises in many applications. For example, the heat equation in the plane is $\frac{\partial u}{\partial t} = \frac{\partial^2 u}{\partial x^2} + \frac{\partial^2 u}{\partial y^2}$, where $u(x, y, t)$ denotes the temperature at the point (x, y) belonging in a planar region $U \subset \mathbb{R}^2$ at time $t \in \mathbb{R}$. Then the steady-state temperature (i.e., u for which $\frac{\partial u}{\partial t} \equiv 0$) satisfies the Laplace equation. The Laplace equation also has a link with stochastic processes, described very roughly below.

Consider $\mathbb{D} := \{z \in \mathbb{C} : |z| < 1\}$, and imagine a particle starting at some point $z \in \mathbb{D}$ and undergoing Brownian motion (e.g., a pollen grain in water, bombarded by many tiny water molecules producing 'random' motion). Intuitively, we expect, since the motion is random, that eventually the particle will leave the boundary $\mathbb{T} := \{z \in \mathbb{C} : |z| = 1\}$ of $\mathbb{D}$. Denote by ζ_z the point on $\mathbb{T}$ where the particle first exits the unit circle $\mathbb{T}$, having started at $z \in \mathbb{D}$. So we get a random variable ζ_z which lives on the unit circle.

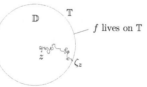

Now let $f : \mathbb{T} \to \mathbb{R}$ be a given continuous function. Then we can think of $f(\zeta_z)$ as being a real-valued random variable on $\mathbb{T}$. Let us denote its expectation by $\mathbb{E}(f(\zeta_z))$. This depends on where one starts initially, that is, it depends on z. Define $u : \mathbb{D} \to \mathbb{R}$ by $u(z) = \mathbb{E}(f(\zeta_z))$ for all $z \in \mathbb{D}$. It turns out that u is then harmonic, and in fact it is a solution to the 'Dirichlet problem', which is the boundary value problem, where given $f : \mathbb{T} \to \mathbb{R}$ on the boundary $\mathbb{T}$, we have to find a function u, solving the Laplace equation in the interior $\mathbb{D}$ of $\mathbb{T}$, such that it has a continuous extension to $\mathbb{T}$, matching with the given data f:

$$\begin{cases} \Delta u = 0 \text{ in } \mathbb{D}, \\ u|_{\mathbb{T}} = f. \end{cases}$$

5.2. Link between harmonic and holomorphic functions

It might appear that harmonic functions should belong just to the realm of real analysis. In this section, we will now learn about two results, which justifies their study in complex analysis. Roughly, a function is harmonic in an open set if and only if locally, it is the real part of some holomorphic function.

Theorem 5.1. *If $U \subset \mathbb{C}$ is an open set and $f \in \mathcal{O}(U)$, then $u := \operatorname{Re} f$ and $v := \operatorname{Im} f$ are harmonic functions in U.*

Vice versa, we will also learn the following converse to this.

Theorem 5.2. *If $U \subset \mathbb{C}$ is simply connected and $u : U \to \mathbb{R}$ is harmonic in U, then there exists a function $v : U \to \mathbb{R}$, such that $f := u + iv \in \mathcal{O}(U)$.*

For the f in the conclusion of the above result, we have $\operatorname{Re} f = u$, and $\operatorname{Im} f = v$. So every harmonic function in a simply connected domain is the real part of some holomorphic function defined there. Since a disc is simply connected, in particular every harmonic function is *locally* the real part of a holomorphic function defined (at least in that disc). We will see later on in Exercise 5.4 that the assumption of simply connectedness is not superfluous, and that given a harmonic function in a *non*-simply connected domain, there may fail to exist a *globally* defined holomorphic function in the whole domain whose real part is the given harmonic function. Before proving Theorem 5.1, let us revisit Example 5.1.

Example 5.3. In Example 5.1, we had seen that $u = x^2 - y^2$ is harmonic in $U = \mathbb{R}^2$. With $z = (x, y) \in \mathbb{R}^2$, $u = \operatorname{Re}(z^2) = \operatorname{Re}(x^2 - y^2 + 2xyi)$. As $z \mapsto z^2$ is entire, Theorem 5.1 delivers again the earlier observation that u is harmonic in $\mathbb{R}^2$. From the calculation above, Theorem 5.1 also shows that $v := 2xy = \operatorname{Im}(z^2)$ is harmonic. (Check: $\frac{\partial u}{\partial x} = 2y$, $\frac{\partial^2 u}{\partial x^2} = 0$, $\frac{\partial v}{\partial x} = 2x$, $\frac{\partial^2 v}{\partial y^2} = 0$, and so $\frac{\partial^2 u}{\partial x^2} + \frac{\partial^2 v}{\partial y^2} = 0 + 0 = 0$.) Also, it follows from Example 5.2 and Theorem 5.1 that for any open subset U of $\mathbb{C}$, $\tilde{u} := x^2 + y^2$ is not the real part of any holomorphic function defined in U. $\diamond$

Proof of Theorem 5.1. We have $f(x + iy) = u(x, y) + iv(x, y)$ for $(x, y) \in U$. Since f is infinitely many times complex differentiable, u, v have continuous partial derivatives of all orders (see Exercise 3.39). Using the Cauchy-Riemann equations, $\frac{\partial^2 u}{\partial x^2} = \frac{\partial}{\partial x}\left(\frac{\partial u}{\partial x}\right) \overset{\text{C-R}}{=} \frac{\partial}{\partial x}\left(\frac{\partial v}{\partial y}\right) \overset{u \in C^2}{=} \frac{\partial}{\partial y}\left(\frac{\partial v}{\partial x}\right) \overset{\text{C-R}}{=} \frac{\partial}{\partial y}\left(-\frac{\partial u}{\partial y}\right) = -\frac{\partial^2 u}{\partial y^2}$, and so u is harmonic. As $v = \operatorname{Re}(-if)$, v is harmonic as well. $\square$

Now we show Theorem 5.2, which gives a converse to the above result when the open set in question is a simply connected domain. As mentioned earlier, for more general domains, it can happen that there are harmonic functions which are not globally the real part of a holomorphic function; see Exercise 5.4, where we take

- $U := \mathbb{R}^2 \backslash \{(0,0)\}$ (which is not simply connected) and
- $u := \log(x^2 + y^2)$ (which is harmonic in U).

Then there is no holomorphic f in $\mathbb{C} \backslash \{0\}$ such that $u = \operatorname{Re} f$ in U.

Proof of of Theorem 5.2. We will construct a holomorphic f with real part u, and then $v := \operatorname{Im} f$ will serve as the required (harmonic) function. Set $g = \frac{\partial u}{\partial x} - i\frac{\partial u}{\partial y}$.

We will prove that g is holomorphic, and then construct a primitive of g, necessarily holomorphic, which will be (up to an additive constant) the f we seek. To show that g is holomorphic, we will use the sufficiency of the Cauchy-Riemann equations (Theorem 2.5). Since u is harmonic, $\operatorname{Re} g = \frac{\partial u}{\partial x}$ and $\operatorname{Im} g = -\frac{\partial u}{\partial y}$ have continuous partial derivatives. Moreover, as u satisfies the Laplace equation, the real and imaginary parts of g satisfy the Cauchy-Riemann equations:

$$\frac{\partial}{\partial x}(\operatorname{Re} g) = \frac{\partial}{\partial x}\left(\frac{\partial u}{\partial x}\right) = \frac{\partial^2 u}{\partial x^2} = -\frac{\partial^2 u}{\partial y^2} = \frac{\partial}{\partial y}\left(-\frac{\partial u}{\partial y}\right) = \frac{\partial}{\partial y}(\operatorname{Im} g),$$

$$\frac{\partial}{\partial y}(\operatorname{Re} g) = \frac{\partial}{\partial y}\left(\frac{\partial u}{\partial x}\right) = \frac{\partial}{\partial x}\left(\frac{\partial u}{\partial y}\right) = -\frac{\partial}{\partial x}\left(-\frac{\partial u}{\partial y}\right) = -\frac{\partial}{\partial x}(\operatorname{Im} g).$$

So $g \in \mathcal{O}(U)$. By Theorem 3.12, as U is simply connected, g has a primitive $G \in \mathcal{O}(U)$. Decompose $G = \tilde{u} + i\tilde{v}$ into its real and imaginary parts $\tilde{u}, \tilde{v}$. Then

$$\frac{\partial u}{\partial x} - i\frac{\partial u}{\partial y} = g = G' = \frac{\partial \tilde{u}}{\partial x} + i\frac{\partial \tilde{v}}{\partial x}, \qquad (*)$$

and so $\frac{\partial(\tilde{u}-u)}{\partial x} = 0$. Hence, by the Fundamental Theorem of Calculus, $\tilde{u}-u$ is locally constant along horizontal lines. Also, $(*)$ gives

$$-\frac{\partial u}{\partial y} = \frac{\partial \tilde{v}}{\partial x} = -\frac{\partial \tilde{u}}{\partial y},$$

where we used the Cauchy-Riemann equations to obtain the last equality. Hence $\frac{\partial(\tilde{u}-u)}{\partial y} = 0$, showing that $\tilde{u}-u$ is locally constant along vertical lines as well. Since U is a domain, any two points in U can be joined by a stepwise path, and so $\tilde{u}-u$ must be a constant, say C ($\in \mathbb{R}$), in U. Consequently, $f := G - C = (\tilde{u}-C)+i\tilde{v} = u+i\tilde{v}$. We have $f \in \mathcal{O}(U)$, $u = \operatorname{Re} f$, $v := \tilde{v} = \operatorname{Im} G$, and $f = u + iv$. $\qquad \square$

Definition 5.2. Let $U \subset \mathbb{C}$ be open and $u : U \to \mathbb{R}$ be harmonic in U. A *harmonic conjugate* of u is any $v : U \to \mathbb{R}$ such that $f := u + iv \in \mathcal{O}(U)$.

Then as v is the imaginary part of the holomorphic function f, v is harmonic too.

Example 5.4. $v := 2xy$ is a harmonic conjugate of the harmonic $u := x^2 - y^2$ considered in Example 5.1, since $f := u + iv = x^2 - y^2 + 2xyi = (x + iy)^2 = z^2$, (where $z = x + iy$, $x, y \in \mathbb{R}$), is entire. $\qquad \diamond$

Harmonic conjugates are obviously not unique since we can just add a constant to a harmonic conjugate and get a new harmonic conjugate. In a domain D, this is the only indeterminacy: If $v, \tilde{v}$ are both harmonic conjugates to a harmonic u in D, then $u + iv, u + i\tilde{v} \in \mathcal{O}(D)$, so that $i(\tilde{v}-v) \in \mathcal{O}(D)$. Hence by the Cauchy-Riemann equations, $\frac{\partial(\tilde{v}-v)}{\partial x} = -\frac{\partial 0}{\partial y} = 0$ and $\frac{\partial(\tilde{v}-v)}{\partial y} = \frac{\partial 0}{\partial x} = 0$, showing that $\tilde{v} - v$ is constant along horizontal line segments and along vertical line segments. As D is a domain, it follows that $\tilde{v} - v$ is constant in D, that is, $\tilde{v} = v + C$ for some $C \in \mathbb{R}$.

In a simply connected domain, for each harmonic function, Theorem 5.2 guarantees the existence of a harmonic conjugate, but it is not particularly useful for finding a harmonic conjugate (from the proof, we see that it relies on the construction of a primitive G). A more direct way is to use the Cauchy-Riemann equations, as shown below.

Example 5.5. Let $u(x, y) = -(\sin x) \sinh y$ for all $(x, y) \in \mathbb{R}^2$. Then $u \in C^2(\mathbb{R}^2)$. Also, we have $\frac{\partial^2 u}{\partial x^2} = (\sin x) \sinh y$ and $\frac{\partial^2 u}{\partial y^2} = -(\sin x) \sinh y$, so that $\frac{\partial^2 u}{\partial x^2} + \frac{\partial^2 u}{\partial y^2} = 0$ in $\mathbb{R}^2$. Thus u is harmonic in $\mathbb{R}^2$. As $\mathbb{R}^2$ is simply connected, a harmonic conjugate exists on $\mathbb{R}^2$. If v is a harmonic conjugate, then $u + iv$ is entire, and so u, v satisfy the Cauchy-Riemann equations. Thus v satisfies

$$\frac{\partial v}{\partial x} = -\frac{\partial u}{\partial y} = (\sin x) \cosh y, \qquad (\star)$$

$$\frac{\partial v}{\partial y} = \frac{\partial u}{\partial x} = -(\cos x) \sinh y. \qquad (\star\star)$$

Integrating $(\star)$ with respect to x from 0 to x while keeping y fixed, we obtain that $v(x, y) - v(0, y) = -(\cos x)(\cosh y) + \cosh y$, and so

$$v(x, y) = -(\cos x)(\cosh y) + C(y), \qquad (*)$$

where $C(y) := v(0, y) + \cosh y$. Differentiating with respect to y, we get

$$-(\cos x)(\sinh y) \overset{(\star\star)}{=} \frac{\partial v}{\partial y} \overset{(*)}{=} -(\cos x)(\sinh y) + \frac{dC}{dy}(y).$$

Consequently $\frac{dC}{dy}(y) = 0$, and so $C(y) = C$ for all y. Thus we could take any constant C, and in particular we may choose $C = 0$. Hence based on the above, we take $v = -(\cos x)(\cosh y)$ as a candidate for a harmonic conjugate of u. In order to verify this guess, we may note that u, v are real differentiable in $\mathbb{R}^2$, and the Cauchy-Riemann equations are satisfied by the pair (u, v), showing that $f := u + iv$ is holomorphic. But instead, let us use an old calculation from Exercise 1.57 to check directly that $f := u + iv$ is holomorphic by giving f explicitly:

$$f = u + iv = -(\sin x)(\sinh y) - i(\cos x)(\cosh y)$$
$$= -i((\cos x)(\cosh y) - i(\sin x)(\sinh y)) = -i \cos z,$$

which is entire. $\diamond$

Exercise 5.3. Find harmonic conjugates for the following harmonic functions in $\mathbb{R}^2$: $e^x \sin y$, $x^3 - 3xy^2 - 2y$, $x(1 + 2y)$.

Exercise 5.4. Show that there is no holomorphic function f defined in $\mathbb{C}\backslash\{0\}$ whose real part is the harmonic function u defined by $u(x, y) = \log(x^2 + y^2)$, $(x, y) \in \mathbb{R}^2\backslash\{(0, 0)\}$. *Hint:* If v is a harmonic conjugate of u, then show that $f \in \mathcal{O}(\mathbb{C}\backslash\{0\})$ satisfies $f'(z) = \frac{2}{z}$.

Exercise 5.5. Is it possible to find a function $v : \mathbb{R}^2 \to \mathbb{R}$ so that the function f defined by $f(x + iy) = x^3 + y^3 + iv(x, y)$, $(x, y) \in \mathbb{R}^2$, is entire?

Exercise 5.6. Let $U \subset \mathbb{C}$ be open, u be harmonic in U, v be a harmonic conjugate of u. Suppose that $u^2 + v^2$ never vanishes in U. Prove that $\frac{u \frac{\partial u}{\partial x} + v \frac{\partial v}{\partial x}}{u^2 + v^2}$ is harmonic in U.

Exercise 5.7. Let $U \subset \mathbb{C}$ be open, and $v : U \to \mathbb{R}$ be a harmonic conjugate of a harmonic function $u : U \to \mathbb{R}$. Prove that the product uv is harmonic in U.

5.3. Consequences of the two way traffic

In the previous section, the two results given in Theorems 5.1 and 5.2 show that there is a two way traffic between the real analysis world of harmonic functions and the complex analysis world of holomorphic functions, allowing a fruitful interaction between the two worlds.

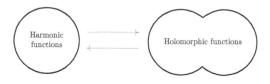

In the previous three chapters we have learnt many pleasant properties possessed by holomorphic functions. Let us now use these to derive some important properties of harmonic functions. In particular, we will now show the following results:

- If u is harmonic, then it is C^∞.
- The Mean Value Property for harmonic functions.
- The Maximum Principle for harmonic functions.
- Uniqueness of solutions to the Dirichlet Problem.

Harmonic functions are smooth.

Corollary 5.3. *Harmonic functions are infinitely many times real differentiable.*

The definition of a harmonic function demands only *twice* continuous differentiability. The remarkable result here says that thanks to the fact that the Laplace equation is satisfied, in fact the function has got to be infinitely differentiable. A result of this type (where some initial assumption of smoothness together with the satisfaction of some differential equation implies extra regularity) is called a *regularity result* in PDE theory.

Proof. Let u be harmonic in an open set $U \subset \mathbb{C}$. Let $z_0 = (x_0, y_0) \in U$. Then there is a $r > 0$ such that $D(z_0, r) \subset U$. As $u|_{D(z_0,r)}$ is harmonic and $D(z_0, r)$ is simply connected, there exists an $f \in \mathcal{O}(D(z_0, r))$ such that $\mathrm{Re}\, f = u$ in $D(z_0, r)$. But f is infinitely many times complex differentiable in $D(z_0, r)$. Consequently, u is infinitely many times real differentiable in $D(z_0, r)$ by Exercise 3.39. As the choice of $z_0 \in D$ was arbitrary, the result follows.   □

Exercise 5.8. Show that all the partial derivatives of a harmonic function are harmonic.

Mean value property. Using the Cauchy Integral Formula, we immediately obtain the following 'mean value property' of harmonic functions, which says that the value of a harmonic function at a point is the average (or mean) of the values on a circle with that point as the centre.

Theorem 5.4 (Mean-value property of harmonic functions). *Let $U \subset \mathbb{C}$ be open, $u : U \to \mathbb{R}$ be harmonic in U, $z_0 \in U$, $R > 0$ be such that $D(z_0, R) \subset U$. Then $u(z_0) = \frac{1}{2\pi} \int_0^{2\pi} u(z_0 + re^{it})dt$ for all $r \in (0, R)$.*

Proof. As $D(z_0, R)$ is simply connected, there exists an $f \in \mathcal{O}(D(z_0, R))$ such that $u = \operatorname{Re} f$. By the Cauchy Integral Formula, with $C(t) := z_0 + re^{it}$, $t \in [0, 2\pi]$,

$$f(z_0) = \frac{1}{2\pi i} \int_C \frac{f(z)}{z - z_0} dz = \frac{1}{2\pi i} \int_0^{2\pi} \frac{f(z_0 + re^{it})}{re^{it}} ire^{it} dt = \frac{1}{2\pi} \int_0^{2\pi} f(z_0 + re^{it}) dt.$$

Equating real parts, the claim is proved. $\qquad\square$

Exercise 5.9. Let $z_0 \in \mathbb{C}$, $R > 0$, and u be a harmonic function in $D(z_0, R)$. Show that $u(z_0) = \frac{1}{\pi R^2} \iint_{D(z_0, R)} u \, dA$, where $dA = dx dy$ is the area measure in $\mathbb{R}^2$.

Exercise 5.10 (Harnack's inequality). Let $z_0 \in \mathbb{C}$, $R > 0$, and u be a pointwise nonnegative harmonic function in $D(z_0, R)$. Show that for all $z \in D(z_0, R)$, $u(z) \leqslant \frac{1}{(1 - \frac{|z - z_0|}{R})^2} u(z_0)$.

Exercise 5.11 (An analogue of Liouville's theorem). Let u be nonnegative and harmonic in $\mathbb{R}^2$. Show that u is constant. Show that the condition of 'nonnegativeness of u' can be relaxed to 'u is bounded above or bounded below'. *Hint:* Pass to the limit $R \to \infty$ in Exercise 5.10. (See Exercise 5.14 for a different proof using Liouville's theorem.)

Intuitively, the mean-value property of harmonic functions rules out the possibility of them having a local maximum or local minimum (unless they are constant). To see this, suppose z_0 is such a would-be (say) local maximum, and take a small enough $r > 0$. The maximum $\max_C u$ on C, is at least as big as the mean of u on C, and by the mean-value property at least as big as $u(z_0)$. But as z_0 is a maximiser, we cannot have a strict inequality, and so $\max_C u$ equals the mean of u on C. This implies u is constant on C. But as this happens with each $r > 0$ small enough, u must be a constant. We give a rigorous proof below.

Maximum Principle. From the Maximum Modulus Theorem (see p. 93), we obtain the following.

Theorem 5.5 (Maximum Principle). *If $U \subset \mathbb{C}$ is a simply connected domain, $u : U \to \mathbb{R}$ is harmonic in U, $z_0 \in U$ is such that $u(z_0) \geqslant u(z)$ for all $z \in D$, then u is constant in U.*

Proof. There exists an $f \in \mathcal{O}(U)$ such that $u = \operatorname{Re} f$. If $g(z) := \exp(f(z))$ ($z \in U$), then $g \in \mathcal{O}(U)$. We have $|g(z_0)| = |\exp(f(z_0))| = e^{\operatorname{Re} z_0} = e^{u(z_0)} \geqslant e^{u(z)} = |g(z)|$ for all $z \in U$. By the Maximum Modulus Theorem applied to g, g is constant in U. Thus $|g|$ is also constant in U, i.e., $|g| = e^{\operatorname{Re} f} = e^u$ is a constant in U. Taking the (real) logarithm, it follows that u is constant in U. $\qquad\square$

Exercise 5.12 (Comparison Principle). Let $\mathbb{D} = \{z \in \mathbb{C} : |z| < 1\}$, $\mathbb{T} := \{z \in \mathbb{C} : |z| = 1\}$. Let $u, v : \mathbb{D} \cup \mathbb{T} \to \mathbb{R}$ be two continuous functions that are harmonic in $\mathbb{D}$. If $u(z) \geqslant v(z)$ for all $z \in \mathbb{T}$, then show that $u(z) \geqslant v(z)$ in $\mathbb{D}$ too.

Uniqueness of solution for the Dirichlet problem. We will now use the Maximum Principle to show the uniqueness of solutions in the Dirichlet problem. Let $\mathbb{D} = \{z \in \mathbb{C} : |z| < 1\}$, $\mathbb{T} := \{z \in \mathbb{C} : |z| = 1\}$. The *Dirichlet problem* is the following:

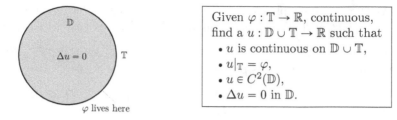

Given $\varphi : \mathbb{T} \to \mathbb{R}$, continuous,
find a $u : \mathbb{D} \cup \mathbb{T} \to \mathbb{R}$ such that
- u is continuous on $\mathbb{D} \cup \mathbb{T}$,
- $u|_{\mathbb{T}} = \varphi$,
- $u \in C^2(\mathbb{D})$,
- $\Delta u = 0$ in $\mathbb{D}$.

φ is called the *boundary data*. The Dirichlet problem arises in applications, for example in heat conduction. Using the Maximum Principle, we have the following.

Proposition 5.6. *The solution to the Dirichlet problem is unique.*

Proof. Let u_1, u_2 be two distinct solutions corresponding to the boundary data φ. In particular, $u_1 = u_2$ on $\mathbb{T}$. So u_1 differs from u_2 somewhere inside $\mathbb{D}$. Without loss of generality, let $w \in \mathbb{D}$ be such that $u_1(w) > u_2(w)$. (Otherwise swap u_1, u_2.) Then $u := u_1 - u_2$ is such that $u \equiv 0$ on $\mathbb{T}$, $u(w) > 0$, and u is harmonic in $\mathbb{D}$. Let $z_0 \in \mathbb{D} \cup \mathbb{T}$ be the maximiser for the real-valued continuous function u on the compact set $\mathbb{D} \cup \mathbb{T}$. As $u \equiv 0$ on $\mathbb{T}$, and $u(w) > 0$, we have $z_0 \notin \mathbb{T}$. So $z_0 \in \mathbb{D}$. Since $u(z_0) \geqslant u(z)$ for all $z \in \mathbb{D}$, the Maximum Principle implies that u must be constant in $\mathbb{D}$. As u is continuous on $\mathbb{D} \cup \mathbb{T}$, and since u is 0 on $\mathbb{T}$, the constant value of u must be 0 everywhere in $\mathbb{D} \cup \mathbb{T}$. Hence $u_1 = u_2$, a contradiction. $\square$

Remark 5.1. Let $u(re^{it}) := \frac{1}{2\pi} \int_0^{2\pi} \frac{1-r^2}{1-2r\cos(\theta-t)+r^2} \varphi(e^{i\theta}) d\theta$, $0 \leqslant r < 1$, $t \in (-\pi, \pi]$. This expression for u is called the *Poisson Integral Formula*, and it can be shown that it solves the Dirichlet problem with boundary data φ. We will not prove this here, but refer the interested reader to e.g. [10] for a proof.   ✳

Exercise 5.13 (Half-plane Dirichlet problems). Given a continuous function $\varphi : \mathbb{R} \to \mathbb{R}$, we seek a continuous, real-valued function h defined in $\{(x,y) \in \mathbb{R}^2 : y \geqslant 0\}$, such that h is harmonic in $\{(x,y) \in \mathbb{R}^2 : y > 0\}$ and $h(x,0) = \varphi(x)$ for all $x \in \mathbb{R}$.

(1) If $\varphi \in \mathbb{R}[x]$, then show that we can take h given by $h(x,y) = \mathrm{Re}(\varphi(x+iy))$.

(2) If $\varphi(x) = \frac{1}{1+x^2}$, then $(x,y) \mapsto \mathrm{Re}(\varphi(x+iy))$ is *not* a solution (as its not defined at i). Show that $h(x,y) := \mathrm{Re}\frac{i}{z+i} = \frac{y+1}{x^2+(y+1)^2}$ gives a solution.

Exercise 5.14 (An analogue of Liouville's theorem redux). Let u be a pointwise nonnegative, harmonic function in $\mathbb{R}^2$. Show that u is constant. *Hint:* If $f \in \mathcal{O}(\mathbb{C})$ and $u = \mathrm{Re} f$, consider $\exp(-f)$. (In Exercise 5.11, we have seen a proof using Harnack's inequality.)

Exercise 5.15. The regularity of Laplace equation solutions is not completely for free, i.e., the solution should be at least C^2 to guarantee that it is then C^∞. Here is an example to show that a *discontinuous* function may satisfy the Laplace equation! Define $u : \mathbb{R}^2 \to \mathbb{R}$ by $u(0,0) = 0$, and for $z = (x,y) \in \mathbb{C}\backslash\{(0,0)\}$, by $u(x,y) = \mathrm{Re}(\exp(-\frac{1}{z^4}))$.

(1) Verify that u is discontinuous at $(0,0)$.

(2) Check that $u(x,0) = e^{-\frac{1}{x^4}}$, $u(0,y) = e^{-\frac{1}{y^4}}$.

(3) As u is the real part of a holomorphic function in $\mathbb{C}\backslash\{0\}$, u satisfies the Laplace equation in $\mathbb{R}^2\backslash\{(0,0)\}$. Show that $\frac{\partial^2 u}{\partial x^2}(0,0)$ and $\frac{\partial^2 u}{\partial y^2}(0,0)$ exist, and $\frac{\partial^2 u}{\partial x^2}(0,0) + \frac{\partial^2 u}{\partial y^2}(0,0) = 0$.

Exercise 5.16. Let D_1, D_2 be domains in $\mathbb{C}$. Let $\varphi : D_1 \to D_2$ be holomorphic and $\varphi(D_1) \subset D_2$. Show that if $h : D_2 \to \mathbb{R}$ is harmonic, then $h \circ \varphi : D_1 \to \mathbb{R}$ is harmonic as well.

Let $\varphi : D_1 \to D_2$ be holomorphic and a bijection, and let $\varphi^{-1} : D_2 \to D_1$ be holomorphic. We call such a map φ a *biholomorphism*. Conclude that $h : D_2 \to \mathbb{R}$ is harmonic if and only if $h \circ \varphi : D_1 \to \mathbb{R}$ is harmonic.

Thus the existence of a biholomorphism between two domains allows one to transplant harmonic (or even holomorphic) functions from one domain to the other. This mobility has the advantage that if D_1 is 'nice' (like a half plane or a disc), while D_2 is complicated, then problems (like the Dirichlet Problem) in D_2 can be solved by first moving over to D_1, solving it there, and then transplanting the solution to D_2.

A natural question is then the following: Given domains D_1 and D_2, is there a biholomorphism between them? An answer is provided by the Riemann Mapping Theorem, a proof of which is beyond the scope of this introductory course, but can be found e.g. in [3].

Riemann Mapping Theorem. *Let $D \neq \mathbb{C}$ be a simply connected domain in $\mathbb{C}$. Then there exists a biholomorphism $\varphi : D \to \mathbb{D} := \{z \in \mathbb{C} : |z| < 1\}$.*

Thus the above result guarantees a biholomorphism between any two proper simply connected domains (by a passage through $\mathbb{D}$). Unfortunately, the proof does not give a practical algorithm for finding the biholomorphism. We also note that the condition $D \neq \mathbb{C}$ is needed, since otherwise φ would be a bounded entire function, and hence a constant by Liouville's Theorem, a contradiction to its purported bijectivity.

Show that the Möbius transformation $\varphi : \mathbb{H} \to \mathbb{D}$, where $\mathbb{H} := \{s \in \mathbb{C} : \mathrm{Re}(s) > 0\}$, given by $\varphi(s) = \frac{s-1}{s+1}$, $s \in \mathbb{H}$, is a biholomorphism between the right half plane $\mathbb{H}$ and the disc $\mathbb{D}$.

Solutions

Solutions to the exercises from the Introduction

Solution to Exercise 0.1. Let the derivative of g at 0 be L. Taking $\epsilon := 1 > 0$, there exists a $\delta > 0$ such that whenever $0 < |x - 0| < \delta$, we have

$$\left| \frac{g(x) - g(0)}{x - 0} - L \right| < \epsilon.$$

In particular, with $x := \frac{\delta}{2}$, we have $0 < |x - 0| = \frac{\delta}{2} < \delta$, and so

$$\left| \frac{g(x) - g(0)}{x - 0} - L \right| = \left| \frac{2\frac{\delta}{2} - 0}{\frac{\delta}{2} - 0} - L \right| = |2 - L| < \epsilon. \tag{6.1}$$

On the other hand, with $x := -\frac{\delta}{2}$, we have $0 < |x - 0| = \frac{\delta}{2} < \delta$, and so

$$\left| \frac{g(x) - g(0)}{x - 0} - L \right| = \left| \frac{-2(-\frac{\delta}{2}) - 0}{-\frac{\delta}{2} - 0} - L \right| = |2 + L| < \epsilon. \tag{6.2}$$

From (6.1) and (6.2) it follows, using the triangle inequality for the real absolute value, that $4 = |2 + L + 2 - L| \leqslant |2 + L| + |2 - L| < \epsilon + \epsilon = 2\epsilon = 2$, a contradiction. Hence g cannot be real differentiable at 0.

Solutions to the exercises from Chapter 1

Solution to Exercise 1.1. Since $(x, y) \neq 0$, at least one among x, y is nonzero, and so $x^2 + y^2 \neq 0$. Thus $(\frac{x}{x^2+y^2}, \frac{-y}{x^2+y^2}) \in \mathbb{R}^2$. Moreover,

$$(x, y) \cdot (\tfrac{x}{x^2+y^2}, \tfrac{-y}{x^2+y^2}) = (x\tfrac{x}{x^2+y^2} - y(\tfrac{-y}{x^2+y^2}), \; x(\tfrac{-y}{x^2+y^2}) + y\tfrac{x}{x^2+y^2})$$
$$= (\tfrac{x^2+y^2}{x^2+y^2}, \tfrac{-xy+xy}{x^2+y^2}) = (1, 0).$$

Hence for $(x, y) \neq (0, 0)$, we have $(x, y)^{-1} = (\frac{x}{x^2+y^2}, \frac{-y}{x^2+y^2})$ in $\mathbb{C}$.

Solution to Exercise 1.2. Since $\theta \in (-\frac{\pi}{2}, \frac{\pi}{2})$, $\tan \theta \in \mathbb{R}$. We have

$$\frac{1}{1-i\tan\theta} = \frac{1}{1^2+(\tan\theta)^2} + i\frac{\tan\theta}{1^2+(\tan\theta)^2} = \frac{(\cos\theta)^2}{(\cos\theta)^2+(\sin\theta)^2} + i\frac{\frac{\sin\theta}{\cos\theta}(\cos\theta)^2}{(\cos\theta)^2+(\sin\theta)^2}$$
$$= \frac{(\cos\theta)^2}{1} + i\frac{(\sin\theta)\cos\theta}{1} = (\cos\theta)^2 + i(\sin\theta)\cos\theta.$$

Hence

$$\frac{1+i\tan\theta}{1-i\tan\theta} = (1 + i\tan\theta)((\cos\theta)^2 + i(\sin\theta)\cos\theta)$$
$$= (\cos\theta)^2 - \frac{\sin\theta}{\cos\theta}(\sin\theta)(\cos\theta) + i((\sin\theta)(\cos\theta) + \frac{\sin\theta}{\cos\theta}(\cos\theta)^2)$$
$$= (\cos\theta)^2 - (\sin\theta)^2 + i2(\sin\theta)\cos\theta = \cos(2\theta) + i\sin(2\theta).$$

Solution to Exercise 1.3. Let $P \subset \mathbb{C}$ be a set of positive elements of $\mathbb{C}$. Then since $i \neq 0$, by (P3), either $i \in P$ or ($i \notin P$ and $-i \in P$). By (P2), we have

$$-1 = i \cdot i = (-i) \cdot (-i) \in P, \text{ and so } 1 = (-1) \cdot (-1) \in P. \qquad (\star)$$

But $1 \neq 0$, and $(\star)$ contradicts (P3) for $x = 1$.

Solution to Exercise 1.4.

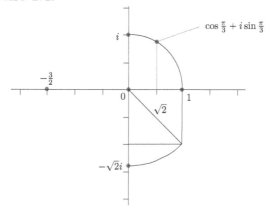

Solution to Exercise 1.5. Let z, w have polar coordinates (r, α) and (ρ, β), respectively. Then we have $\angle POS = \alpha$, $\angle QOS = \beta$, and $\angle ROS = \alpha + \beta$. Thus $\angle ROQ = \angle ROS - \angle QOS = (\alpha + \beta) - \beta = \alpha = \angle POS$. Also, $\frac{OP}{OR} = \frac{r}{r\rho} = \frac{1}{\rho} = \frac{OS}{OQ}$. As the ratio of a pair of sides is equal, and the included angle is equal too, $\triangle POS$ is similar to $\triangle ROQ$.

Solution to Exercise 1.6. For $\theta \in \mathbb{R}$, $(\cos\theta + i\sin\theta)^3 = \cos(3\theta) + i\sin(3\theta)$. But

$$(\cos\theta + i\sin\theta)^3 = (\cos\theta + i\sin\theta)((\cos\theta)^2 - (\sin\theta)^2 + i2(\cos\theta)(\sin\theta))$$
$$= (\cos\theta)((\cos\theta)^2 - (\sin\theta)^2) - (\sin\theta)2(\cos\theta)(\sin\theta) + i(\cdots).$$

Taking the real parts, we obtain

$$\cos(3\theta) = \mathrm{Re}(\cos(3\theta) + i\sin(3\theta)) = \mathrm{Re}((\cos\theta + i\sin\theta)^3)$$
$$= (\cos\theta)((\cos\theta)^2 - (\sin\theta)^2) - 2(\cos\theta)(\sin\theta)^2$$
$$= (\cos\theta)((\cos\theta)^2 - 1 + (\cos\theta)^2) - 2(\cos\theta)(1 - (\cos\theta)^2)$$
$$= (\cos\theta)^3 - \cos\theta + (\cos\theta)^3 - 2\cos\theta + 2(\cos\theta)^3$$
$$= 4(\cos\theta)^3 - 3\cos\theta.$$

Solution to Exercise 1.7. As $1 + i = \sqrt{2}(\frac{1}{\sqrt{2}} + i\frac{1}{\sqrt{2}}) = \sqrt{2}(\cos\frac{\pi}{4} + i\sin\frac{\pi}{4})$,

$$(1 + i)^{10} = (\sqrt{2})^{10}(\cos\frac{\pi}{4} + i\sin\frac{\pi}{4})^{10} = 2^5(\cos(10\frac{\pi}{4}) + i\sin(10\frac{\pi}{4}))$$
$$= 32(\cos(2\pi + \frac{\pi}{2}) + i\sin(2\pi + \frac{\pi}{2}))$$
$$= 32(\cos\frac{\pi}{2} + i\sin\frac{\pi}{2}) = 32(0 + i1) = 32i.$$

Solution to Exercise 1.8. The angle made by $2 + i$ with the positive real axis is $\tan^{-1}\frac{1}{2}$, and the angle made by $3 + i$ with the positive real axis is $\tan^{-1}\frac{1}{3}$. Thus the angle made by $(2 + i)(3 + i)$ with the positive real axis is $\tan^{-1}\frac{1}{2} + \tan^{-1}\frac{1}{3}$. As $(2 + i)(3 + i) = 6 - 1 + i(2 + 3) = 5 + 5i$, the angle made by $(2 + i)(3 + i)$ with the positive real axis is $\tan^{-1}\frac{5}{5} = \tan^{-1}1 = \frac{\pi}{4}$. Consequently, $\frac{\pi}{4} = \tan^{-1}\frac{1}{2} + \tan^{-1}\frac{1}{3}$.

Solution to Exercise 1.9. Suppose that the vertices A, B, C of the equilateral triangle are at the complex numbers z_A, z_B, z_C, and that they are labelled in the anticlockwise fashion. Since $AC = AB$ and $\angle CAB = \frac{\pi}{3}$, we have

$$z_C - z_A = (\cos\frac{\pi}{3} + i\sin\frac{\pi}{3})(z_B - z_A). \qquad (*)$$

We argue by contradiction. Let $p, q, m, n \in \mathbb{Z}$ be such that $z_C - z_A = p + iq$, and $z_B - z_A = m + in$. Then $(*)$ becomes $p + iq = (\frac{1}{2} + \frac{\sqrt{3}}{2}i)(m + in)$, i.e., $p = \frac{m}{2} - \frac{\sqrt{3}}{2}n$, and $q = \frac{m\sqrt{3}}{2} + \frac{n}{2}$. Multiplying the expression for p by $-n$, the one for q by m, and adding, we obtain $qm - pn = \frac{\sqrt{3}}{2}(m^2 + n^2)$. But $m^2 + n^2 \neq 0$ (as $z_B \neq z_A$!), and so $\sqrt{3} = \frac{2(qm-pn)}{m^2+n^2} \in \mathbb{Q}$, a contradiction.

Solution to Exercise 1.10. Let $w = \rho(\cos\alpha + i\sin\alpha)$, where $\rho \geqslant 0$ and $\alpha \in \mathbb{R}$. Then $\rho^4(\cos(4\alpha) + i\sin(4\alpha)) = w^4 = -1 = 1(\cos\pi + i\sin\pi)$. Hence $\rho^4 = 1$, and as $\rho \geqslant 0$, $\rho = 1$. Also, $4\alpha \in \{\pi, \pi\pm 2\pi, \pi\pm 4\pi, \cdots\}$, and so $\alpha \in \{\frac{\pi}{4}, \frac{\pi}{4}\pm\frac{\pi}{2}, \frac{\pi}{4}\pm\pi, \cdots\}$. So $w = \rho(\cos\alpha + i\sin\alpha) = 1(\cos\alpha + i\sin\alpha)$ belongs to the set

$\{\cos\frac{\pi}{4}+i\sin\frac{\pi}{4}, \cos\frac{3\pi}{4}+i\sin\frac{3\pi}{4}, \cos\frac{5\pi}{4}+i\sin\frac{5\pi}{4}, \cos\frac{7\pi}{4}+i\sin\frac{7\pi}{4}\} = \{\frac{1+i}{\sqrt{2}}, \frac{-1+i}{\sqrt{2}}, \frac{-1-i}{\sqrt{2}}, \frac{1-i}{\sqrt{2}}\}$.

The location of the four fourth roots of -1 in the complex plane is shown below.

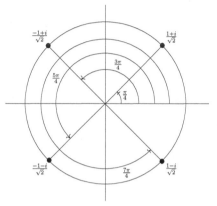

Solution to Exercise 1.11. $0 = z^6 - z^3 - 2 = (z^3)^2 - 2z^3 + z^3 - 2 = (z^3 - 2)(z^3 + 1)$. So $z^3 = 2$ or $z^3 = -1$. We have $z^3 = 2$ if and only if

$z \in \{\sqrt[3]{2}(\cos\frac{2\pi}{3}+i\sin\frac{2\pi}{3}), \sqrt[3]{2}(\cos\frac{4\pi}{3}+i\sin\frac{4\pi}{3}), \sqrt[3]{2}\} = \{\sqrt[3]{2}(-\frac{1}{2}+i\frac{\sqrt{3}}{2}), \sqrt[3]{2}(-\frac{1}{2}-i\frac{\sqrt{3}}{2}), \sqrt[3]{2}\}$.

We have $z^3 = -1$ if and only if

$z \in \{\cos\frac{\pi}{3}+i\sin\frac{\pi}{3}, \cos\pi+i\sin\pi, \cos\frac{5\pi}{3}+i\sin\frac{5\pi}{3}\} = \{\frac{1}{2}+i\frac{\sqrt{3}}{2}, -1, \frac{1}{2}-i\frac{\sqrt{3}}{2}\}$.

So $z^6 - z^3 - 2 = 0$ if and only if $[(z^3 = 2)$ or $(z^3 = -1)]$, that is, if and only if

$$z \in \{\sqrt[3]{2}(-\tfrac{1}{2}+i\tfrac{\sqrt{3}}{2}), \sqrt[3]{2}(-\tfrac{1}{2}-i\tfrac{\sqrt{3}}{2}), \sqrt[3]{2}\} \cup \{\tfrac{1}{2}+i\tfrac{\sqrt{3}}{2}, -1, \tfrac{1}{2}-i\tfrac{\sqrt{3}}{2}\}$$
$$= \{\sqrt[3]{2}(-\tfrac{1}{2}+i\tfrac{\sqrt{3}}{2}), \sqrt[3]{2}(-\tfrac{1}{2}-i\tfrac{\sqrt{3}}{2}), \sqrt[3]{2}, \tfrac{1}{2}+i\tfrac{\sqrt{3}}{2}, -1, \tfrac{1}{2}-i\tfrac{\sqrt{3}}{2}\}.$$

Solution to Exercise 1.12. Suppose that $\omega \in \mathbb{C}\backslash\mathbb{R}$ is such that $\omega^3 = 1$. Then $0 = \omega^3 - 1 = (\omega - 1)(\omega^2 + \omega + 1)$, and since $\omega \neq 1$, we have $\omega^2 + \omega + 1 = 0$. Hence

$$((b-a)\omega+(b-c))((b-a)\omega^2+(b-c)) = (b-a)^2\omega^3+(b-a)(b-c)(-1)+(b-c)^2$$
$$= (b-a)^2 1+(b-a)(b-c)(-1)+(b-c)^2$$
$$= (b-a)(b-a-b+c) + (b-c)^2$$
$$= (b-a)(c-a) + (b-c)^2$$
$$= bc - ca - ab + a^2 + b^2 - 2bc + c^2$$
$$= a^2 + b^2 + c^2 - ab - bc - ca = 0.$$

Hence $(b-a)\omega = c - b$ or $(b-a)\omega^2 = c - b$. But the latter case is the same as $(b-a)\omega^3 = (c-b)\omega$, that is, $(c-b)\omega = b - a$. In either of these cases, it follows that the distance of $b - a$ to the origin is the same as the distance of $c - b$ to the origin, and moreover, the angle between the line segments joining a to b and c to b is $\frac{\pi}{3}$. So the triangle formed by a, b, c is equilateral.

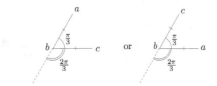

If a, b, c are all real, then the equilateral triangle must degenerate to a point $r \in \mathbb{R}$, and so $a = b = c\ (= r)$. Thus we recover the real case result.

Solution to Exercise 1.13.
Let $\omega \in \mathbb{C}\backslash\mathbb{R}$ satisfy $\omega^3 = 1$. Then $0 = 1 - \omega^3 = (1 - \omega)(1 + \omega + \omega^2) = 0$. As $\omega \neq 1$, $1 + \omega + \omega^2 = 0$. Also, $1 + \omega^2 + \omega^4 = 1 + \omega^2 + \omega\omega^3 = 1 + \omega^2 + \omega = 0$. We have
$(1 + 1)^{3n} + (1 + \omega)^{3n} + (1 + \omega^2)^{3n} = \sum_{k=0}^{3n} \binom{3n}{k}(1 + \omega^k + \omega^{2k})$. But

$$(1 + \omega^k + \omega^{2k}) = \begin{cases} 1 + 1 + 1 & \text{if } k \equiv 0 \bmod 3 \\ 1 + \omega + \omega^2 & \text{if } k \equiv 1 \bmod 3 \\ 1 + \omega^2 + \omega^4 & \text{if } k \equiv 2 \bmod 3 \end{cases} = \begin{cases} 3 & \text{if } k \equiv 0 \bmod 3 \\ 0 & \text{if } k \equiv 1 \bmod 3 \\ 0 & \text{if } k \equiv 2 \bmod 3. \end{cases}$$

Thus $(1 + 1)^{3n} + (1 + \omega)^{3n} + (1 + \omega^2)^{3n} = 3(\binom{3n}{0} + \binom{3n}{3} + \cdots + \binom{3n}{3n}))$. But also

$$\begin{aligned} (1 + 1)^{3n} + (1 + \omega)^{3n} + (1 + \omega^2)^{3n} &= 2^{3n} + (-\omega^2)^{3n} + (-\omega)^{3n} \\ &= 2^{3n} + (-1)^n + (-1)^n = 2^{3n} + 2(-1)^n. \end{aligned}$$

Solution to Exercise 1.14. Let A, B, C, D correspond to the complex numbers a, b, c, d, respectively. Since AB' is obtained from AB by rotating AB about A in an clockwise fashion by $90°$, B' corresponds to $a - i(b - a)$. Since P is the midpoint of BB', P corresponds to $\frac{b+a-i(b-a)}{2}$. Similarly, Q, R, S correspond to $\frac{c+b-i(c-b)}{2}$, $\frac{d+c-i(d-c)}{2}$, $\frac{a+d-i(a-d)}{2}$, respectively.

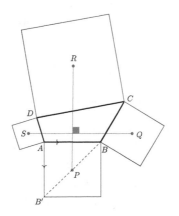

Denote the complex numbers corresponding to P, Q, R, S, by p, q, r, s, respectively,

$$i(q - s) = i(\tfrac{c+b-i(c-b)}{2} - \tfrac{a+d-i(a-d)}{2}) = \tfrac{c-b-a+d+i(c+b-a-d)}{2}$$

$$= \tfrac{-b-a+i(b-a)}{2} + \tfrac{d+c-i(d-c)}{2} = -p + r.$$

Hence the distance of $q - s$ to the origin is the same as that of $p - r$ to the origin, showing that $QS = PR$. Also, as multiplication by i produces a rotation about the origin by $90°$, it follows from $i(q - s) = r - p$ that $PR \perp QS$.

Solution to Exercise 1.15. For $k \in \mathbb{N}$, $(z - z_0)(z^{k-1} + z^{k-2}z_0 + \cdots + zz_0^{k-2} + z_0^{k-1})$ when expanded 'telescopes' to give $z^k - z_0^k$. We have:

$$
\begin{aligned}
p(z) &= c_0 + c_1 z + c_2 z^2 + \cdots + c_d z^d \\
p(z_0) &= c_0 + c_1 z_0 + c_2 z_0^2 + \cdots + c_d z_0^d \\
\hline
p(z) - p(z_0) &= c_1(z - z_0) + c_2(z^2 - z_0^2) + \cdots + c_d(z^d - z_0^d).
\end{aligned}
$$

With $q(z) := c_1 + c_2(z + z_0) + \cdots + c_d(z^{d-1} + z^{k-2}z_0 + \cdots + zz_0^{k-2} + z_0^{k-1})$, we have $p(z) - p(z_0) = (z - z_0)q(z)$, i.e., $p(z) = (z - z_0)q(z) + p(z_0)$. As $c_d \neq 0$, the polynomial q has degree $d - 1$. If $p(z_0) = 0$, then $p(z) = (z - z_0)q(z) + 0 = (z - z_0)q(z)$.

If $p(z) := z^n - 1$, then since $p(z_1) = 0$, we have $p(z) = (z - z_1)q_1(z)$ for some polynomial q_1 of degree $n - 1$. But $0 = p(z_2) = (z_2 - z_1)q_1(z_2)$, and since $z_2 \neq z_1$, $q_1(z_2) = 0$, and so $q_1(z) = (z - z_2)q_2(z)$ for some polynomial q_2 of degree $n - 2$. Hence $p(z) = (z - z_1)(z - z_2)q_2(z)$. Proceeding in this manner, we eventually get $p(z) = (z - z_1)(z - z_2) \cdots (z - z_{n-1})q_{n-1}(z)$ for a polynomial q_{n-1} of degree 1. Moreover, $q_{n-1}(z_n) = 0$, so that $q_{n-1}(z) = C(z - z_n)$ for some constant C. But

$$-1 = 0^n - 1 = p(0) = C(0 - z_1) \cdots (0 - z_n) = C(-1)^n z_1 \cdots z_n$$

$$= C(-1)^n \prod_{k=0}^{n-1}(\cos(\tfrac{2\pi}{n} + k\tfrac{2\pi}{n}) + i\sin(\tfrac{2\pi}{n} + k\tfrac{2\pi}{n}))$$

$$= C(-1)^n(\cos(2\pi + (n-1)\pi) + \sin(2\pi + (n-1)\pi)) = C(-1)^n(-1)^{n-1} = -C,$$

and so $C = 1$.

Solution to Exercise 1.16. In particular, for all $x \in \mathbb{R}$, $c_0 + c_1 x + \cdots + c_d x^d = 0$. Setting $x = 0$, we obtain $c_0 = 0$. Suppose that for some $n \in \{0, 1, \cdots, d\}$ we have shown $c_0 = \cdots = c_n = 0$. Hence for all $x \in \mathbb{R}$, $c_{n+1}x^{n+1} + \cdots + c_d x^d = 0$. Differentiating $n + 1$ times with respect to x, and then setting $x = 0$, we obtain $c_{n+1}(n+1)! + 0 = 0$, and so $c_{n+1} = 0$. By induction, it follows that $c_0 = \cdots = c_d = 0$.

Solution to Exercise 1.17. First we show uniqueness. Let $p = qh + r = \tilde{q}h + \tilde{r}$ for polynomials $q, \tilde{q}, r, \tilde{r}$, where $r, \tilde{r}$ each have a degree strictly less than d. Then $(q - \tilde{q})h = \tilde{r} - r$. If $q \neq \tilde{q}$, then the left-hand side has degree at least d, while the right-hand side has degree strictly less than d, a contradiction. So $q = \tilde{q}$. This then yields $\tilde{r} - r = 0h = 0$, and so $\tilde{r} = r$ as well.

We use induction on the degree of p to show existence. If p is constant, then with $q := 0$ and $r := p(0)$, we have $p(z) = p(0) = 0 \cdot h(z) + p(0) = q(z)h(z) + r(z)$.

Suppose the claim is true for all polynomials of degree $\leqslant n$, for some $n \in \mathbb{N} \cup \{0\}$. Let p be a polynomial of degree $n + 1$ ($\geqslant 0 + 1 = 1$). By the Division Algorithm for *integers*, $n + 1 = d \cdot k + \ell$, for nonnegative integers k, ℓ, where $0 \leqslant \ell < d$. Let $p(z) = c_{n+1}z^{n+1} + \cdots + c_0$, and $h(z) = a_d z^d + \cdots + a_0$. Then the coefficient of the leading term, namely z^{n+1}, of the polynomial $\frac{c_{n+1}}{a_d^k}h^k$ is c_{n+1}, and so $p - \frac{c_{n+1}}{a_d^k}h^k$ has degree $\leqslant n$. By the induction hypothesis, there exist polynomials $\tilde{q}, \tilde{r}$ such that $p - \frac{c_{n+1}}{a_d^k}h^k = h\tilde{q} + \tilde{r}$, and such that the degree of $\tilde{r}$ is strictly less than d.

- If $k = 0$, then $q := \tilde{q}$, and absorb the constant $\frac{c_{n+1}}{a_d^k}h^k$ in $\tilde{r}$, i.e., $r := \tilde{r} + \frac{c_{n+1}}{a_d^k}h^k$.

- If $k > 0$, then $r := \tilde{r}$, and $q := \tilde{q} + \frac{c_{n+1}}{a_d^k}h^{k-1}$.

In either case, $p = qh + r$, with r having degree $< d$, completing the induction step.

Solution to Exercise 1.18. Let $z_1 = x_1 + iy_1$, $z_2 = x_2 + iy_2$, where x_1, x_2, y_1, y_2 belong to $\mathbb{R}$. Then $z_1 z_2 = x_1 x_2 - y_1 y_2 + i(x_1 y_2 + y_1 x_2)$, and

$$
\begin{aligned}
|z_1 z_2|^2 &= (x_1 x_2 - y_1 y_2)^2 + (x_1 y_2 + y_1 x_2)^2 \\
&= x_1^2 x_2^2 - 2x_1 x_2 y_1 y_2 + y_1^2 y_2^2 + x_1^2 y_2^2 + 2x_1 y_2 y_1 x_2 + y_1^2 x_2^2 \\
&= x_1^2(x_2^2 + y_2^2) + y_1^2(y_2^2 + x_2^2) = (x_1^2 + y_1^2)(x_2^2 + y_2^2) = |z_1|^2 |z_2|^2.
\end{aligned}
$$

Since $|z_1|, |z_2|, |z_1 z_2|$ are all nonnegative, it follows that $|z_1 z_2| = |z_1| |z_2|$.

Solution to Exercise 1.19. Let $z = x + iy$, where $x, y \in \mathbb{R}$. Then

$$\overline{(\overline{z})} = \overline{x - iy} = x - i(-y) = x + iy = z.$$

Also, $z\overline{z} = (x + iy)(x - iy) = x^2 + y^2 + i(-xy + xy) = x^2 + y^2 = |z|^2$.

Finally, $\frac{z + \overline{z}}{2} = \frac{x + iy + x - iy}{2} = x = \operatorname{Re} z$, and $\frac{z - \overline{z}}{2i} = \frac{x + iy - x + iy}{2i} = \frac{2iy}{2i} = y = \operatorname{Im} z$.

Solution to Exercise 1.20. Let $z = x + iy$, where $x, y \in \mathbb{R}$. Then

$$
\begin{aligned}
|z| &= |x + iy| = \sqrt{x^2 + y^2} = \sqrt{x^2 + (-y)^2} = |x - iy| = |\overline{z}|, \\
|\operatorname{Re} z| &= |x| = \sqrt{x^2} = \sqrt{x^2 + 0} \leqslant \sqrt{x^2 + y^2} = |x + iy| = |z|, \\
|\operatorname{Im} z| &= |y| = \sqrt{y^2} = \sqrt{0 + y^2} \leqslant \sqrt{x^2 + y^2} = |x + iy| = |z|.
\end{aligned}
$$

Since $\overline{z}$ is the reflection of z in the real axis, and as $0 \in \mathbb{R}$, the distance of z to 0 is equal to the distance of $\overline{z}$ to 0, i.e., $|z| = |\overline{z}|$. The inequalities $|\operatorname{Re} z| \leqslant |z|$ and $|\operatorname{Im} z| \leqslant |z|$ express the fact that the length of any side in a right-angled triangle is at most the length of the hypotenuse.

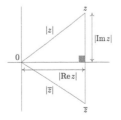

Solution to Exercise 1.21. $|\bar{a}z| = |\bar{a}||z| = |a||z| < 1(1) = 1$. So $\bar{a}z \neq 1$. We have

$$|\tfrac{z-a}{1-\bar{a}z}|^2 = \tfrac{z-a}{1-\bar{a}z}\overline{\left(\tfrac{z-a}{1-\bar{a}z}\right)} = \tfrac{(z-a)}{(1-\bar{a}z)}\tfrac{(\bar{z}-\bar{a})}{(1-a\bar{z})} = \tfrac{z\bar{z}-a\bar{z}-\bar{a}z+a\bar{a}}{1-a\bar{z}-\bar{a}z+a\bar{a}z\bar{z}} = \tfrac{|z|^2-a\bar{z}-\bar{a}z+|a|^2}{1-a\bar{z}-\bar{a}z+|a|^2|z|^2}$$

$$= \tfrac{1-a\bar{z}-\bar{a}z+|a|^2|z|^2+|z|^2+|a|^2-1-|a|^2|z|^2}{1-a\bar{z}-\bar{a}z+|a|^2|z|^2} = 1 + \tfrac{|z|^2+|a|^2-1-|a|^2|z|^2}{1-a\bar{z}-\bar{a}z+|a|^2|z|^2}$$

$$= 1 + \tfrac{|z|^2+|a|^2-1-|a|^2|z|^2}{|1-\bar{a}z|^2} = 1 - \tfrac{(1-|z|^2)(1-|a|^2)}{|1-\bar{a}z|^2}.$$

$\tfrac{(1-|z|^2)(1-|a|^2)}{|1-\bar{a}z|^2} \geqslant 0$ as $|z| \leqslant 1$, $|a| < 1$. So $|\tfrac{z-a}{1-\bar{a}z}|^2 = 1 - \tfrac{(1-|z|^2)(1-|a|^2)}{|1-\bar{a}z|^2} \leqslant 1 - 0 = 1$.

Solution to Exercise 1.22. In the picture below, $0, z, w, w - z$ forms a parallelogram. So when we form the parallelogram determined by $0, z, t(w - z)$, the fourth vertex, namely $z + t(w - z)$, lies on the segment joining z and w (as the line joining 0 and $t(w - z)$ is parallel to the line joining z and w). As the t increases from 0 to 1, $z + t(w - z)$ moves from z to w along the segment joining them.

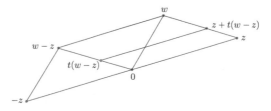

Solution to Exercise 1.23. As $p(w) = 0$, $\overline{c_0 + c_1w + \cdots + c_dw^d} = \overline{p(w)} = \overline{0} = 0$. We have $0 = \overline{c_0 + c_1w + \cdots + c_dw^d} = \overline{c_0} + \overline{c_1w} + \cdots + \overline{c_dw^d} = c_0 + c_1\overline{w} + \cdots + c_d(\overline{w})^d$ (using $c_k \in \mathbb{R}$, and $\overline{w^k} = (\overline{w})^k$, $1 \leqslant k \leqslant d$). So $0 = c_0 + c_1\overline{w} + \cdots + c_d(\overline{w})^d = p(\overline{w})$.

Solution to Exercise 1.24. Let $a = |a|(\cos\alpha + i\sin\alpha)$, and $b = |b|(\cos\beta + i\sin\beta)$, where $\alpha, \beta \in [0, 2\pi)$. Then $a\bar{b} = |a|(\cos\alpha + i\sin\alpha)|b|(\cos\beta - i\sin\beta)$ and so we have $\text{Im}(a\bar{b}) = |a||b|((\cos\alpha)(-\sin\beta) + (\sin\alpha)(\cos\beta)) = |a||b|\sin(\alpha - \beta)$.

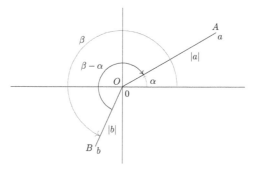

The area of the triangle OAB formed by 0, a, b (where $O \equiv 0$, $A \equiv a$, $B \equiv b$) is given by $\tfrac{1}{2}OA \cdot OB \cdot \sin\angle AOB = \tfrac{1}{2}|a||b||\sin(\alpha - \beta)| = \tfrac{1}{2}|\text{Im}(a\bar{b})| = |\tfrac{\text{Im}(a\bar{b})}{2}|$.

Solution to Exercise 1.25. In $\triangle ABC$, the points C', A', B' on AB, BC, CA, respectively, are such that $AC':C'B = BA':A'C = CB':B'A = 2:1$. Join AA', BB', CC', and label the intersection points in the line segment pairs (CC', AA'), (AA', BB'), (BB', CC') as X, Y, Z, respectively. We need the ratio of the areas of the internal triangle $\triangle XYZ$ to that of $\triangle ABC$.

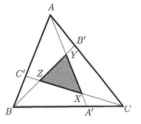

Let A, B, C correspond to $z, 0, w \in \mathbb{C}$, respectively. Then the points C', A', B' correspond to $\frac{z}{3}$, $\frac{2w}{3}$, $\frac{2z+w}{3}$, respectively. As Y lies on the intersection of AA' and BB', for some $s, t \in [0, 1]$, $t\frac{2z+w}{3} = (1-s)z + s\frac{2w}{3}$, i.e., $(2t - 3(1-s))z = (2s - t)w$. Since $0, z, w$ are not collinear, it follows that $2t = 3(1-s)$ and $t = 2s$. Solving, we get $s = \frac{3}{7}$. Hence Y corresponds to the complex number $\frac{4z+2w}{7}$. Noticing the symmetry of the situation (that is, by 'shifting the origin' to z and then to w), we can read off from the above that X, Z correspond to the complex numbers $\frac{4(w-z)+2(-z)}{7} + z = \frac{4w+z}{2}$, and $\frac{4(-w)+2(z-w)}{7} + w = \frac{w+2z}{7}$, respectively. Shifting the origin to say Z, we have that X, Y then correspond to $a = \frac{-z+3w}{7}$, $b = \frac{2z+w}{7}$, and so the area of the internal triangle, $\triangle XYZ$, will be

$$\left|\tfrac{1}{2}\mathrm{Im}\left(\tfrac{(2z+w)}{7}\tfrac{(-\overline{z}+3\overline{w})}{7}\right)\right| = \left|\tfrac{1}{2}\tfrac{1}{7(7)}\mathrm{Im}(-2|z|^2 + 3|w|^2 - w\overline{z} + 6z\overline{w})\right|$$

$$= \left|\tfrac{1}{2}\tfrac{1}{7(7)}\mathrm{Im}(-w\overline{z} - \overline{w}z + 7z\overline{w})\right| = \tfrac{1}{7}\tfrac{|\mathrm{Im}(z\overline{w})|}{2}.$$

The area of $\triangle ABC$ is $\frac{|\mathrm{Im}(z\overline{w})|}{2}$, and so the above shows that the area of $\triangle XYZ$ is one-seventh the area of $\triangle ABC$.

Solution to Exercise 1.26. For $z_1, z_2, z_3 \in \mathbb{C}$, $w := i \det \begin{bmatrix} 1 & z_1 & \overline{z_1} \\ 1 & z_2 & \overline{z_2} \\ 1 & z_3 & \overline{z_3} \end{bmatrix} = -i \det \begin{bmatrix} 1 & z_1 & \overline{z_1} \\ 1 & z_2 & \overline{z_2} \\ 1 & z_3 & \overline{z_3} \end{bmatrix}$.

Let S_n denote the set of all permutations on $\{1, \cdots, n\}$. For $M = [m_{i,j}] \in \mathbb{C}^{n \times n}$,

$$\overline{\det M} = \overline{\sum_{\sigma \in S_n} (\mathrm{sign}\ \sigma)m_{1,\sigma(1)} \cdots m_{n,\sigma(n)}} = \sum_{\sigma \in S_n} (\mathrm{sign}\ \sigma)\overline{m_{1,\sigma(1)}} \cdots \overline{m_{n,\sigma(n)}} = \det \overline{M},$$

where $\overline{M}$ is obtained from the matrix M by taking entrywise complex conjugates. So

$$\overline{\det \begin{bmatrix} 1 & z_1 & \overline{z_1} \\ 1 & z_2 & \overline{z_2} \\ 1 & z_3 & \overline{z_3} \end{bmatrix}} = \det \begin{bmatrix} 1 & \overline{z_1} & z_1 \\ 1 & \overline{z_2} & z_2 \\ 1 & \overline{z_3} & z_3 \end{bmatrix} = -\det \begin{bmatrix} 1 & z_1 & \overline{z_1} \\ 1 & z_2 & \overline{z_2} \\ 1 & z_3 & \overline{z_3} \end{bmatrix},$$

where the last equality follows by swapping the second and third columns. Thus,

$$\overline{i \det \begin{bmatrix} 1 & z_1 & \overline{z_1} \\ 1 & z_2 & \overline{z_2} \\ 1 & z_3 & \overline{z_3} \end{bmatrix}} = -i \overline{\det \begin{bmatrix} 1 & z_1 & \overline{z_1} \\ 1 & z_2 & \overline{z_2} \\ 1 & z_3 & \overline{z_3} \end{bmatrix}} = -i\left(-\det \begin{bmatrix} 1 & z_1 & \overline{z_1} \\ 1 & z_2 & \overline{z_2} \\ 1 & z_3 & \overline{z_3} \end{bmatrix}\right) = i \det \begin{bmatrix} 1 & z_1 & \overline{z_1} \\ 1 & z_2 & \overline{z_2} \\ 1 & z_3 & \overline{z_3} \end{bmatrix}.$$

So w is its own complex conjugate, and hence it is real.

Solution to Exercise 1.27. The claim is a consequence of the following:

$$\left|\tfrac{|w|}{|z|}z + \tfrac{|z|}{|w|}w\right|^2 = \left(\tfrac{|w|}{|z|}z + \tfrac{|z|}{|w|}w\right)\left(\tfrac{|w|}{|z|}\overline{z} + \tfrac{|z|}{|w|}\overline{w}\right)$$

$$= \tfrac{|w|^2}{|z|^2}|z|^2 + \tfrac{|z|^2}{|w|^2}|w|^2 + \tfrac{|w|}{|z|}z\tfrac{|z|}{|w|}\overline{w} + \tfrac{|z|}{|w|}w\tfrac{|w|}{|z|}\overline{z}$$

$$= |w|^2 + |z|^2 + z\overline{w} + w\overline{z} = (w+z)(\overline{w}+\overline{z}) = |w+z|^2.$$

As $\tfrac{|w|}{|z|}z$ is just a scaled version of z, it lies along the ray starting from 0 and passing through z, and the distance of $\tfrac{|w|}{|z|}z$ from the origin is $|w|$, which is one of the lengths of the sides of the parallelogram drawn with solid lines below. Similarly, $\tfrac{|z|}{|w|}w$ lies along the ray passing through 0 and w, and is at a distance of $|z|$ from 0, which is the other length of the side of the parallelogram drawn with solid lines. Also the included angle between $\tfrac{|w|}{|z|}z$ and $\tfrac{|z|}{|w|}w$ is the same as the angle between z and w. So the parallelogram drawn with solid lines is congruent to the parallelogram drawn with dashed lines, and in particular, the lengths of the corresponding diagonals are the same, giving $|z+w| = \left|\tfrac{|w|}{|z|}z + \tfrac{|z|}{|w|}w\right|$.

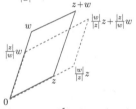

Solution to Exercise 1.28. If $z, w \in S^1$, then $|zw| = |z||w| = 1(1) = 1$, and so $zw \in S^1$. Complex multiplication is associative and commutative. Also, $1 \in S^1$, and $z1 = z = 1z$ for all $z \in S^1$. If $z \in S^1$, then $z \neq 0$ and $\tfrac{1}{z} \in S^1$ too, because $\left|\tfrac{1}{z}\right| = \tfrac{1}{|z|} = \tfrac{1}{1} = 1$, and so every element in S^1 has a multiplicative inverse in S^1. Hence S^1 forms an Abelian group.

For $n \in \mathbb{N}$, let $\zeta = \cos\tfrac{2\pi}{n} + i\sin\tfrac{2\pi}{n}$. If for some $m \in \mathbb{N}$, $\zeta^m = 1$, then in particular, $\cos\tfrac{2\pi m}{n} = 1$, so that $(0 \neq)\tfrac{2\pi m}{n} = 2\pi k$ for some $k \in \mathbb{Z}\backslash\{0\}$, showing that n divides m. This means that $m \geqslant n$. So $\zeta, \cdots, \zeta^{n-1}$ cannot be 1. Also, $\zeta^n = 1$. Thus the order of $\zeta \in S^1$ is n, and $H_n = \{1, \zeta, \cdots, \zeta^{n-1}\}$ is a subgroup of S^1 of order n.

Let $H := \{\cos(\sqrt{2}\pi m) + i\sin(\sqrt{2}\pi m) : m \in \mathbb{Z}\} \subset S^1$, and $1 = \cos 0 + i\sin 0 \in H$. Also, H is closed under multiplication because if $z, w \in H$, then there exist $m, n \in \mathbb{Z}$ such that $z = \cos(\sqrt{2}\pi m) + i\sin(\sqrt{2}\pi m)$ and $w = \cos(\sqrt{2}\pi n) + i\sin(\sqrt{2}\pi n)$, giving $zw = \cos(\sqrt{2}\pi(m+n)) + i\sin(\sqrt{2}\pi(m+n)) \in H$, as $m+n \in \mathbb{Z}$. Finally, H is closed under taking inverses, since if $z = \cos(\sqrt{2}\pi m) + i\sin(\sqrt{2}\pi m) \in H$, where $m \in \mathbb{Z}$, then $z^{-1} = \cos(\sqrt{2}\pi(-m)) + i\sin(\sqrt{2}\pi(-m)) \in H$ too. The map $\varphi : \mathbb{Z} \to H$ given by $\mathbb{Z} \ni m \mapsto \cos(\sqrt{2}\pi m) + i\sin(\sqrt{2}\pi m) \in H$ is surjective. It is also injective, since if for some distinct $m, n \in \mathbb{Z}$, $\cos(\sqrt{2}\pi m) + i\sin(\sqrt{2}\pi m) = \cos(\sqrt{2}\pi n) + i\sin(\sqrt{2}\pi n)$, then it follows that $\sqrt{2}\pi m = \sqrt{2}\pi n + 2\pi k$ for some integer k, giving $\sqrt{2} = \tfrac{2k}{m-n} \in \mathbb{Q}$, a contradiction. Thus the map φ is a bijection. As $\mathbb{Z}$ is countable, so is H. (There are many other examples of infinite countable subgroups too.)

Solution to Exercise 1.29. Let $S = z_1 + \cdots + z_n$. With $\zeta := \cos\frac{2\pi}{n} + i\sin\frac{2\pi}{n}$, the set $\{\zeta z_1, \cdots, \zeta z_n\} = \{z_1, \cdots, z_n\}$. Thus $S = z_1 + \cdots + z_n = \zeta z_1 + \cdots + \zeta z_n = \zeta S$, i.e., $(\zeta - 1)S = 0$. But as $n \geqslant 2$, $\zeta \neq 1$ (otherwise $\frac{2\pi}{n} = 2\pi k$ for some integer k, giving $1 = nk$ for integers n, k with $n \geqslant 2$, which is absurd). So $S = 0$.

Alternatively, by Exercise 1.15, $z^n - 1 = (z - z_1) \cdots (z - z_n)$, and by expanding the right-hand side, $z^n - 1 = z^n - (z_1 + \cdots + z_n)z^{n-1} + \cdots + (-1)^n z_1 \cdots z_n$. 'Comparing coefficients' of z^{n-1} on both sides (see Exercise 1.16), $z_1 + \cdots + z_n = 0$.

Solution to Exercise 1.30. For $z_1, z_2 \in \mathbb{C}$, $|z_1| = |z_1 - z_2 + z_2| \leqslant |z_1 - z_2| + |z_2|$. So $|z_1| - |z_2| \leqslant |z_1 - z_2|$. As this holds for *all* $z_1, z_2 \in \mathbb{C}$, by swapping z_1 and z_2, we also obtain $|z_2| - |z_1| \leqslant |z_2 - z_1| = |-(z_1 - z_2)| = |-1||z_1 - z_2| = |z_1 - z_2|$. Thus $||z_1| - |z_2|| \leqslant |z_1 - z_2|$.

Solution to Exercise 1.31. We have $z \in C$ if and only if

$$
\begin{aligned}
r^2 &= |z|^2 - 2\,\mathrm{Re}(\overline{a}z) + |a|^2 \\
&= z\overline{z} - (\overline{a}z + a\overline{z}) + a\overline{a} \\
&= z(\overline{z} - \overline{a}) - a(\overline{z} - \overline{a}) = (z - a)(\overline{z} - \overline{a}) \\
&= |z - a|^2.
\end{aligned}
$$

So C is the set of all points $z \in \mathbb{C}$, whose distance $|z - a|$ to a is r. Thus C is a circle with centre a and radius r.

Solution to Exercise 1.32. We have

$$
\begin{aligned}
|z_1 + z_2|^2 + |z_1 - z_2|^2 &= (z_1 + z_2)(\overline{z_1} + \overline{z_2}) + (z_1 - z_2)(\overline{z_1} - \overline{z_2}) \\
&= |z_1|^2 + z_1\overline{z_2} + z_2\overline{z_1} + |z_2|^2 + |z_1|^2 - z_1\overline{z_2} - z_2\overline{z_1} + |z_2|^2 \\
&= 2(|z_1|^2 + |z_2|^2).
\end{aligned}
$$

Consider the parallelogram P with vertices at $0, z_1, z_2, z_1 + z_2$ in the complex plane. Then $|z_1 + z_2|$ denotes the length of one diagonal of P, while $|z_1 - z_2|$ is the length of the other diagonal of P. Also, $|z_1|, |z_2|$ are the lengths of the sides of P. So the above equality says:

> In a parallelogram, the sum of the squares of the lengths of the diagonals equals the sum of the squares of the lengths of all the sides.

Solution to Exercise 1.33. From Exercise 1.32, with $z_1 = z$ and $z_2 = r$, we obtain $|z - r|^2 + |z + r|^2 = 2(|z|^2 + |r|^2) = 2(r^2 + r^2) = 4r^2 = (2r)^2$. Thus in the triangle formed by $z, (-r, 0), (0, r)$, the sum of the squares of the lengths of the two sides (namely, $|z - r|^2 + |z + r|^2$) equals the square of the length of the third side (i.e., $(2r)^2$). By the converse of the Pythagoras Theorem, the triangle formed by $z, (-r, 0), (0, r)$ is a right angled triangle. Also, the right angle is opposite the longest side (hypotenuse), and the longest side in our case is the diameter.

Solution to Exercise 1.34. (1),(2),(3) are depicted below (left to right):

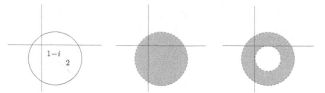

(4) If $z = x + iy$ $(x, y \in \mathbb{R})$, then $\mathrm{Re}(z - (1 - i)) = 3 \Leftrightarrow x - 1 = 3$, i.e., $x = 4$. See the picture on the left below.

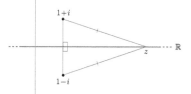

(5) If $z = x + iy$ $(x, y \in \mathbb{R})$, then $|\mathrm{Im}(z - (1 - i))| < 3 \Leftrightarrow |y + 1| < 3$, i.e., $-4 < y < 2$. See the picture on the right above.

(6) $\{z \in \mathbb{C} : |z - (1 - i)| = |z - (1 + i)|\}$ is the set of all complex numbers whose distance to $1 - i$ is equal to its distance to $1 + i$. So it is the set of all points lying on the perpendicular bisector of the line segment joining $1 - i$ to $1 + i$. Thus the set is the real line $\mathbb{R}$.

(7) $|z - (1 - i)| + |z - (1 + i)| = 2$ means the sum of the distances of z to $1 + i$ and to $1 - i$ is 2. As the distance between $1 - i$ and $1 + i$ is 2, z lies on the line segment joining $1 - i$ and $1 - i$.

(8) The set of z such that $|z - (1 - i)| + |z - (1 + i)| = 3$ lies on an ellipse E with foci at $1 + i$ and $1 - i$, and so $\{z \in \mathbb{C} : |z - (1 - i)| + |z - (1 + i)| < 3\}$ is the region in the interior of the ellipse E. See the picture on the right above.

Solution to Exercise 1.35. Let z, w make angles α, β, respectively, with the positive x-axis in the counterclockwise direction. Then

$$\bar{z} = \overline{|z|(\cos\alpha + i\sin\alpha)} = |z|(\cos\alpha - i\sin\alpha) = |z|(\cos(-\alpha) + i\sin(-\alpha)).$$

So $\bar{z}w = |z|(\cos(-\alpha)+i\sin(-\alpha))|w|(\cos\beta+i\sin\beta) = |z||w|(\cos(\beta-\alpha)+i\sin(\beta-\alpha)).$
Thus $\operatorname{Re}(\bar{z}w) = |z||w|\cos(\beta-\alpha) = (OA)(OB)\cos\angle AOB$. Hence

$$AB^2 = |z - w|^2 = (z - w)(\bar{z} - \bar{w}) = |z|^2 + |w|^2 - \bar{z}w - z\bar{w}$$
$$= |z|^2 + |w|^2 - 2\operatorname{Re}(\bar{z}w) = OA^2 + OB^2 - 2(OA)(OB)\cos\angle AOB.$$

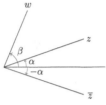

Solution to Exercise 1.36. Let $(u, v) \in \mathbb{R}^2$. We want $p = (x, y, z) \in S^2 \setminus \{\mathbf{n}\}$ such that $\varphi(p) = (u, v)$, i.e., $(\frac{x}{1-z}, \frac{y}{1-z}) = (u, v)$. For such a triple (x, y, z), we have that

$$u^2 + v^2 = \frac{x^2}{(1-z)^2} + \frac{y^2}{(1-z)^2} = \frac{x^2+y^2}{(1-z)^2} = \frac{1-z^2}{(1-z)^2} = \frac{1+z}{1-z}, \text{ and so } z = \frac{u^2+v^2-1}{u^2+v^2+1}. \text{ Also,}$$

$$x = u(1 - z) = u(1 - \tfrac{u^2+v^2-1}{u^2+v^2+1}) = \frac{2u}{u^2+v^2+1} \text{ and } y = v(1 - z) = \frac{2v}{u^2+v^2+1}.$$

So we define $p := (x, y, z) = (\frac{2u}{u^2+v^2+1}, \frac{2v}{u^2+v^2+1}, \frac{u^2+v^2-1}{u^2+v^2+1})$. Then

$$x^2 + y^2 + z^2 = \frac{4u^2+4v^2+u^4+v^4+1+2u^2v^2-2u^2-2v^2}{(u^2+v^2+1)^2} = \frac{(u^2+v^2+1)^2}{(u^2+v^2+1)^2} = 1.$$

Thus $p \in S^2$. Also $p \neq \mathbf{n}$, since otherwise $(\frac{2u}{u^2+v^2+1}, \frac{2v}{u^2+v^2+1}, \frac{u^2+v^2-1}{u^2+v^2+1}) = (0, 0, 1)$, and so from the first two entries, $u = 0$ and $v = 0$, implying $\frac{u^2+v^2-1}{u^2+v^2+1} = -1 \neq 1$, a contradiction. So $p \in S^2 \setminus \{\mathbf{n}\}$. Finally, $\varphi(p) = \frac{1}{1-\frac{u^2+v^2-1}{u^2+v^2+1}}(\frac{2u}{u^2+v^2+1}, \frac{2v}{u^2+v^2+1}) = (u, v)$.

Hence $\varphi^{-1}(u, v) = (\frac{2u}{u^2+v^2+1}, \frac{2v}{u^2+v^2+1}, \frac{u^2+v^2-1}{u^2+v^2+1})$.

Let $w = (u, v)$ and $w' = (u', v')$. Let A, A', a, a' be the lengths in $\mathbb{R}^3$ of the segments joining $\mathbf{n}$ to $w, w', \varphi^{-1}w, \varphi^{-1}w'$, respectively. Then $A^2 = u^2 + v^2 + 1 = |w|^2 + 1$ and $A'^2 = u'^2 + v'^2 + 1 = |w'|^2 + 1$.

Considering the similar triangles shown in the picture on page 16, $\frac{a}{A} = \frac{1-z}{1}$, where $z = \frac{u^2+v^2-1}{u^2+v^2+1} = \frac{|w|^2-1}{|w|^2+1}$ is the z coordinate of $\varphi^{-1}w \in S^2 \setminus \{\mathbf{n}\}$. Similarly, $\frac{a'}{A'} = \frac{1-z'}{1}$, where $z' = \frac{u'^2+v'^2-1}{u'^2+v'^2+1} = \frac{|w'|^2-1}{|w'|^2+1}$. These yield $a = \frac{2}{\sqrt{1+|w|^2}}$ and $a' = \frac{2}{\sqrt{1+|w'|^2}}$.

The triangles formed by $\mathbf{n}, \varphi^{-1}w, \varphi^{-1}w'$ and by $\mathbf{n}, w, w'$ share the angle θ between the ray from $\mathbf{n}$ to w, and the ray from $\mathbf{n}$ to w'. The Cosine Formula in each triangle gives $|w - w'|^2 = A^2 + A'^2 - 2AA' \cos\theta$ and $d_c(w, w')^2 = a^2 + a'^2 - 2aa' \cos\theta$. Thus

$$d_c(w, w')^2 = a^2 + a'^2 - 2aa' \frac{A^2 + A'^2 - |w - w'|^2}{2AA'}$$

$$= \frac{4}{|w|^2 + 1} + \frac{4}{|w'|^2 + 1} - 2 \frac{2}{\sqrt{|w|^2 + 1}} \frac{2}{\sqrt{|w'|^2 + 1}} \frac{|w|^2 + 1 + |w'|^2 + 1 - |w - w'|^2}{2\sqrt{|w|^2 + 1}\sqrt{|w'|^2 + 1}}$$

$$= \frac{4(|w|^2 + 1) + 4(|w'|^2 + 1) - 4(|w|^2 + 1 + |w'|^2 + 1 - |w - w'|^2)}{(1 + |w|^2)(1 + |w'|^2)}$$

$$= \frac{4|w - w'|^2}{(1 + |w|^2)(1 + |w'|^2)}.$$

Taking square roots, the result follows. As $d_c(w, w')$ is at most the diameter of the sphere, $d_c(w, w') \leqslant 2$. If $w, w' \in \mathbb{C}\backslash\{0\}$, then

$$d_c(w^{-1}, w'^{-1}) = \frac{2|\frac{1}{w} - \frac{1}{w'}|}{\sqrt{(1 + \frac{1}{|w|^2})(1 + \frac{1}{|w'|^2})}} = \frac{2\frac{|w' - w|}{|w||w'|}}{\frac{\sqrt{(|w|^2 + 1)(|w'|^2 + 1)}}{|w||w'|}} = \frac{2|w - w'|}{\sqrt{(1 + |w|^2)(1 + |w'|^2)}} = d_c(w, w').$$

Solution to Exercise 1.37. For $R \geqslant 1$, $k \in \{0, 1, \cdots, d\}$, and $z \in \mathbb{C}$ satisfying $|z| > R$, we have $|z|^k = |z|^k 1 \leqslant |z|^k R^{d-k} \leqslant |z|^k |z|^{d-k} = |z|^d$. Hence for any $R \geqslant 1$ and all $z \in \mathbb{C}$ satisfying $|z| > R$, we have

$$|p(z)| = |c_0 + c_1 z + \cdots + c_d z^d| \leqslant |c_0| + |c_1||z| + \cdots + |c_d||z|^d$$

$$\leqslant |c_0||z|^d + |c_1||z|^d + \cdots + |c_d||z|^d = M|z|^d,$$

where $M := |c_0| + |c_1| + \cdots + |c_d| > 0$.

For $z \neq 0$, $p(z) = z^d(c_d + \frac{c_{d-1}}{z} + \cdots + \frac{c_1}{z^{d-1}} + \frac{c_0}{z^d})$. As $\lim_{n\to\infty}(\frac{|c_{d-1}|}{n} + \cdots + \frac{|c_1|}{n^{d-1}} + \frac{|c_0|}{n^d}) = 0$, there exists an N large enough so that $\frac{|c_{d-1}|}{N} + \cdots + \frac{|c_1|}{N^{d-1}} + \frac{|c_0|}{N^d} < \frac{|c_d|}{2}$. Hence for $|z| > R := N \ (\geqslant 1)$, with $m := \frac{|c_d|}{2}$, we have

$$|p(z)| = |z|^d |c_d + \frac{c_{d-1}}{z} + \cdots + \frac{c_1}{z^{d-1}} + \frac{c_0}{z^d}| \geqslant |z|^d(|c_d| - |\frac{c_{d-1}}{z} + \cdots + \frac{c_1}{z^{d-1}} + \frac{c_0}{z^d}|)$$

$$\geqslant |z|^d(|c_d| - (\frac{|c_{d-1}|}{|z|} + \cdots + \frac{|c_1|}{|z|^{d-1}} + \frac{|c_0|}{|z|^d}))$$

$$\geqslant |z|^d(|c_d| - (\frac{|c_{d-1}|}{N} + \cdots + \frac{|c_1|}{N^{d-1}} + \frac{|c_0|}{N^d}))$$

$$\geqslant |z|^d(|c_d| - \frac{|c_d|}{2}) = \frac{|c_d|}{2}|z|^d = m|z|^d.$$

Solution to Exercise 1.38. ('If' part) Let $(\operatorname{Re} z_n)_{n\in\mathbb{N}}$, $(\operatorname{Im} z_n)_{n\in\mathbb{N}}$ be convergent in $\mathbb{R}$ to $\operatorname{Re} L$, $\operatorname{Im} L$ respectively. Given $\epsilon > 0$, let $N \in \mathbb{N}$ be such that for all $n > N$, $|\operatorname{Re} z_n - \operatorname{Re} L| < \frac{\epsilon}{\sqrt{2}}$ and $|\operatorname{Im} z_n - \operatorname{Im} L| < \frac{\epsilon}{\sqrt{2}}$. Then for all $n > N$,

$$|z_n - L| = \sqrt{(\operatorname{Re} z_n - \operatorname{Re} L)^2 + (\operatorname{Im} z_n - \operatorname{Im} L)^2} < \sqrt{\frac{\epsilon^2}{2} + \frac{\epsilon^2}{2}} = \epsilon.$$

So $(z_n)_{n\in\mathbb{N}}$ converges to L.

('Only if' part) Let $(z_n)_{n\in\mathbb{N}}$ converge to L. Given $\epsilon > 0$, let $N \in \mathbb{N}$ be such that for all $n > N$, $|z - L| < \epsilon$. For $n > N$, $|\operatorname{Re} z_n - \operatorname{Re} L| = |\operatorname{Re}(z_n - L)| \leqslant |z_n - L| < \epsilon$, and $|\operatorname{Im} z_n - \operatorname{Im} L| = |\operatorname{Im}(z_n - L)| \leqslant |z_n - L| < \epsilon$. Hence $(\operatorname{Re} z_n)_{n\in\mathbb{N}}$, $(\operatorname{Im} z_n)_{n\in\mathbb{N}}$ are convergent in $\mathbb{R}$ to $\operatorname{Re} L$, $\operatorname{Im} L$, respectively.

Solution to Exercise 1.39. ('Only if' part) Suppose $(z_n)_{n\in\mathbb{N}}$ converges to L. Then $(\text{Re}\, z_n)_{n\in\mathbb{N}}$ and $(\text{Im}\, z_n)_{n\in\mathbb{N}}$ converge to $\text{Re}\, L$ and $\text{Im}\, L$, respectively. Hence $(\text{Re}\, z_n)_{n\in\mathbb{N}}$ and $(-\text{Im}\, z_n)_{n\in\mathbb{N}}$ converge to $\text{Re}\, L$ and $-\text{Im}\, L$, respectively, that is, $(\text{Re}\,\overline{z_n})_{n\in\mathbb{N}}$ and $(\text{Im}\,\overline{z_n})_{n\in\mathbb{N}}$ converge to $\text{Re}\,\overline{L}$ and $\text{Im}\,\overline{L}$, respectively. Consequently $(\overline{z_n})_{n\in\mathbb{N}}$ converges to $\overline{L}$.

('If' part) Suppose $(\overline{z_n})_{n\in\mathbb{N}}$ converges to $\overline{L}$. By the previous part, $((\overline{(z_n)}))_{n\in\mathbb{N}}$ converges to $(\overline{L})$, that is, $(z_n)_{n\in\mathbb{N}}$ converges to L.

Solution to Exercise 1.40. Let $(z_n)_{n\in\mathbb{N}}$ be a Cauchy sequence in $\mathbb{C}$. Then

$$|\text{Re}\, z_n - \text{Re}\, z_m| = |\text{Re}(z_n - z_m)| \leqslant |z_n - z_m| \text{ and}$$
$$|\text{Im}\, z_n - \text{Im}\, z_m| = |\text{Im}(z_n - z_m)| \leqslant |z_n - z_m|$$

show that the two real sequences $(\text{Re}\, z_n)_{n\in\mathbb{N}}$ and $(\text{Im}\, z_n)_{n\in\mathbb{N}}$ are then also Cauchy sequences in $\mathbb{R}$. Since $\mathbb{R}$ is complete, they are convergent to, say, $a, b \in \mathbb{R}$, respectively. But then $(z_n)_{n\in\mathbb{N}}$ converges in $\mathbb{C}$ to $a + ib$. Hence $\mathbb{C}$ is complete.

Solution to Exercise 1.41.
(1) Given $R > 0$, let $N \in \mathbb{N}$ be such that $N \cdot 1 > R$ (such an N exists by the Archimedean Property). For $n > N$, $|z_n| = |ni^n| = n > N > R$. So $(ni^n)_{n\in\mathbb{N}}$ tends to ∞.

(2) Let $z_n = x_n + iy_n$, where $x_n, y_n \in \mathbb{R}$, $n \in \mathbb{N}$. From Exercise 1.36,
$$d_2(\varphi^{-1}z_n, \mathbf{n})^2 = \big(\tfrac{2x_n}{x_n^2+y_n^2+1}\big)^2 + \big(\tfrac{2y_n}{x_n^2+y_n^2+1}\big)^2 + \big(\tfrac{x_n^2+y_n^2-1}{x_n^2+y_n^2+1}-1\big)^2 = \tfrac{4}{x_n^2+y_n^2+1} = \tfrac{4}{|z_n|^2+1}.$$

('Only if' part) Given $\epsilon > 0$, let $R := \tfrac{2}{\epsilon} > 0$. As $(z_n)_{n\in\mathbb{N}}$ tends to ∞, there exists an $N \in \mathbb{N}$ such that for all $n > N$, we have $|z_n| > R$. So for all $n > N$, $|d_2(\varphi^{-1}z_n, \mathbf{n}) - 0| = \tfrac{2}{\sqrt{|z_n|^2+1}} < \tfrac{2}{|z_n|} < \tfrac{2}{R} = \epsilon$. So $\lim\limits_{n\to\infty} d_2(\varphi^{-1}z_n, \mathbf{n}) = 0$.

('If' part) Let $R > 0$. Set $\epsilon = \tfrac{2}{\sqrt{1+R^2}} > 0$. As the sequence $(d_2(\varphi^{-1}z_n, \mathbf{n}))_{n\in\mathbb{N}}$ converges to 0, there exists an index $N \in \mathbb{N}$ such that for all $n > N$, we have $\tfrac{2}{\sqrt{|z_n|^2+1}} = |d_2(\varphi^{-1}z_n, \mathbf{n}) - 0| < \tfrac{2}{\sqrt{1+R^2}}$, i.e., $|z_n| > R$. So $(z_n)_{n\in\mathbb{N}}$ tends to ∞.

Solution to Exercise 1.42. Let $z_0 \in \mathbb{C}$ and $\epsilon > 0$. Set $\delta = \epsilon > 0$. If $|z - z_0| < \delta$, then $|\text{Re}\, z - \text{Re}\, z_0| = |\text{Re}(z - z_0)| \leqslant |z - z_0| < \delta = \epsilon$, and by the reverse triangle inequality (Exercise 1.30), $||z| - |z_0|| \leqslant |z - z_0| < \delta = \epsilon$. So $z \mapsto \text{Re}\, z, |z|$ are continuous at z_0. As $z_0 \in \mathbb{C}$ was arbitrary, $z \mapsto \text{Re}\, z, |z|$ are continuous on $\mathbb{C}$.

Solution to Exercise 1.43. First consider the case of the unit circle; $r = 1$. There is no loss of generality in assuming that $P_1, \cdots, P_n$ are the n distinct n^{th} roots of unity, say $1, z_1, \cdots, z_{n-1}$, and these are the distinct roots of the polynomial $z^n - 1$. So $(z-1)(z^{n-1} + z^{n-2} + \cdots + z + 1) = z^n - 1 = (z-1)(z-z_1)\cdots(z-z_{n-1})$. Thus for all $z \neq 1$, $(z - z_1)\cdots(z - z_{n-1}) = z^{n-1} + z^{n-2} + \cdots + z + 1$. As both sides are continuous functions on $\mathbb{C}$ that match everywhere on $\mathbb{C}\backslash\{1\}$, their values must also match at 1 (this can be seen e.g. via the sequential characterisation of continuity at a point). So $(1 - z_1) \cdots (1 - z_{n-1}) = 1 + 1^1 + \cdots + 1^{n-1} = n$. Hence $|1 - z_1| \cdots |1 - z_{n-1}| = n$. So in the unit circle case, we have shown that $(P_1 P_2)(P_1 P_3) \cdots (P_1 P_n) = n$. Next, if we scale the unit circle by $r > 0$, then each diagonal length of the inscribed n-gon gets amplified by a factor of r, and so $(P_1 P_2)(P_1 P_3) \cdots (P_1 P_n) = n r^{n-1}$.

Solution to Exercise 1.44. ('If' part) Suppose f is not continuous at z_0, i.e., $\neg[\forall \epsilon > 0 \ \exists \delta > 0$ such that $\forall z \in S$ satisfying $|z - z_0| < \delta$, $|f(z) - f(z_0)| < \epsilon]$. Thus $\exists \epsilon > 0$ such that $\forall \delta > 0 \ \exists z \in S$ such that $|z - z_0| < \delta$, but $|f(z) - f(z_0)| \geq \epsilon$. For $n \in \mathbb{N}$, taking $\delta = \frac{1}{n}$, there exists a $z_n \in S$ such that $|z_n - z_0| < \delta = \frac{1}{n}$, but $|f(z_n) - f(z_0)| \geq \epsilon$. The sequence $(z_n)_{n \in \mathbb{N}}$ is contained in S and is convergent with limit z_0: Indeed, given $\varepsilon > 0$, by the Archimedean Property, there exists an $N \in \mathbb{N}$ such that $N > \varepsilon^{-1}$, and for all $n > N$, $|z_n - z_0| < \frac{1}{n} < \frac{1}{N} < \varepsilon$. But the sequence $(f(z_n))_{n \in \mathbb{N}}$ does not converge to $f(z_0)$: E.g. $\frac{\epsilon}{2} > 0$, and if $N \in \mathbb{N}$ is such that for all $n > N$, $|f(z_n) - f(z_0)| < \frac{\epsilon}{2}$, then $\epsilon \leq |f(z_{N+1}) - f(z_0)| < \frac{\epsilon}{2}$, a contradiction.

('Only if' part) Let f be continuous at z_0, and $(z_n)_{n \in \mathbb{N}}$ be a sequence contained in S with limit z_0. As f is continuous at z_0, given $\epsilon > 0$, there exists a $\delta > 0$ such that for all $z \in S$ satisfying $|z - z_0| < \delta$, $|f(z) - f(z_0)| < \epsilon$. As $(z_n)_{n \in \mathbb{N}}$ is convergent with limit z_0, there exists an $N \in \mathbb{N}$ such that for all $n > N$, $|z_n - z_0| < \delta$. Consequently, for $n > N$, $|f(z_n) - f(z_0)| < \epsilon$. So $(f(z_n))_{n \in \mathbb{N}}$ is convergent with limit $f(z_0)$.

Solution to Exercise 1.45. ('If' part) Suppose u and v are continuous at z_0. Let $\epsilon > 0$. Then there exists a $\delta_1 > 0$ such that whenever $z \in S$ satisfies $|z - z_0| < \delta_1$, $|u(z) - u(z_0)| < \epsilon/2$. Also, there exists a $\delta_2 > 0$ such that whenever $z \in S$ satisfies $|z - z_0| < \delta_2$, $|v(z) - v(z_0)| < \epsilon/2$. So with $\delta := \min\{\delta_1, \delta_2\} > 0$, if $|z - z_0| < \delta$, then $|f(z) - f(z_0)| = |u(z) + iv(z) - u(z_0) - iv(z_0)| \leq |u(z) - u(z_0)| + |i||v(z) - v(z_0)| < \frac{\epsilon}{2} + \frac{\epsilon}{2} = \epsilon$. So f is continuous at z_0.

('Only if' part) Let f be continuous at z_0. Let $\epsilon > 0$. There exists a $\delta > 0$ such that if $z \in S$ satisfies $|z - z_0| < \delta$, then $|f(z) - f(z_0)| < \epsilon$. So if $|z - z_0| < \delta$, then $|u(z) - u(z_0)| = |\text{Re } f(z) - \text{Re } f(z_0)| = |\text{Re}(f(z) - f(z_0))| \leq |f(z) - f(z_0)| < \epsilon$, and $|v(z) - v(z_0)| = |\text{Im } f(z) - \text{Im } f(z_0)| = |\text{Im}(f(z) - f(z_0))| \leq |f(z) - f(z_0)| < \epsilon$. Thus both u and v are continuous at z_0.

Solution to Exercise 1.46. Let $U := \{z \in \mathbb{C} : (\text{Re } z)(\text{Im } z) > 1\}$, and define $F := \complement U$ (the complement of U). If $(z_n)_{n \in \mathbb{N}}$ is a sequence in F such that $(z_n)_{n \in \mathbb{N}}$ converges to L in $\mathbb{C}$, then for all $n \in \mathbb{N}$, $(\text{Re } z_n)(\text{Im } z_n) \leq 1$, and $(\text{Re } z_n)_{n \in \mathbb{N}}$, $(\text{Im } z_n)_{n \in \mathbb{N}}$ converge respectively to $\text{Re } L$ and $\text{Im } L$. So their termwise product $((\text{Re } z_n)(\text{Im } z_n))_{n \in \mathbb{N}}$ converges $(\text{Re } L)(\text{Im } L)$. The inequalities $(\text{Re } z_n)(\text{Im } z_n) \leq 1$ $(n \in \mathbb{N})$ yield $(\text{Re } L)(\text{Im } L) \leq 1$, i.e., $L \in F$. So F is closed, i.e., $\complement F = U$ is open. (Alternatively, as U is the inverse image $\psi^{-1}((1, \infty))$ of the open set $(1, \infty) \subset \mathbb{R}$ under the continuous map $\mathbb{R}^2 \ni (x, y) \mapsto \psi(x, y) := xy \in \mathbb{R}$, U is open.)

Next we will show that U is not a domain. Let U be path-connected. Then there is a (stepwise) path $\gamma : [a, b] \to U$ that joins $\gamma(a) = 2 + 2i \in U$ to $\gamma(b) = -2 - 2i \in U$.

As $\mathbb{C} \ni z \mapsto \operatorname{Re} z \in \mathbb{R}$ is continuous, $[a,b] \ni t \xrightarrow{\varphi} \operatorname{Re} \gamma(t) \in \mathbb{R}$ is continuous too. We have $\varphi(a) = \operatorname{Re}\gamma(a) = \operatorname{Re}(2+2i) = 2$ and $\varphi(b) = \operatorname{Re}\gamma(b) = \operatorname{Re}(-2-2i) = -2$. As $\varphi(a) = 2 > 0 > -2 = \varphi(b)$, by the Intermediate Value Theorem, there is a $t_* \in [a,b]$ such that $0 = \varphi(t_*) = \operatorname{Re}\gamma(t_*)$. So $(\operatorname{Re}\gamma(t_*))(\operatorname{Im}\gamma(t_*)) = (0)(\operatorname{Im}\gamma(t_*)) = 0 \not> 1$, showing that $\gamma(t_*) \notin U$, a contradiction.

Solution to Exercise 1.47. Since D is open, it is clear that its reflection in the real axis, $\tilde{D}$, is also open. Let $w_1, w_2 \in \tilde{D}$. Then $\overline{w_1}, \overline{w_2} \in D$. As D is a domain, there exists a stepwise path $\gamma : [a,b] \to \mathbb{C}$ such that $\gamma(a) = \overline{w_1}$, $\gamma(b) = \overline{w_2}$ and for all $t \in [a,b]$, $\gamma(t) \in D$. Now define $\tilde{\gamma} : [a,b] \to \mathbb{C}$ by $\tilde{\gamma}(t) = \overline{\gamma(t)}$, $t \in [a,b]$. Then $\tilde{\gamma}(a) = \overline{\overline{w_1}} = w_1$, $\tilde{\gamma}(b) = \overline{\overline{w_2}} = w_2$, and for all $t \in [a,b]$, $\tilde{\gamma}(t) = \overline{\gamma(t)} \in \tilde{D}$. As $\tilde{\gamma}$ is the composition of the continuous functions γ and $z \mapsto \bar{z}$, $\tilde{\gamma}$ is continuous. Also, since γ is a stepwise path, there exist points $t_0 = a < t_1 < \cdots < t_n < t_{n+1} = b$ such that for each $k = 0, 1, \cdots, n$, the restriction $\gamma|_{[t_k, t_{k+1}]}$ has either constant real part or has a constant imaginary part. Consequently $\tilde{\gamma}|_{[t_k, t_{k+1}]}$ (which has the same real part as $\gamma|_{[t_k, t_{k+1}]}$ and has imaginary part which is minus the imaginary part of $\gamma|_{[t_k, t_{k+1}]}$) also has either a constant real part or has a constant imaginary part. So $\tilde{\gamma}$ is a stepwise path too. Hence $\tilde{D}$ is path-connected.

Solution to Exercise 1.48. $\exp(3 + \pi i) = e^3(\cos \pi + i \sin \pi) = e^3(-1 + i0) = -e^3$ and $\exp(i\frac{9\pi}{2}) = \exp(i(4\pi + \frac{\pi}{2})) = e^0(\cos(4\pi + \frac{\pi}{2}) + i \sin(4\pi + \frac{\pi}{2})) = 1(0 + i1) = i$.

Solution to Exercise 1.49. With $z = x + iy$ $(x, y \in \mathbb{R})$, $e^x(\cos y + i \sin y) = \pi i$. Taking absolute values, $e^x = \pi$, and so $x = \log \pi$. Then $\cos y + i \sin y = i$, so that $\sin y = 1$ and $\cos y = 0$. Thus $y = \frac{\pi}{2} + 2\pi k$, $k \in \mathbb{Z}$. Hence if $\exp z = \pi i$, then $z \in \{\log \pi + i(\frac{\pi}{2} + 2\pi k), \ k \in \mathbb{Z}\}$.

Vice versa, if $z = \log \pi + i(\frac{\pi}{2} + 2\pi k)$ for some $k \in \mathbb{Z}$, then

$$\exp z = e^{\log \pi}(\cos(\tfrac{\pi}{2} + 2\pi k) + i \sin(\tfrac{\pi}{2} + 2\pi k)) = \pi(0 + i1) = \pi i.$$

Consequently, $\exp z = \pi i$ if and only if $z \in \{\log \pi + i(\frac{\pi}{2} + 2\pi k), \ k \in \mathbb{Z}\}$.

Solution to Exercise 1.50. Let $\gamma(t) := \exp(it)$, $t \in [0, 2\pi]$. Then

$$\gamma(t) = \exp(it) = e^0(\cos t + i \sin t) = \cos t + i \sin t.$$

The point $(\cos t, \sin t)$ lies on a circle of radius 1 and centre $(0,0)$, and with increasing t, this point moves anticlockwise. Hence the curve $t \mapsto \gamma(t)$ is the circle traversed once in the anticlockwise direction, as shown below.

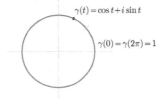

$\gamma(t) = \cos t + i \sin t$

$\gamma(0) = \gamma(2\pi) = 1$

Solution to Exercise 1.51. If $z = x + iy$ ($x, y \in \mathbb{R}$), then $|\exp(-iz)| < 1$ if and only if $|\exp(-ix + y)| < 1$, i.e., $e^y < 1$, i.e., $y < 0$. So $D = \{z \in \mathbb{C} : \operatorname{Im} z < 0\}$, i.e., D is the half-plane below the real axis in the complex plane. As D is open and path-connected, it is a domain.

Solution to Exercise 1.52. As $\exp(t + it) = e^t(\cos t + i \sin t)$, the curve is given by $t \mapsto (e^t \cos t, e^t \sin t)$. The curve is a spiral, and as $t \searrow -\infty$, $e^t(\cos t + i \sin t)$ converges to 0, while the curve spirals outwards as $t \nearrow \infty$.

Solution to Exercise 1.53. If $z = x + iy$, where $x, y \in \mathbb{R}$, then
$$\exp(z^2) = \exp((x+iy)^2) = \exp(x^2 - y^2 + 2xyi) = e^{x^2-y^2}(\cos(2xy) + i\sin(2xy)).$$
So $|\exp(z^2)| = e^{x^2-y^2}$, $\operatorname{Re} \exp(z^2) = e^{x^2-y^2}\cos(2xy)$, $\operatorname{Im} \exp(z^2) = e^{x^2-y^2}\sin(2xy)$.
For $z \neq 0$, $\exp\frac{1}{z} = \exp\frac{1}{x+iy} = \exp\frac{x-iy}{x^2+y^2} = e^{\frac{x}{x^2+y^2}}(\cos\frac{-y}{x^2+y^2} + i\sin\frac{-y}{x^2+y^2})$. Thus
$|\exp\frac{1}{z}| = e^{\frac{x}{x^2+y^2}}$, $\operatorname{Re}\exp\frac{1}{z} = e^{\frac{x}{x^2+y^2}}\cos\frac{y}{x^2+y^2}$, $\operatorname{Im}\exp\frac{1}{z} = -e^{\frac{x}{x^2+y^2}}\sin\frac{y}{x^2+y^2}$.

Solution to Exercise 1.54. Let $z = x + iy$, where $x, y \in \mathbb{R}$. Then
$$\overline{\exp z} = \overline{e^x\cos y + ie^x\sin y} = e^x\cos y - ie^x\sin y = e^x(\cos(-y) + i\sin(-y))$$
$$= \exp(x - iy) = \exp\overline{z}.$$

Solution to Exercise 1.55. For $x, y \in \mathbb{R}$, $\exp(x + iy) = e^x(\cos y + i\sin y)$, and so the image of vertical lines are circles and the image of horizontal lines are rays from 0. Thus the image of the square is a portion of an annulus, as shown below.

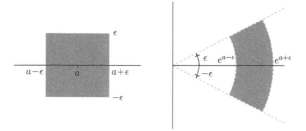

The area of the sector with radius r and opening angle θ is $\frac{1}{2}\theta r^2$, so that the ratio $\rho(\epsilon)$ of the areas is $\rho(\epsilon) = \frac{\frac{1}{2}(2\epsilon)((e^{a+\epsilon})^2 - (e^{a-\epsilon})^2)}{(2\epsilon)^2} = e^{2(a-\epsilon)}\frac{e^{4\epsilon}-1}{4\epsilon}$. Hence we obtain

$$\lim_{\epsilon \to 0} \rho(\epsilon) = \lim_{\epsilon \to 0} e^{2(a-\epsilon)}\frac{e^{4\epsilon}-1}{4\epsilon} = e^{2(a-0)}\frac{de^x}{dx}\Big|_{x=0} = e^{2a}e^0 = e^{2a}.$$

Solution to Exercise 1.56. For $z_1, z_2 \in \mathbb{C}$, we have

$$(\sin z_1)(\cos z_2) + (\cos z_1)(\sin z_2)$$

$$= \frac{(\exp(iz_1) - \exp(-iz_1))}{2i}\frac{(\exp(iz_2) + \exp(-iz_2))}{2} + \frac{(\exp(iz_1) + \exp(-iz_1))}{2}\frac{(\exp(iz_2) - \exp(-iz_2))}{2i}$$

$$= \frac{2\exp(i(z_1+z_2)) - 2\exp(-i(z_1+z_2))}{4i} = \sin(z_1 + z_2).$$

Solution to Exercise 1.57. For $y \in \mathbb{R}$, we have

$$(\cosh y)^2 - (\sinh y)^2 = \left(\frac{e^y + e^{-y}}{2}\right)^2 - \left(\frac{e^y - e^{-y}}{2}\right)^2 = \frac{e^{2y} + 2 + e^{-2y} - (e^{2y} - 2 + e^{-2y})}{4} = \frac{4}{4} = 1.$$

Let $z = x + iy$, where $x, y \in \mathbb{R}$. Then using the addition formula on page 24,

$$\cos z = \cos(x + iy) = (\cos x)(\cos(iy)) - (\sin x)(\sin(iy))$$

$$= (\cos x)\frac{e^{-y} + e^y}{2} - (\sin x)\frac{e^{-y} - e^y}{2i} = (\cos x)(\cosh y) - (\sin x)\left(-\frac{\sinh y}{i}\right)$$

$$= (\cos x)(\cosh y) - i(\sin x)(\sinh y).$$

Thus

$$|\cos z|^2 = (\cos x)^2(\cosh y)^2 + (\sin x)^2(\sinh y)^2$$
$$= (1 - (\sin x)^2)(\cosh y)^2 + (\sin x)^2((\cosh y)^2 - 1)$$
$$= (\cosh y)^2 - (\sin x)^2.$$

Solution to Exercise 1.58. Let $z = x + iy$, where $x, y \in \mathbb{R}$. Then $\cos z = 3$ gives

$$(\cos x)(\cosh y) = 3 \qquad (*)$$
$$(\sin x)(\sinh y) = 0. \qquad (**)$$

Now $\sinh y = 0$ if and only if $y = 0$. If $y = 0$, then $z = x + iy = x$ is real, and $[-1, 1] \ni \cos x = \cos z = 3$, a contradiction. Thus $\sinh y \neq 0$, and so $(**)$ implies that $\sin x = 0$. Hence $x \in \{n\pi : n \in \mathbb{Z}\}$. So $\cos x = \pm 1$. As $\cosh y = \frac{e^y + e^{-y}}{2} > 0$ for all $y \in \mathbb{R}$, $(*)$ implies $\cos x$ cannot be -1. So $\cos x = 1$ and $x \in \{2\pi n : n \in \mathbb{Z}\}$. Then $\cosh y = 3$ yields $\frac{e^y + e^{-y}}{2} = 3$, that is, $(e^y)^2 - 6e^y + 1 = 0$. Thus $e^y = 3 \pm 2\sqrt{2}$, and so $y = \log(3 + 2\sqrt{2})$ or $y = \log(3 - 2\sqrt{2}) = \log\frac{9-8}{3+2\sqrt{2}} = \log\frac{1}{3+2\sqrt{2}} = -\log(3 + 2\sqrt{2})$. Hence $z \in \{2\pi n \pm i\log(3 + 2\sqrt{2}), n \in \mathbb{Z}\}$.

Vice versa, if $z = 2\pi n \pm i\log(3 + 2\sqrt{2})$ for some $n \in \mathbb{Z}$, then

$$\cos z = (\cos(2\pi n))\cosh(\pm\log(3 \pm 2\sqrt{2})) - i(\sin(2\pi n))(\sinh \cdots)$$

$$= 1\cdot\cosh(\pm\log(3 \pm 2\sqrt{2})) - i0(\sinh \cdots) = \cosh(\pm\log(3 \pm 2\sqrt{2}))$$

$$= \frac{e^{\log(3+2\sqrt{2})} + e^{-\log(3+2\sqrt{2})}}{2} = \frac{3 + 2\sqrt{2} + (3 + 2\sqrt{2})^{-1}}{2} = \frac{3 + 2\sqrt{2} + 3 - 2\sqrt{2}}{2} = 3.$$

Consequently, $\cos z = 3$ if and only if $z \in \{2\pi n \pm i\log(3 + 2\sqrt{2}), n \in \mathbb{Z}\}$.

Solution to Exercise 1.59. If $\cos z + i \sin z = e$, then we obtain

$$\frac{\exp(iz)+\exp(-iz)}{2} + i\frac{\exp(iz)-\exp(-iz)}{2i} = e,$$

giving $\exp(iz) = e$. Let $z = x + iy$, where $x, y \in \mathbb{R}$. Taking complex absolute values in $\exp(iz) = e$, $e^{-y} = |\exp(-y+ix)| = |\exp(i(x+iy))| = |\exp(iz)| = |e| = e$. So $y = -1$. Thus $e \exp(ix) = e^{-y} \exp(ix) = \exp(-y+ix) = \exp(iz) = e$, that is, $\exp(ix) = 1$. Hence $\cos x + i \sin x = 1$, so that $\cos x = 1$ and $\sin x = 0$. Thus $x \in 2\pi\mathbb{Z}$. So if $\cos z + i \sin z = e$, then $z \in \{2\pi n - i : n \in \mathbb{Z}\}$. Vice versa, if $z = 2\pi n - i$ $(n \in \mathbb{Z})$, then $\cos z + i \sin z = \exp(iz) = \exp(i(2\pi n - i)) = e^1 \exp(i2\pi n) = e(1 + i0) = e$.

Solution to Exercise 1.60.

Solution to Exercise 1.61. We have

$$\mathrm{Log}\,(1+i) = \mathrm{Log}(\sqrt{2}(\tfrac{1}{\sqrt{2}} + i\tfrac{1}{\sqrt{2}})) = \mathrm{Log}\,(\sqrt{2}(\cos\tfrac{\pi}{4} + i\sin\tfrac{\pi}{4})) = \log\sqrt{2} + i\tfrac{\pi}{4}.$$

Solution to Exercise 1.62. We have

$$\mathrm{Log}\,(-1) = \mathrm{Log}\,(1(\cos\pi + i\sin\pi)) = \log 1 + i\pi = 0 + i\pi = i\pi,$$
$$\mathrm{Log}\,1 = \mathrm{Log}\,(1(\cos 0 + i\sin 0)) = \log 1 + i0 = 0 + i0 = 0.$$

With $z = -1$, $\mathrm{Log}\,(z^2) = \mathrm{Log}\,((-1)^2) = \mathrm{Log}\,1 = 0$, and $2\mathrm{Log}\,z = 2\mathrm{Log}\,(-1) = 2\pi i$. So when $z = -1$, $\mathrm{Log}\,(z^2) = 0 \neq 2\pi i = 2\mathrm{Log}\,z$.

For $z \in \mathbb{C}\backslash\{0\}$, $\mathrm{Log}(z^n) = n\,\mathrm{Log}\,z \Leftrightarrow \log|z^n| + i\,\mathrm{Arg}(z^n) = n\log|z| + in\mathrm{Arg}\,z$. But $\log|z^n| = \log(|z|^n) = n\log|z|$. So $\mathrm{Log}(z^n) = n\,\mathrm{Log}\,z \Leftrightarrow \mathrm{Arg}(z^n) = n\mathrm{Arg}\,z$. If $\mathrm{Log}(z^n) = n\,\mathrm{Log}\,z$, then $n\mathrm{Arg}\,z = \mathrm{Arg}(z^n) \in (-\pi, \pi]$, so that $\mathrm{Arg}\,z \in (-\tfrac{\pi}{n}, \tfrac{\pi}{n}]$, that is, z belongs to the 'sector' $\Sigma := \{z \in \mathbb{C} : -\tfrac{\pi}{n} < \mathrm{Arg}\,z \leqslant \tfrac{\pi}{n}\}$. Vice versa, if $z \in \Sigma$, and $\theta := \mathrm{Arg}\,z \in (-\tfrac{\pi}{n}, \tfrac{\pi}{n}]$, then we have $z = |z|(\cos\theta + i\sin\theta)$, so that $z^n = |z|^n(\cos\theta + i\sin\theta)^n = |z^n|(\cos(n\theta) + i\sin(n\theta))$, with $n\theta \in (-\pi, \pi]$, showing that $\mathrm{Arg}(z^n) = n\theta = n\mathrm{Arg}\,z$, and $\mathrm{Log}(z^n) = n\,\mathrm{Log}\,z$.

Solution to Exercise 1.63.
For $w \in \mathbb{A} = \{r(\cos\theta + i\sin\theta) : 1 < r < e,\ \theta \in (-\pi, \pi]\}$, $\mathrm{Log}\,w = \log r + i\mathrm{Arg}\,w$, and so $\mathrm{Re}(\mathrm{Log}\,w) = \log r \in (0,1)$, while $\mathrm{Im}(\mathrm{Log}\,w) \in (-\pi, \pi]$. So we have that the image $\mathrm{Log}\,\mathbb{A} \subset \mathbb{I} := \{x + iy : 0 < x < 1,\ -\pi < y \leqslant \pi\}$.

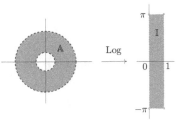

We claim $\mathbb{I} \subset \mathrm{Log}\,\mathbb{A}$. Let $z \in \mathbb{I}$. So $x := \mathrm{Re}\,z \in (0,1)$, $y := \mathrm{Im}\,z \in (-\pi, \pi]$. Then $w := \exp(x + iy) \in \mathbb{A}$, since $|w| = e^x \in (1, e)$. Also,

$$\mathrm{Log}\,w = \mathrm{Log}(e^x(\cos y + i \sin y)) = (\log e^x) + iy = x + iy = z.$$

Hence $\mathrm{Log}\,\mathbb{A} = \mathbb{I}$.

Solution to Exercise 1.64. $\mathrm{Log}(1 + i) = \mathrm{Log}(\sqrt{2}(\cos\frac{\pi}{4} + i\sin\frac{\pi}{4}) = \log\sqrt{2} + i\frac{\pi}{4}$. Thus the principal value of $(1 + i)^{1-i}$ is

$$\exp((1-i)\mathrm{Log}(1+i)) = \exp((1-i)(\log\sqrt{2} + i\tfrac{\pi}{4})) = e^{\log\sqrt{2} + \frac{\pi}{4}}\exp(i(\tfrac{\pi}{4} - \log\sqrt{2}))$$
$$= \sqrt{2}\,e^{\frac{\pi}{4}}\tfrac{(1+i)}{\sqrt{2}}\exp(-i\log\sqrt{2}) = e^{\frac{\pi}{4}}(1 + i)(\cos(\log\sqrt{2}) - i\sin(\log\sqrt{2})).$$

Thus we have that the real and imaginary parts of $(1 + i)^{1-i}$ are given respectively by $e^{\frac{\pi}{4}}(\cos(\log\sqrt{2}) + \sin(\log\sqrt{2}))$ and $e^{\frac{\pi}{4}}(\cos(\log\sqrt{2}) - \sin(\log\sqrt{2}))$.

Solution to Exercise 1.65. Since $1 = \exp 0 = \exp(z - z) = (\exp z)(\exp(-z))$, $\exp(-z) = \frac{1}{\exp z}$ for all $z \in \mathbb{C}$. We have

$$w^{-n} = \exp(-n\,\mathrm{Log}\,w) = \exp(n(-\mathrm{Log}\,w)) = (\exp(-\mathrm{Log}\,w))^n = \left(\frac{1}{\exp(\mathrm{Log}\,w)}\right)^n = \left(\frac{1}{w}\right)^n.$$

For $m \in \mathbb{N}$, $(e^{it})^m = (\cos t + i\sin t)^m = \cos(mt) + i\sin(mt) = e^{imt}$. Let $m = -n$, $n \in \mathbb{N}$. With $w := e^{it} = \cos t + i\sin t$, we have

$$\frac{1}{w} = \frac{1}{\cos t + i\sin t} = \cos t - i\sin t = \cos(-t) + i\sin(-t) = e^{i(-t)},$$

and so $(e^{it})^m = (e^{it})^{-n} = (e^{i(-t)})^n = e^{in(-t)} = e^{imt}$.

Solution to Exercise 1.66. (1) Suppose first that $c \neq 0$. We note that if $z \in \mathbb{C}$ and $\frac{az+b}{cz+d} = \frac{a}{c}$, then we get $caz + bc = acz + ad$, giving $ac - bd = 0$, which is a contradiction. So, if $z \in \widehat{\mathbb{C}}$ and $f(z) = \frac{a}{c}$, then $z \notin \mathbb{C}$, that is, $z = \infty$.

$1°$ If $z, z' \in \widehat{\mathbb{C}}$ are such that $f(z) = f(z') \in \mathbb{C}\backslash\{\frac{a}{c}\}$, then $z, z' \in \mathbb{C}$ and $\frac{az+b}{cz+d} = \frac{az'+b}{cz'+d}$. This yields upon simplifying that $(ad - bc)(z - z') = 0$, and so $z = z'$.

$2°$ If $z, z' \in \widehat{\mathbb{C}}$ are such that $f(z) = f(z') = \frac{a}{c}$, then $z = z' = \infty$.

$3°$ If $z, z' \in \widehat{\mathbb{C}}$ and $f(z) = f(z') = \infty$, then $z = z' = -\frac{d}{c}$.

Thus f is injective when $c \neq 0$.

Next we show surjectivity. (Solving for z in $w = \frac{az+b}{cz+d}$, we arrive at $w = \frac{-dw+b}{cw-a}$.)

For $w \in \mathbb{C}\backslash\{\frac{a}{c}\}$, $f(\frac{-dw+b}{cw-a}) = \dfrac{a(\frac{-dw+b}{cw-a}) + b}{c(\frac{-dw+b}{cw-a}) + d} = \dfrac{-adw + ab + bcw - ab}{-cdw + bc + cdw - ad} = \dfrac{-(ad-bc)w}{-(ad-bc)} = w$.

Also, $f(\infty) = \frac{a}{c}$ and $f(-\frac{d}{c}) = \infty$. Hence every point of $\widehat{\mathbb{C}}$ is in the range of f, showing that f is surjective.

To show the inverse is given by $g(w) = \frac{-dw+b}{cw-a}$ for $w \in \mathbb{C}\backslash\{\frac{a}{c}\}$, with the understanding that $g(\frac{a}{c}) = \infty$ and $g(\infty) = \frac{-d}{c}$, it is enough to show $f \circ g = \mathrm{id}$, where id is the identity map on $\widehat{\mathbb{C}}$ (as then $g = \mathrm{id} \circ g = (f^{-1} \circ f) \circ g = f^{-1} \circ (f \circ g) = f^{-1} \circ \mathrm{id} = f^{-1}$).

$1°$ For $w \in \mathbb{C}\backslash\{\frac{a}{c}\}$, we have $\mathrm{id}(w) = w = f(\frac{-dw+b}{cw-a}) = f(g(w)) = (f \circ g)(w)$.

$2°$ For $w = \frac{a}{c}$, $(f \circ g)(w) = f(g(\frac{a}{c})) = f(\infty) = \frac{a}{c} = w = \mathrm{id}(w)$.

$3°$ For $w = \infty$, then $(f \circ g)(w) = f(g(\infty)) = f(\frac{-d}{c}) = \infty = w = \mathrm{id}(w)$.

This completes the proof when $c \neq 0$.

Next we suppose that $c = 0$. Then $0 \neq ad - bc = ad$ implies that $a \neq 0$ and $d \neq 0$. Thus $\alpha := \frac{a}{d} \neq 0$. With $\beta := \frac{b}{d}$, we have $f(z) = \frac{a}{d}z + \frac{b}{d} = \alpha z + \beta \in \mathbb{C}$ for all $z \in \mathbb{C}$. Also $f(\infty) = \infty$. If $z, z' \in \widehat{\mathbb{C}}$ are such that $f(z) = f(z') \in \mathbb{C}$, then we have $\alpha z + \beta = \alpha z' + \beta$, and so $\alpha(z - z') = 0$, giving $z = z'$ (as $\alpha \neq 0$). Also, if $z, z' \in \widehat{\mathbb{C}}$ and $f(z) = f(z') = \infty$, then $z = z' = \infty$. Thus f is injective.

For any $w \in \mathbb{C}$, we have $f(\frac{w-\beta}{\alpha}) = \alpha\frac{(w-\beta)}{\alpha} + \beta = w$. Also $f(\infty) = \infty$. Hence every point of $\widehat{\mathbb{C}}$ belongs to the range of f, showing that f is surjective.

Finally, we will check that the inverse is given by g, where $g(w) := \frac{-dw+b}{0-a} = \frac{w}{\alpha} - \frac{\beta}{\alpha}$ for $w \in \mathbb{C}$, and $g(\infty) := \infty$.

$1°$ For all $w \in \mathbb{C}$, $(f \circ g)(w) = f(\frac{w}{\alpha} - \frac{\beta}{\alpha}) = \alpha\frac{(w-\beta)}{\alpha} + \beta = w = \mathrm{id}(w)$.

$2°$ Also, $(f \circ g)(\infty) = f(\infty) = \infty = \mathrm{id}(\infty)$.

(2) With the Möbius transformations $f(z) = \frac{az+b}{cz+d}$ and $g(z) = \frac{a'z+b'}{c'z+d'}$, we associate the matrices $F = \begin{bmatrix} a & b \\ c & d \end{bmatrix}$ and $G = \begin{bmatrix} a' & b' \\ c' & d' \end{bmatrix}$. Then their determinants are nonzero.

We have $(f \circ g)(z) = f(\frac{a'z+b'}{c'z+d'}) = \frac{a(\frac{a'z+b'}{c'z+d'})+b}{c(\frac{a'z+b'}{c'z+d'})+d} = \frac{(aa'+bc')z+ab'+bd'}{(ca'+dc')z+(cb'+dd')} = \frac{Az+B}{Cz+D}$, where

$H := \begin{bmatrix} A & B \\ C & D \end{bmatrix} := \begin{bmatrix} aa'+bc' & ab'+bd' \\ ca'+dc' & cb'+dd' \end{bmatrix} = \begin{bmatrix} a & b \\ c & d \end{bmatrix}\begin{bmatrix} a' & b' \\ c' & d' \end{bmatrix} = FG$.

Now to show that $f \circ g$ is a Möbius transformation, it suffices to show $AD - BC \neq 0$. But $AD - BC = \det H = \det(FG) = (\det F)(\det G) = (ad - bc)(a'd' - b'c') \neq 0$.

(3) Let $C = \{z \in \mathbb{C} : |z - z_0| = r\}$ be the circle with centre $z_0 \in \mathbb{C}$ and radius $r > 0$. For a translation $f(z) = z + b$, $f(C) = \{w \in \mathbb{C} : |w - (z_0 + b)| = r\}$, which is a circle with centre $z_0 + b$ and radius $r > 0$.

For a dilation $f(z) = az$ ($a > 0$), $f(C) = \{w \in \mathbb{C} : |w - az_0| = ar\}$, which is a circle with centre az_0 and radius $ar > 0$.

For a rotation $f(z) = e^{i\theta}z$ ($\theta \in \mathbb{R}$), $f(C) = \{w \in \mathbb{C} : |w - e^{i\theta}z_0| = r\}$, which is a circle with centre $e^{i\theta}z_0$ and radius $r > 0$.

So for translations, dilations, and rotations, the images of circles are circles.

With inversions, we will now see that the image of a circle is either a circle or a straight line (which can be viewed as a circle with an infinite radius). If $f(z) = \frac{1}{z}$, then $f(C) \cap \mathbb{C} = \{w \in \mathbb{C} : |\frac{1}{w} - z_0| = r\} = \{w \in \mathbb{C} : |1 - wz_0| = r|w|\}$. So:

$$w \in f(C) \cap \mathbb{C} \Leftrightarrow 0 = |1 - wz_0|^2 - r^2|w|^2 \Leftrightarrow 0 = (1 - wz_0)(1 - \overline{w}\,\overline{z_0}) - r^2|w|^2$$
$$\Leftrightarrow (|z_0|^2 - r^2)|w|^2 - z_0 w - \overline{z_0}\,\overline{w} + 1 = 0.$$

$1°$ $|z_0| = r > 0$. Then $w \in f(C) \cap \mathbb{C}$ if and only if $2\operatorname{Re}(z_0 w) = 1$. If $z_0 = (x_0, y_0) \in \mathbb{R}^2$, we have with $w = (x, y) \in \mathbb{R}^2$ that $w \in f(C)$ if and only if $x_0 x - y_0 y = \frac{1}{2}$. Thus $f(C) = \{(x, y) \in \mathbb{R}^2 : x_0 x - y_0 y = \frac{1}{2}\} \cup \{\infty\}$, which is a line in the complex plane. (Since $|z_0| = r > 0$, $(x_0, y_0) \neq (0,0)$. Also, $0 \in C$, and so $\infty \in f(C)$.)

$2°$ $|z_0| > r$. Let $k > 0$ be such that $k^2 = |z_0|^2 - r^2 > 0$. Also, let $a = \frac{\overline{z_0}}{k^2} \in \mathbb{C}$. Then $w \in f(C)$ if and only if $|w|^2 - 2\operatorname{Re}(\overline{a}w) + |a|^2 = (\frac{r}{k^2})^2$. Thus (see e.g. Exercise 1.31) $f(C) = \{w \in \mathbb{C} : |w - a| = \frac{r}{k^2}\}$, which is a circle in the complex plane with centre a and radius $\frac{r}{k^2}$.

$3°$ $|z_0| < r$. Let $k > 0$ be such that $k^2 = r^2 - |z_0|^2 > 0$. Also, let $a = -\frac{\overline{z_0}}{k^2} \in \mathbb{C}$. Then $w \in f(C)$ if and only if $|w|^2 - 2\operatorname{Re}(\overline{a}w) + |a|^2 = (\frac{r}{k^2})^2$. It follows that $f(C) = \{w \in \mathbb{C} : |w - a| = \frac{r}{k^2}\}$, which is a circle with centre a and radius $\frac{r}{k^2}$.

Next we consider the case of lines. From Exercise 1.22, a line passing through z_0, z_1, $z_0 \neq z_1$, is $L = \{tz_1 + (1 - t)z_0 : t \in \mathbb{R}\}$. So if $d := z_1 - z_0 \neq 0$, then $L = \{z_0 + td : t \in \mathbb{R}\}$ is the line passing through z_0 and $z_0 + d$ ($\neq z_0$).

For a translation $f(z) = z + b$, $f(L) = \{z_0 + b + td : t \in \mathbb{R}\}$, which is a line passing through $z_0 + b$ and $z_0 + b + d$.

For a dilation $f(z) = az$ ($a > 0$), $f(L) = \{az_0 + tad : t \in \mathbb{R}\}$, which is a line passing through az_0 and $az_0 + ad$.

For a rotation $f(z) = e^{i\theta}z$ ($\theta \in \mathbb{R}$), $f(L) = \{e^{i\theta}z_0 + te^{i\theta}d : t \in \mathbb{R}\}$, which is a line passing through $e^{i\theta}z_0$ and $e^{i\theta}z_0 + e^{i\theta}d$.

Thus in each of the cases above, the images of lines are lines again.

With inversions $f(z) = \frac{1}{z}$, we will now see that the image of a line L is either a line (if L passes through the origin) or a circle (if L does not pass through the origin).

$1°$ If $L = \{td : t \in \mathbb{R}\}$, then using $f(\infty) = 0$ and $f(0) = \infty$, it can be seen that $f(L \cup \{\infty\}) = \{s\frac{1}{d} : s \in \mathbb{R}\} \cup \{\infty\}$. The set $\{s\frac{1}{d} : s \in \mathbb{R}\}$ describes a line passing through 0 (and is obtained by reflecting L in the real axis).

$2°$ If L does not pass through the origin, then there is a nonzero complex number $z_0 \in L$ closest to the origin, so that the segment joining 0 to z_0 is perpendicular to L. So we may take $d = iz_0$, and $L = \{z_0 + tiz_0 : t \in \mathbb{R}\}$. For $w := \frac{1}{z_0(1+it)}$,

$$|w - \frac{1}{2z_0}| = \frac{1}{2|z_0|}|\frac{2}{1+it} - 1| = \frac{1}{2|z_0|}|\frac{2(1-it)}{1+t^2} - 1| = \frac{1}{2|z_0|}|\frac{1-t^2}{1+t^2} - i\frac{2t}{1+t^2}|$$
$$= \frac{1}{2|z_0|(1+t^2)}\sqrt{(1-t^2)^2 + 4t^2} = \frac{1}{2|z_0|(1+t^2)}(1 + t^2) = \frac{1}{2|z_0|}.$$

So $f(L \cup \{\infty\}) = \{w \in \mathbb{C} : |w - \frac{1}{2z_0}| = \frac{1}{2|z_0|}\}$, which is a circle passing through the origin.

Solutions to the exercises from Chapter 2

Solution to Exercise 2.1. Let L_1, L_2 denote the limits as $z \to z_0$ of h_1, h_2, respectively. Let $\epsilon > 0$. There exists $\delta_1 > 0$ such that whenever $z \in U$ satisfies $0 < |z - z_0| < \delta_1$, $|h_1(z) - L_1| < \epsilon/2$. Also, there exists $\delta_2 > 0$ such that whenever $z \in U$ satisfies $0 < |z - z_0| < \delta_2$, $|h_2(z) - L_2| < \epsilon/2$. Set $\delta = \min\{\delta_1, \delta_2\} > 0$. If $z \in U$ and $0 < |z - z_0| < \delta$, then $|(h_1 + h_2)(z) - (L_1 + L_2)| \leqslant |h_1(z) - L_1| + |h_2(z) - L_2| < \epsilon$.
Next we consider the case of pointwise product. Let $\epsilon > 0$. There exists $\delta_1 > 0$ such that whenever $z \in U$ satisfies $0 < |z - z_0| < \delta_1$, $|h_1(z) - L_1| < \frac{\epsilon}{2(|L_2|+1)} \leqslant \frac{\epsilon}{2}$. Also, there exists $\delta_2 > 0$ such that whenever $z \in U$ satisfies $0 < |z - z_0| < \delta_2$, $|h_2(z) - L_2| < \frac{\epsilon}{\epsilon + 2|L_1|}$. Set $\delta = \min\{\delta_1, \delta_2\} > 0$. If $z \in U$ and $0 < |z - z_0| < \delta$,

$$
\begin{aligned}
|(h_1 h_2)(z) - L_1 L_2| &= |h_1(z) h_2(z) - L_2 h_1(z) + L_2 h_1(z) - L_1 L_2| \\
&= |h_1(z)||h_2(z) - L_2| + |L_2||h_1(z) - L_1| \\
&\leqslant (|h_1(z) - L_1| + |L_1|)|h_2(z) - L_2| + |L_2||h_1(z) - L_1| \\
&< (\tfrac{\epsilon}{2} + |L_1|)\tfrac{\epsilon}{\epsilon + 2|L_1|} + |L_2|\tfrac{\epsilon}{2(|L_2|+1)} = \epsilon.
\end{aligned}
$$

Solution to Exercise 2.2. Rough work: Let L be the limit of f as $z \to z_0$. For $z, z_0 \in U$, $|\frac{1}{f(z)} - \frac{1}{L}| = \frac{|f(z) - L|}{|f(z)||L|}$. For z close to z_0, $|f(z)|$ is close to $|L| > 0$, and $|f(z)|$ stays away from zero as follows: if $|f(z) - L| < \frac{|L|}{2}$, then using the reverse triangle inequality, $\frac{|L|}{2} > |f(z) - L| \geqslant ||f(z)| - |L|| \geqslant |L| - |f(z)|$, giving $|f(z)| > \frac{|L|}{2}$.
Given $\epsilon > 0$, $\epsilon' := \min\{\frac{\epsilon|L|^2}{2}, \frac{|L|}{2}\} > 0$. Let $\delta > 0$ be such that whenever $z \in U$ satisfies $0 < |z - z_0| < \delta$, we have $|f(z) - L| < \epsilon'$. Thus, if $0 < |z - z_0| < \delta$, then we have $\frac{|L|}{2} > |f(z) - L| \geqslant ||f(z)| - |L|| \geqslant |L| - |f(z)|$, giving $|f(z)| > \frac{|L|}{2}$. So if $0 < |z - z_0| < \delta$, $|\frac{1}{f(z)} - \frac{1}{L}| = \frac{|f(z) - L|}{|f(z)||L|} < \epsilon' \frac{2}{|L|^2} \leqslant \frac{\epsilon|L|^2}{2} \frac{2}{|L|^2} = \epsilon$.

Solution to Exercise 2.3. For $z \neq 0$, $\frac{f(z) - f(0)}{z - 0} = \frac{|z|^2 - 0}{z - 0} = \frac{|z|^2}{z} = \bar{z}$. So we guess that $f'(0) = 0$. Given $\epsilon > 0$, set $\delta := \epsilon > 0$. If $0 < |z - 0| = |z| < \delta$, then $|\frac{f(z) - f(0)}{z - 0} - 0| = |\bar{z}| = |z| < \delta = \epsilon$. So f is complex differentiable at 0, and $f'(0) = 0$.

Solution to Exercise 2.4. Let $w_0 \in \widetilde{D}$. Then $\overline{w_0} \in D$. As f is holomorphic on D, given $\epsilon > 0$, there exists a $\delta > 0$ such that whenever $z \in D$ and $0 < |z - \overline{w_0}| < \delta$,

$$
|\tfrac{f(z) - f(\overline{w_0})}{z - \overline{w_0}} - f'(\overline{w_0})| < \epsilon. \qquad (\star)
$$

If $w \in \widetilde{D}$ satisfies $0 < |w - w_0| < \delta$, then firstly, $\overline{w} \in D$, and moreover we have that $0 < |\overline{w} - \overline{w_0}| = |\overline{w - w_0}| = |w - w_0| < \delta$. So

$$
\begin{aligned}
|\tfrac{\widetilde{f}(w) - \widetilde{f}(w_0)}{w - w_0} - \overline{f'(\overline{w_0})}| &= |\overline{\tfrac{f(\overline{w}) - f(\overline{w_0})}{\overline{w} - \overline{w_0}}} - \overline{f'(\overline{w_0})}| = |\overline{\tfrac{f(\overline{w}) - f(\overline{w_0})}{\overline{w} - \overline{w_0}} - f'(\overline{w_0})}| \\
&= |\tfrac{f(\overline{w}) - f(\overline{w_0})}{\overline{w} - \overline{w_0}} - f'(\overline{w_0})| < \epsilon \quad \text{(using $(\star)$).}
\end{aligned}
$$

So $\widetilde{f}$ is complex differentiable at w_0 and $(\widetilde{f})'(w_0) = \overline{f'(\overline{w_0})}$. As $w_0 \in \widetilde{D}$ was arbitrary, $\widetilde{f}$ is holomorphic in $\widetilde{D}$.

Solution to Exercise 2.5. Let $z_0 \in \mathbb{C}\backslash\{0\}$. For $z, z_0 \in \mathbb{C}\backslash\{0\}$, $z \neq z_0$, we have

$$|\frac{\frac{1}{z} - \frac{1}{z_0}}{z - z_0} - (-\frac{1}{z_0^2})| = |-\frac{(z-z_0)}{zz_0(z-z_0)} + \frac{1}{z_0^2}| = |-\frac{1}{zz_0} + \frac{1}{z_0^2}| = \frac{|z-z_0|}{|z||z_0|^2}. \quad (\star)$$

For z close to z_0, the numerator is small. Also, the denominator term $|z|$ is close to $|z_0| > 0$, and hence can be made to stay away from zero (so that $1/|z|$ can be bounded), as follows: If (the to-be-chosen) $\delta > 0$ is such that $\delta \leq |z_0|/2$, then whenever $0 < |z - z_0| < \delta$, we have (using the reverse triangle inequality) $|z_0|/2 \geq \delta > |z - z_0| \geq ||z| - |z_0|| \geq |z_0| - |z|$, which yields $|z| > |z_0|/2$. Then the right-hand estimate in $(\star)$ can be bounded above: $\frac{|z-z_0|}{|z||z_0|^2} < \delta \frac{2}{|z_0|} \frac{1}{|z_0|^2}$, and so to make this latter combination smaller than ϵ, we can further choose $\delta \leq \epsilon|z_0|^3/2$. So, given $\epsilon > 0$, setting $\delta := \min\{\frac{|z_0|}{2}, \frac{\epsilon|z_0|^3}{2}\}$, for all $z \in \mathbb{C}\backslash\{0\}$ such that $0 < |z-z_0| < \delta$,

$$|\frac{\frac{1}{z} - \frac{1}{z_0}}{z - z_0} - (-\frac{1}{z_0^2})| = \frac{|z-z_0|}{|z||z_0|^2} < \delta \frac{2}{|z_0|} \frac{1}{|z_0|^2} \leq \frac{\epsilon|z_0|^3}{2} \frac{2}{|z_0|} \frac{1}{|z_0|^2} = \epsilon.$$

Thus $f'(z_0) = -\frac{1}{z_0^2}$.

Solution to Exercise 2.6. Since f is complex differentiable at z_0, there exist $r > 0$ and $h : D(z_0, r)(\subset U) \to \mathbb{C}$ such that $f(z) = f(z_0) + (f'(z_0) + h(z))(z - z_0)$ for all $z \in D(z_0, r)$, and $\lim_{z \to z_0} h(z) = 0$.

Let $r' \in (0, r)$ be such that for all $z \in D_*(z_0, r') := \{z \in \mathbb{C} : 0 < |z - z_0| < r'\}$, we have $|h(z)| < 1$. Given $\epsilon > 0$, set $\delta = \min\{\frac{\epsilon}{|f'(z_0)|+1}, r'\}$. Then whenever $0 < |z - z_0| < \delta$, we have $z \in D(z_0, r')(\subset D(z_0, r) \subset U)$ and so

$$|f(z) - f(z_0)| = |f'(z_0) + h(z)||z - z_0| \leq (|f'(z_0)| + |h(z)|)\frac{\epsilon}{|f'(z_0)|+1}$$
$$< (|f'(z_0)| + 1)\frac{\epsilon}{|f'(z_0)|+1} = \epsilon.$$

Solution to Exercise 2.7. As $f, g : U \to \mathbb{C}$ are complex differentiable at $z_0 \in U$, by Lemma 2.1 there exist $r > 0$ and $h_f, h_g : D(z_0, r) \to \mathbb{C}$, such that for $z \in D(z_0, r)$,

$$f(z) = f(z_0) + (f'(z_0) + h_f(z))(z - z_0), \quad (*)$$
$$g(z) = g(z_0) + (g'(z_0) + h_g(z))(z - z_0), \quad (**)$$

and $\lim_{z \to z_0} h_f(z) = 0 = \lim_{z \to z_0} h_g(z)$.

(1) Adding $(*)$ and $(**)$, $(f+g)(z) = (f+g)(z_0) + (f'(z_0)+g'(z_0)+h_{f+g}(z))(z-z_0)$, for $z \in D(z_0, r)$, where $h_{f+g}(z) := h_f(z) + h_g(z)$ for all $z \in D(z_0, r)$. Moreover, $\lim_{z \to z_0} h_{f+g}(z) = \lim_{z \to z_0}(h_f(z) + h_g(z)) = \lim_{z \to z_0} h_f(z) + \lim_{z \to z_0} h_g(z) = 0+0 = 0$. By Lemma 2.1, $f + g$ is complex differentiable and $(f + g)'(z_0) = f'(z_0) + g'(z_0)$.

(2) Multiplying $(*)$ by α, $(\alpha \cdot f)(z) = (\alpha \cdot f)(z_0) + (\alpha f'(z_0) + h_{\alpha \cdot f}(z))(z - z_0)$, for $z \in D(z_0, r)$, where $h_{\alpha \cdot f}(z) := \alpha h_f(z)$ for all $z \in D(z_0, r)$. Moreover, we have $\lim_{z \to z_0} h_{\alpha \cdot f}(z) = \lim_{z \to z_0}(\alpha h_f(z)) = \alpha \lim_{z \to z_0} h_f(z) = \alpha 0 = 0$. By Lemma 2.1, $\alpha \cdot f$ is complex differentiable and $(\alpha \cdot f)'(z_0) = \alpha f'(z_0)$.

(3) We have $(fg)(z) = (fg)(z_0) + \big(f'(z_0)g(z_0) + f(z_0)g'(z_0) + h_{fg}(z)\big)(z - z_0)$, for $z \in D(z_0, r)$ (obtained by multiplying $(*)$ and $(**)$), where h_{fg} is given by $h_{fg}(z) := f(z_0)h_g(z) + g(z_0)h_f(z) + (z - z_0)(f'(z_0) + h_f(z))(g'(z_0) + h_g(z))$, for $z \in D(z_0, r)$. Then $\lim_{z \to z_0} h_{fg}(z) = f(z_0) \, 0 + g(z_0) \, 0 + 0 \, (f'(z_0) + 0)(g'(z_0) + 0) = 0$. So fg is complex differentiable and $(fg)'(z_0) = f'(z_0)g(z_0) + f(z_0)g'(z_0)$ by Lemma 2.1.

Solution to Exercise 2.8. No. (Let $\mathcal{O}(\mathbb{C})$ be finite-dimensional with dimension $d \in \mathbb{N}$. As $\mathcal{O}(\mathbb{C})$ contains the nonzero vector $\mathbb{C} \ni z \mapsto 1$, we have $d \neq 0$. The $d + 1$ vectors $1, z, \cdots, z^d \in \mathcal{O}(\mathbb{C})$ are linearly dependent. So there exist $\alpha_0, \cdots, \alpha_d \in \mathbb{C}$, not all zeroes, such that for all $z \in \mathbb{C}$, we have $\alpha_0 \cdot 1 + \alpha_1 \cdot z + \cdots + \alpha_d \cdot z^d = 0$. Let $k \in \{0, 1, \cdots, d\}$ be the smallest index such that $\alpha_k \neq 0$. Differentiating k times, and evaluating at $0 \in \mathbb{C}$, $0 + \alpha_k \cdot k! + 0 = 0$, and so $\alpha_k = 0$, a contradiction.)

Solution to Exercise 2.9. Let $z_0 \in U$. As f is complex differentiable at z_0, there exist $r > 0$ and $h : D(z_0, r) (\subset U) \to \mathbb{C}$ such that $f(z) = f(z_0) + (f'(z_0) + h(z))(z - z_0)$ for all $z \in D(z_0, r)$, and $\lim_{z \to z_0} h(z) = 0$. With $g := \frac{1}{f}$,

$$\tfrac{1}{g(z)} = \tfrac{1}{g(z_0)} + (f'(z_0) + h(z))(z - z_0),$$

and so $g(z_0) = g(z) + (f'(z_0) + h(z))g(z_0)g(z)(z - z_0)$. Rearranging,

$$g(z) = g(z_0) + \big(-f'(z_0)g(z_0)g(z) - h(z)g(z_0)g(z)\big)(z - z_0)$$
$$= g(z_0) + \Big(-\tfrac{f'(z_0)}{(f(z_0))^2} + \tfrac{f'(z_0)}{(f(z_0))^2} - \tfrac{f'(z_0)}{f(z_0)f(z)} - \tfrac{h(z)}{f(z_0)f(z)}\Big)(z - z_0)$$
$$= g(z_0) + \Big(-\tfrac{f'(z_0)}{(f(z_0))^2} + \varphi(z)\Big)(z - z_0),$$

where $\varphi(z) := \frac{f'(z_0)}{(f(z_0))^2} - \frac{f'(z_0)}{f(z_0)f(z)} - \frac{h(z)}{f(z_0)f(z)}$ for all $z \in D(z_0, r)$. As f is continuous at z_0 and $\lim_{z \to z_0} h(z) = 0$, $\lim_{z \to z_0} \varphi(z) = \frac{f'(z_0)}{(f(z_0))^2} - \frac{f'(z_0)}{f(z_0)f(z_0)} - \frac{0}{f(z_0)f(z_0)} = 0$. Hence g is complex differentiable at z_0 and $g'(z_0) = -\frac{f'(z_0)}{(f(z_0))^2}$.

Solution to Exercise 2.10. For $m \geqslant 0$, this was shown in Example 2.3. Let $m = -n$ for some $n \in \mathbb{N}$. Set $f(z) = z^n$ for $z \in \mathbb{C} \backslash \{0\}$. Consider the map $z \mapsto z^m = z^{-n} = \frac{1}{z^n} = \frac{1}{f(z)}$. As $f \in \mathcal{O}(\mathbb{C} \backslash \{0\})$ and f is pointwise nonzero, $\frac{1}{f}$ is holomorphic, and $(\frac{1}{f})'(z) = -\frac{f'(z)}{(f(z))^2} = -\frac{n z^{n-1}}{(z^n)^2} = -n \frac{1}{z^{n+1}} = m \frac{1}{z^{-m+1}} = m z^{m-1}$.

Solution to Exercise 2.11. Define $f : \mathbb{D} \to \mathbb{C}$ by $f(z) = -\frac{1+z}{1-z}$ for all $z \in \mathbb{D}$, and $g : \mathbb{C} \to \mathbb{C}$ by $g(z) = \exp z$, for all $z \in \mathbb{C}$. Then $f(\mathbb{D}) \subset \mathbb{C} = U_g$. By the chain rule, $g \circ f$ is holomorphic in $\mathbb{D}$, and

$$(g \circ f)'(z) = g'(f(z))f'(z) = \exp(-\tfrac{1+z}{1-z})\tfrac{d}{dz}(-\tfrac{1+z}{1-z})$$
$$= \exp(-\tfrac{1+z}{1-z})\big(-(1+z)\tfrac{d}{dz}(\tfrac{1}{1-z}) - \tfrac{1}{1-z}\tfrac{d}{dz}(1+z)\big)$$
$$= \exp(-\tfrac{1+z}{1-z})\big(-\tfrac{(1+z)}{(1-z)^2} - \tfrac{1}{1-z}\big) = -\tfrac{2}{(1-z)^2}\exp(-\tfrac{1+z}{1-z}).$$

Thus $\frac{d}{dz}\exp(-\frac{1+z}{1-z}) = -\frac{2}{(1-z)^2}\exp(-\frac{1+z}{1-z})$ in $\mathbb{D}$.

Solution to Exercise 2.12. With $z = x + iy$, where $x, y \in \mathbb{R}$, $|z|^2 = x^2 + y^2$. So if u, v denote the real part and the imaginary part of $|z|^2$, then $u = x^2 + y^2$ and $v = 0$. So $\frac{\partial u}{\partial x} = 2x$, $\frac{\partial v}{\partial y} = 0$, $\frac{\partial u}{\partial y} = 2y$, $\frac{\partial v}{\partial x} = 0$. Since $z \neq 0$, it follows that x or y is nonzero, and so at least one of the Cauchy-Riemann equations is violated, that is, $\frac{\partial u}{\partial x} = 2x \neq 0 = \frac{\partial v}{\partial y}$ or $\frac{\partial u}{\partial y} = 2y \neq 0 = -\frac{\partial v}{\partial y}$. So $|z|^2$ is not differentiable at any nonzero complex number.

Solution to Exercise 2.13. 1° Let $y > 0$. If $u := \operatorname{Re}(\operatorname{Log})$ and $v := \operatorname{Im}(\operatorname{Log})$ then $u(x, y) = \log\sqrt{x^2 + y^2} = \frac{1}{2}\log(x^2 + y^2)$, and $v(x, y) = \cos^{-1}\dfrac{x}{\sqrt{x^2+y^2}}$. So

$$\frac{\partial u}{\partial x} = \frac{1}{2}\frac{1}{x^2+y^2}2x = \frac{x}{x^2+y^2}, \qquad \frac{\partial v}{\partial y} = -\frac{1}{\sqrt{1-\frac{x^2}{x^2+y^2}}}\frac{0 - x\frac{1}{2\sqrt{x^2+y^2}}2y}{x^2+y^2} = \frac{x}{x^2+y^2},$$

$$\frac{\partial u}{\partial y} = \frac{1}{2}\frac{1}{x^2+y^2}2y = \frac{y}{x^2+y^2}, \qquad \frac{\partial v}{\partial x} = -\frac{1}{\sqrt{1-\frac{x^2}{x^2+y^2}}}\frac{1\sqrt{x^2+y^2}-x\frac{x}{\sqrt{x^2+y^2}}}{x^2+y^2} = -\frac{y}{x^2+y^2}.$$

These partials are continuous on $\mathbb{R} \times (0, \infty)$, and the Cauchy-Riemann equations are satisfied. Thus Log is holomorphic here, and $\operatorname{Log}' z = \frac{x}{x^2+y^2} - i\frac{y}{x^2+y^2} = \frac{\bar{z}}{z\bar{z}} = \frac{1}{z}$, where $z = (x, y) \in \mathbb{R} \times (0, \infty)$.

2° Let $x > 0$. If $u := \operatorname{Re}(\operatorname{Log})$ and $v := \operatorname{Im}(\operatorname{Log})$ then $u(x, y) = \frac{1}{2}\log(x^2 + y^2)$, and $v(x, y) = \sin^{-1}\dfrac{y}{\sqrt{x^2+y^2}}$. So

$$\frac{\partial u}{\partial x} = \frac{x}{x^2+y^2}, \qquad \frac{\partial v}{\partial x} = \frac{1}{\sqrt{1-\frac{y^2}{x^2+y^2}}}\frac{0 - y\frac{x}{\sqrt{x^2+y^2}}}{x^2+y^2} = -\frac{y}{x^2+y^2},$$

$$\frac{\partial u}{\partial y} = \frac{y}{x^2+y^2}, \qquad \frac{\partial v}{\partial y} = \frac{1}{\sqrt{1-\frac{y^2}{x^2+y^2}}}\frac{1\sqrt{x^2+y^2}-y\frac{y}{\sqrt{x^2+y^2}}}{x^2+y^2} = \frac{x}{x^2+y^2}.$$

These partials are continuous on $(0, \infty) \times \mathbb{R}$, and the Cauchy-Riemann equations are satisfied. Thus Log is holomorphic here, and $\operatorname{Log}' z = \frac{x}{x^2+y^2} - i\frac{y}{x^2+y^2} = \frac{\bar{z}}{z\bar{z}} = \frac{1}{z}$, where where $z = (x, y) \in (0, \infty) \times \mathbb{R}$.

3° Let $y < 0$. If $u := \operatorname{Re}(\operatorname{Log})$ and $v := \operatorname{Im}(\operatorname{Log})$ then $u(x, y) = \frac{1}{2}\log(x^2 + y^2)$, and $v(x, y) = -\cos^{-1}\dfrac{x}{\sqrt{x^2+y^2}}$. Note that $\sqrt{y^2} = -y$ since $y < 0$. We have

$$\frac{\partial u}{\partial x} = \frac{x}{x^2+y^2}, \qquad \frac{\partial v}{\partial y} = \frac{1}{\sqrt{1-\frac{x^2}{x^2+y^2}}}\frac{0 - x\frac{y}{\sqrt{x^2+y^2}}}{x^2+y^2} = \frac{x}{x^2+y^2},$$

$$\frac{\partial u}{\partial y} = \frac{y}{x^2+y^2}, \qquad \frac{\partial v}{\partial x} = \frac{1}{\sqrt{1-\frac{x^2}{x^2+y^2}}}\frac{1\sqrt{x^2+y^2}-x\frac{x}{\sqrt{x^2+y^2}}}{x^2+y^2} = -\frac{y}{x^2+y^2}.$$

These partials are continuous on $\mathbb{R} \times (-\infty, 0)$, and the Cauchy-Riemann equations are satisfied. Thus Log is holomorphic here, and $\operatorname{Log}' z = \frac{x}{x^2+y^2} - i\frac{y}{x^2+y^2} = \frac{\bar{z}}{z\bar{z}} = \frac{1}{z}$, where $z = (x, y) \in \mathbb{R} \times (-\infty, 0)$.

Solution to Exercise 2.14. Let $w_0 := f(z_0)$, $w \in V \backslash \{w_0\}$, and $z := g(w)$. Then $z_0 = g(f(z_0)) = g(w_0)$. Also, $z \neq z_0$ (otherwise, $w = f(g(w)) = f(z) = f(z_0) = w_0$). For all $w \in V \backslash \{w_0\}$,

$$1 = \frac{\mathrm{id}_V(w) - \mathrm{id}_V(w_0)}{w - w_0} = \frac{f(g(w)) - f(g(w_0))}{w - w_0} = \frac{f(g(w)) - f(g(w_0))}{g(w) - g(w_0)} \frac{g(w) - g(w_0)}{w - w_0}.$$

By the continuity of g, $\lim_{w \to w_0} g(w) = g(w_0)$. As f is complex differentiable at z_0,

$$\lim_{w \to w_0} \frac{f(g(w)) - f(g(w_0))}{g(w) - g(w_0)} = f'(g(w_0)) = f'(z_0) \neq 0.$$

Hence by Exercise 2.2, $\lim_{w \to w_0} \frac{g(w) - g(w_0)}{w - w_0} = \lim_{w \to w_0} \frac{1}{\frac{f(g(w)) - f(g(w_0))}{g(w) - g(w_0)}} = \frac{1}{f'(z_0)}$.

With $U = \mathbb{R} \times (-\pi, \pi)$, $V = \mathbb{C} \backslash (-\infty, 0]$, $f = \exp$ and $g = \mathrm{Log}$, f, g are continuous, and $f \circ g = \mathrm{id}_V$, $g \circ f = \mathrm{id}_U$. As $f \in \mathcal{O}(U)$, and for all $z \in U$, $f'(z) = \exp z \neq 0$, we have for all $w \in V$ and with $z = g(w)$, that $\mathrm{Log}'w = g'(w) = \frac{1}{f'(z)} = \frac{1}{\exp z} = \frac{1}{\exp(\mathrm{Log}\, w)} = \frac{1}{w}$.

Solution to Exercise 2.15. Let $z = x + iy$, where $x, y \in \mathbb{R}$. Then we have that $z^3 = (x + iy)^3 = x^3 + 3x^2(iy) + 3x(iy)^2 + (iy)^3 = x^3 - 3xy^2 + i(3x^2y - y^3)$. If u, v denote the real and imaginary parts of z^3, then $u(x, y) = x^3 - 3xy^2$ and $v(x, y) = 3x^2y - y^3$. We have $\frac{\partial u}{\partial x} = 3x^2 - 3y^2 = \frac{\partial v}{\partial y}$, and $\frac{\partial u}{\partial y} = -6xy = -\frac{\partial v}{\partial x}$, which are continuous on $\mathbb{R}^2$. Hence u, v are real differentiable in $\mathbb{R}^2$. Also, the Cauchy-Riemann equations are satisfied in $\mathbb{R}^2$. Thus $z \mapsto z^3$ is entire.

Solution to Exercise 2.16. If $z = x + iy$, where $x, y \in \mathbb{R}$, then $\mathrm{Re}\, z = \mathrm{Re}(x + iy) = x$. The real and imaginary parts u, v of $\mathrm{Re}\, z$ are $u(x, y) = x$ and $v(x, y) = 0$. Thus $\frac{\partial u}{\partial x} = 1 \neq 0 = \frac{\partial v}{\partial y}$ for all $(x, y) \in \mathbb{R}^2$. Hence the Cauchy-Riemann equations are satisfied nowhere in $\mathbb{R}^2$. So at each point of $\mathbb{C}$, $\mathrm{Re}\, z$ is not complex differentiable.

Solution to Exercise 2.17. Let u, v be the real and imaginary parts of f. Then $v = 0$ everywhere in D. Thus $\frac{\partial u}{\partial x} = \frac{\partial v}{\partial y} = 0$ and $\frac{\partial u}{\partial y} = -\frac{\partial v}{\partial x} = 0$ everywhere in D. If the line segment joining $(x, y_0) \in D$ to $(x_0, y_0) \in D$ lies in D, then there exists a ξ in the open interval with endpoints x, x_0 such that $\frac{u(x, y_0) - u(x_0, y_0)}{x - x_0} = \frac{\partial u}{\partial x}(\xi, y_0) = 0$. Also, whenever the line segment joining $(x_0, y_0) \in D$ to $(x_0, y) \in D$ lies in D, then $\frac{u(x_0, y) - u(x_0, y_0)}{y - y_0} = \frac{\partial u}{\partial y}(x_0, \eta) = 0$ for some η in the open interval with endpoints y, y_0. So u is constant along horizontal and vertical line segments lying in D. As D is path-connected, u is constant in D (as any two points in D can be joined by a stepwise path lying in D). So $f = u + i0 = u$ is constant in D.

Solution to Exercise 2.18. Let u, v be the real and imaginary parts of f. Then $f'(z) = \frac{\partial u}{\partial x} + i\frac{\partial v}{\partial x} = 0$ gives $\frac{\partial u}{\partial x} = \frac{\partial v}{\partial x} = 0$ in D, and using the Cauchy-Riemann equations, also that $\frac{\partial u}{\partial y}(= -\frac{\partial v}{\partial x}) = 0$ and $\frac{\partial v}{\partial y}(= \frac{\partial u}{\partial x}) = 0$ everywhere in D. If the line segment joining $(x, y_0) \in D$ to $(x_0, y_0) \in D$ lies in D, then there exists a ξ in the open interval with endpoints x, x_0 such that $\frac{u(x, y_0) - u(x_0, y_0)}{x - x_0} = \frac{\partial u}{\partial x}(\xi, y_0) = 0$. Also, whenever the line segment joining $(x_0, y_0) \in D$ to $(x_0, y) \in D$ lies in D, then $\frac{u(x_0, y) - u(x_0, y_0)}{y - y_0} = \frac{\partial u}{\partial y}(x_0, \eta) = 0$ for some η in the open interval with endpoints y, y_0. So u is constant along horizontal and vertical line segments lying inside D. As D

is path-connected, u is constant in D (because any two points in D can be joined by a stepwise path). Similarly v is constant in D. So $f = u + iv$ is constant in D.

Solution to Exercise 2.19. By the chain rule, $\frac{\partial u}{\partial x}(x, y) = h'(v(x,y))\frac{\partial v}{\partial x}(x, y)$ and $\frac{\partial u}{\partial y}(x, y) = h'(v(x,y))\frac{\partial v}{\partial y}(x, y)$. Using these and the Cauchy-Riemann equations,

$$\frac{\partial u}{\partial y}(x, y) = h'(v(x,y))\frac{\partial v}{\partial y}(x, y) = h'(v(x,y))\frac{\partial u}{\partial x}(x, y)$$
$$= h'(v(x,y))\left(h'(v(x,y))\frac{\partial v}{\partial x}(x, y)\right) = (h'(v(x,y)))^2\frac{\partial v}{\partial x}(x, y)$$
$$= -(h'(v(x,y)))^2\frac{\partial u}{\partial y}(x, y),$$

and so $(1+(h'(v(x,y)))^2)\frac{\partial u}{\partial y}(x, y) = 0$. As $(1+(h'(v(x,y)))^2) \geqslant 1 > 0$, $\frac{\partial u}{\partial y}(x, y) = 0$. Using the Cauchy-Riemann equations, we also obtain $\frac{\partial v}{\partial x}(x, y) = -\frac{\partial u}{\partial y}(x, y) = 0$. Thus $\frac{\partial u}{\partial x}(x, y) = h'(v(x,y))\frac{\partial v}{\partial x}(x, y) = h'(v(x,y))0 = 0$, and using the Cauchy-Riemann equations again, $\frac{\partial v}{\partial y}(x, y) = \frac{\partial u}{\partial x}(x, y) = 0$. So u is constant along horizontal and along vertical line segments. Since every $z \in \mathbb{C}$ can be joined to the origin using a combination of a horizontal and a vertical line segment, u is constant in $\mathbb{C}$. Similarly v is constant in $\mathbb{C}$. So $f = u + iv$ is constant in $\mathbb{C}$.

Solution to Exercise 2.20. Let u, v be the real and imaginary parts of f. Then $u(x, y) = x^2 - y^2$ and $v(x, y) = kxy$ for all $(x, y) \in \mathbb{R}^2$.

('Only if' part) If f is entire, then the Cauchy-Riemann equations are satisfied in $\mathbb{R}^2$, and so $2x = \frac{\partial u}{\partial x} = \frac{\partial v}{\partial y} = kx$ for all $x, y \in \mathbb{R}$. Taking $x = 1$ and any $y \in \mathbb{R}$, $k = 2$.

('If' part) If $k = 2$, then $f(z) = x^2 - y^2 + 2xyi = x^2 + (iy)^2 + 2x(iy) = (x+iy)^2 = z^2$. We have seen in Example 2.1 that $z \mapsto z^2$ is entire.

Solution to Exercise 2.21.

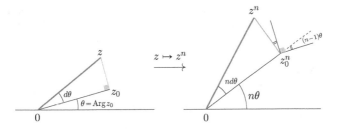

$|z - z_0| = |z_0| \tan d\theta \approx |z_0| d\theta$, while $|z^n - z_0^n| = |z_0|^n \tan(nd\theta) \approx |z_0|^n nd\theta$, and so the magnification produced locally by $z \mapsto z^n$ is $\frac{|z^n - z_0^n|}{|z - z_0|} \approx \frac{|z_0|^n nd\theta}{|z_0| d\theta} = n|z_0|^{n-1}$. Also, from the picture, the anticlockwise local rotation is $\frac{\pi}{2} - (\frac{\pi}{2} - (n-1)\theta) = (n-1)\theta$. So $f'(z_0) = n|z_0|^{n-1}(\cos((n-1)\theta)+i\sin((n-1)\theta)) = n(|z_0|(\cos\theta+i\sin\theta))^{n-1} = nz_0^{n-1}$. Consequently, $\frac{d}{dz}z^n = nz^{n-1}$ in $\mathbb{C}$.

Solution to Exercise 2.22. Let $z \in \mathbb{C}\backslash\{0\}$. For simplicity, we suppose z is in the first quadrant and $|z| > 1$. Displace z radially outwards through a small distance to obtain $\widetilde{z}$. Let $\mathrm{Arg}\, z = \theta = \mathrm{Arg}\, \widetilde{z}$. Then $z = |z|(\cos\theta + i\sin\theta)$, and so

$$\frac{1}{z} = \frac{1}{|z|(\cos\theta + i\sin\theta)} = \frac{1}{|z|}(\cos\theta - i\sin\theta) = \frac{1}{|z|}(\cos(-\theta) + i\sin(-\theta)).$$

Thus $\mathrm{Arg}\, \frac{1}{z} = -\theta = \mathrm{Arg}\, \frac{1}{\widetilde{z}}$. We note that $|\frac{1}{\widetilde{z}}| < |\frac{1}{z}|$. So the local anticlockwise rotation produced is $\pi - 2\theta$.

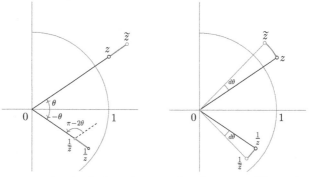

Next, displace z_0 counterclockwise along a circular arc with centre 0 and radius $|z_0|$ through an angle $d\theta$ to obtain $\widetilde{z}$. Then it is clear that $\frac{1}{\widetilde{z}}$ is obtained from $\frac{1}{z}$ by moving clockwise through an angle of $d\theta$ along a circular arc with radius $|\frac{1}{z}|$ centred at 0. So the local magnification is $\frac{1}{|z|}d\theta/(|z|d\theta) = \frac{1}{|z|^2}$.

Consequently, $\frac{d}{dz}(\frac{1}{z}) = \frac{1}{|z|^2}(\cos(\pi - 2\theta) + i\sin(\pi - 2\theta)) = \frac{-1}{|z|^2(\cos\theta + i\sin\theta)^2} = -\frac{1}{z^2}$.

Solution to Exercise 2.23.

From the left picture above, the magnification is $\frac{e^x\delta}{\delta} = e^x$. From the right picture, the anticlockwise local rotation produced is y. Thus the complex derivative at z_0 must be $e^x(\cos y + i\sin y) = \exp(x + iy)$. Consequently, $\exp' z = \exp z$ for all $z \in \mathbb{C}$.

Solution to Exercise 2.24. Let $w \in \mathbb{C}\backslash(-\infty, 0]$. Displace w radially outwards through a distance δ to obtain $\widetilde{w}$. If $\mathrm{Arg}\, w = \theta = \mathrm{Arg}\, \widetilde{w}$, then $\mathrm{Log}\, \widetilde{w}$ is a horizontal rightward displacement of $\mathrm{Log}\, w$ (through a distance $\log(1 + \frac{\delta}{|w|})$). So the local anticlockwise rotation produced is $2\pi - \theta$.

Next, displace w counterclockwise along a circular arc with centre 0 and radius $|w|$ through an angle $d\theta$ to obtain w'. Then it is clear that $\mathrm{Log}\, w'$ is obtained from $\mathrm{Log}\, w$ by a vertical displacement through a distance $d\theta$. So the local magnification is $\frac{d\theta}{|w|d\theta} = \frac{1}{|w|}$. Thus $\mathrm{Log}'\, w = \frac{1}{|w|}(\cos(2\pi - \theta) + i\sin(2\pi - \theta)) = \frac{1}{|w|(\cos\theta + i\sin\theta)} = \frac{1}{w}$.

Solution to Exercise 2.25. Let $z_0 \in \mathbb{C}$. Move z_0 along the line with slope 1 through a distance of δ to get the point z. Similarly move z_0 horizontally to the left by a distance δ to get the point $\tilde{z}$. Suppose that the map $\zeta \mapsto \mathrm{Re}\,\zeta$ is complex differentiable at z_0. In the following picture, by considering the images of $z, \tilde{z}$ under the mapping $\mathrm{Re}\cdot$, we get conflicting values of the local rotation produced as being $45°$ and $0°$, i.e., $\frac{\pi}{4} = \mathrm{Arg}(\mathrm{Re}'\, z_0) = 0$, a contradiction. So the map cannot be complex differentiable at z_0. As the choice of $z_0 \in \mathbb{C}$ was arbitrary, the map is complex differentiable nowhere.

Solution to Exercise 2.26. We have $|\exp'(a+i0)|^2 = |\exp(a+i0)|^2 = |e^a|^2 = e^{2a}$. Thus we do have that $\lim_{\epsilon \to 0} \rho(\epsilon) = e^{2a} = |\exp'(a+i0)|^2$. This is expected, since for small ϵ, the image under $\exp$ of the square region S_ϵ with a side length 2ϵ centred at a will be approximately a square region with a side length $2\epsilon|\exp'(a+0i)|$, and so the ratio of the areas is $\frac{4\epsilon^2|\exp'(a+i0)|^2}{4\epsilon^2} = |\exp'(a+i0)|^2 = |e^a|^2 = e^{2a}$.

Solution to Exercise 2.27. Let $\mathbf{0} := (0,0)$. For $\mathbf{x} = (x,y) \in \mathbb{R}^2 \setminus \{\mathbf{0}\}$, its distance to the origin $\mathbf{0}$ is $\|\mathbf{x}\|_2 = \sqrt{x^2 + y^2} =: r > 0$. Denote by α the angle made by the ray joining $\mathbf{0}$ to $\mathbf{x}$ with the positive real axis in the counterclockwise direction. Then $x = r\cos\alpha$ and $y = r\sin\alpha$. Thus

$$R\mathbf{x} = \begin{bmatrix} \cos\theta & -\sin\theta \\ \sin\theta & \cos\theta \end{bmatrix} \begin{bmatrix} r\cos\alpha \\ r\sin\alpha \end{bmatrix} = \begin{bmatrix} r(\cos\theta\cos\alpha - \sin\theta\sin\alpha) \\ r(\sin\theta\cos\alpha + \cos\theta\sin\alpha) \end{bmatrix} = \begin{bmatrix} r\cos(\theta+\alpha) \\ r\sin(\theta+\alpha) \end{bmatrix}.$$

Hence the angle made by the ray joining $\mathbf{0}$ to $R\mathbf{x}$ with the positive real axis in the counterclockwise direction is $\theta + \alpha$, and the distance of $R\mathbf{x}$ to $\mathbf{0}$ is

$$\|R\mathbf{x}\|_2 = r\sqrt{(\cos(\theta+\alpha))^2 + (\sin(\theta+\alpha))^2} = r = \|\mathbf{x}\|_2 \text{ (the distance of } \mathbf{x} \text{ to } \mathbf{0}\text{).}$$

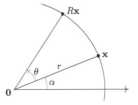

Thus $R\mathbf{x}$ is obtained by rotating $\mathbf{x}$ about the origin $\mathbf{0}$ counterclockwise through θ.

Solution to Exercise 2.28.

(1) We have

$$\tfrac{\partial f}{\partial \bar{z}}(z_0) = \tfrac{\partial u}{\partial \bar{z}}(z_0) + i\tfrac{\partial v}{\partial \bar{z}}(z_0) = \tfrac{1}{2}(\tfrac{\partial u}{\partial x}(z_0) + i\tfrac{\partial u}{\partial y}(z_0)) + i\tfrac{1}{2}(\tfrac{\partial v}{\partial x}(z_0) + i\tfrac{\partial v}{\partial y}(z_0))$$

$$= \tfrac{1}{2}(\tfrac{\partial u}{\partial x}(z_0) - \tfrac{\partial v}{\partial y}(z_0)) + i\tfrac{1}{2}(\tfrac{\partial u}{\partial y}(z_0) + \tfrac{\partial v}{\partial x}(z_0)).$$

So $\tfrac{\partial f}{\partial \bar{z}}(z_0) = 0$ if and only if the Cauchy-Riemann equations hold at z_0. As it is given that u and v are real differentiable at z_0, Theorems 2.5 and 2.4 imply that f is complex differentiable at z_0 if and only if $\tfrac{\partial f}{\partial \bar{z}}(z_0) = 0$.

(2) We have

$$\tfrac{\partial f}{\partial z}(z_0) = \tfrac{\partial u}{\partial z}(z_0) + i\tfrac{\partial v}{\partial z}(z_0) = \tfrac{1}{2}(\tfrac{\partial u}{\partial x}(z_0) - i\tfrac{\partial u}{\partial y}(z_0)) + i\tfrac{1}{2}(\tfrac{\partial v}{\partial x}(z_0) - i\tfrac{\partial v}{\partial y}(z_0))$$

$$= \tfrac{1}{2}(\tfrac{\partial u}{\partial x}(z_0) + \tfrac{\partial v}{\partial y}(z_0)) + i\tfrac{1}{2}(-\tfrac{\partial u}{\partial y}(z_0) + \tfrac{\partial v}{\partial x}(z_0)) = \tfrac{\partial u}{\partial x}(z_0) + i\tfrac{\partial v}{\partial x}(z_0) = f'(z_0).$$

(3) For the map $\mathbb{R}^2 \ni (x, y) = z \mapsto \bar{z}$, we have

$$\tfrac{\partial \bar{z}}{\partial \bar{z}} = \tfrac{1}{2}(\tfrac{\partial x}{\partial x} + i\tfrac{\partial x}{\partial y}) + i\tfrac{1}{2}(\tfrac{\partial(-y)}{\partial x} + i\tfrac{\partial(-y)}{\partial y}) = \tfrac{1}{2} - i\tfrac{1}{2}i = 1 \neq 0$$

everywhere in $\mathbb{R}^2$. (So $\bar{z}$ is complex differentiable nowhere in $\mathbb{C}$.)

For the map $\mathbb{R}^2 \ni (x, y) = z \mapsto |z|^2 = z\bar{z} = x^2 + y^2$, we have

$$\tfrac{\partial |z|^2}{\partial \bar{z}} = \tfrac{\partial(z\bar{z})}{\partial \bar{z}} = \tfrac{\partial(x^2+y^2)}{\partial \bar{z}} = \tfrac{1}{2}(\tfrac{\partial(x^2+y^2)}{\partial x} + i\tfrac{\partial(x^2+y^2)}{\partial y}) = \tfrac{1}{2}(2x + i2y) = x + iy = z.$$

So $|z|^2$ is complex differentiable only at 0.

(4) $\mathbb{R}^2 \ni (x, y) = z \mapsto f(z) = z^2 = x^2 - y^2 + i(2xy)$ is entire, because

$$\tfrac{\partial z^2}{\partial \bar{z}} = \tfrac{1}{2}(\tfrac{\partial(x^2-y^2)}{\partial x} + i\tfrac{\partial(x^2-y^2)}{\partial y}) + i\tfrac{1}{2}(\tfrac{\partial(2xy)}{\partial x} + i\tfrac{\partial(2xy)}{\partial y})$$

$$= \tfrac{1}{2}(2x - i2y) + i\tfrac{1}{2}(2y + i2x) = 0.$$

Moreover,

$$\tfrac{\partial z^2}{\partial z} = \tfrac{1}{2}(\tfrac{\partial(x^2-y^2)}{\partial x} - i\tfrac{\partial(x^2-y^2)}{\partial y}) + i\tfrac{1}{2}(\tfrac{\partial(2xy)}{\partial x} - i\tfrac{\partial(2xy)}{\partial y})$$

$$= \tfrac{1}{2}(2x + i2y) + i\tfrac{1}{2}(2y - i2x) = 2(x + iy) = 2z.$$

Solutions to the exercises from Chapter 3

Solution to Exercise 3.1. For $t \in [0, 2\pi]$, we have $\gamma_1(t) = \exp(it) = \cos t + i \sin t$, $\gamma_2(t) = \exp(2it) = \cos(2t) + i \sin(2t)$, $\gamma_3(t) = \exp(-it) = \cos t - i \sin t$, and in each case, $(\operatorname{Re} \gamma_k(t))^2 + (\operatorname{Im} \gamma_k(t))^2 = 1$, $k = 1, 2, 3$, $t \in [0, 2\pi]$. So the range of γ_k is contained in the unit circle $\mathbb{T}$ with centre 0. Also, if $z = \exp(i\theta)$ with $\theta \in [0, 2\pi)$, then $z = \gamma_1(\theta) = \gamma_2(\theta/2) = \gamma_3(2\pi - \theta)$, and so every $z \in \mathbb{T}$ belongs to the range of each of $\gamma_1, \gamma_2, \gamma_3$. We have $\gamma_1'(t) = -\sin t + i \cos t = i(\cos t + i \sin t) = i \exp(it)$, and similarly, $\gamma_2'(t) = 2i \exp(2it)$, $\gamma_3'(t) = -i \exp(-it)$ for all $t \in [0, 2\pi]$. Thus

$$\int_{\gamma_1} \tfrac{1}{z} dz = \int_0^{2\pi} \tfrac{1}{\exp(it)} i \exp(it) \, dt = 2\pi i,$$
$$\int_{\gamma_2} \tfrac{1}{z} dz = \int_0^{2\pi} \tfrac{1}{\exp(2it)} 2i \exp(2it) \, dt = 4\pi i,$$
$$\int_{\gamma_3} \tfrac{1}{z} dz = \int_0^{2\pi} \tfrac{1}{\exp(-it)} (-i) \exp(-it) \, dt = -2\pi i.$$

Solution to Exercise 3.2. Let $\gamma(t) = x(t) + iy(t)$, $t \in [a, b]$, where x, y are real-valued. Also, let u, v be the real and imaginary parts of f, respectively. Then

$f'(\gamma(t))\gamma'(t)$

$= \big(\tfrac{\partial u}{\partial x}(x(t), y(t)) + i \tfrac{\partial v}{\partial x}(x(t), y(t)) \big)(x'(t) + iy'(t))$

$= \tfrac{\partial u}{\partial x}(x(t), y(t))x'(t) - \tfrac{\partial v}{\partial x}(x(t), y(t))y'(t) + i\big(\tfrac{\partial u}{\partial x}(x(t), y(t))y'(t) + \tfrac{\partial v}{\partial x}(x(t), y(t))x'(t) \big)$

$= \tfrac{\partial u}{\partial x}(x(t), y(t))x'(t) + \tfrac{\partial u}{\partial y}(x(t), y(t))y'(t) + i\big(\tfrac{\partial v}{\partial y}(x(t), y(t))y'(t) + \tfrac{\partial v}{\partial x}(x(t), y(t))x'(t) \big)$

     (using the Cauchy-Riemann equations)

$= \tfrac{d}{dt}u(x(t), y(t)) + i\tfrac{d}{dt}v(x(t), y(t))$ (using the Chain Rule)

$= \tfrac{d}{dt}\big(u(x(t), y(t)) + iv(x(t), y(t)) \big) = \tfrac{d}{dt}f(\gamma(t)).$

Solution to Exercise 3.3. With $u := \operatorname{Re} f$, $v := \operatorname{Im} f$, and $x := \operatorname{Re}\gamma$, $y := \operatorname{Im}\gamma$, we have $\int_0^1 (ux' - vy') \, dt = \operatorname{Re}(\int_\gamma f(z) \, dz) = \int_\gamma \operatorname{Re}(f(z)) \, dz = \int_0^1 u(x' + iy') \, dt$, giving, $\int_0^1 -vy' \, dt = i \int_0^1 uy' \, dt$, and this will not hold in general. To construct a concrete counterexample, take $f \equiv i$ and $\gamma(t) = it$, $t \in [0, 1]$. Then f is entire, γ is smooth, and $\operatorname{Re}(\int_\gamma f(z) \, dz) = -1 \neq 0 = \int_\gamma \operatorname{Re} f(z) \, dz$.

Solution to Exercise 3.4.
- $\int_\gamma (z + \bar{z}) dz = \int_0^{2\pi} (2e^{it} + 2e^{-it}) 2ie^{it} dt = 4i \int_0^{2\pi} (e^{2it} + 1) dt = 4i \, 0 + 4i \, 2\pi = 8\pi i.$
- $\int_\gamma (z^2 - 2z + 3) dz = \int_0^{2\pi} (4e^{2it} - 4e^{it} + 3) 2ie^{it} dt = \int_0^{2\pi} i(8e^{3it} - 8e^{2it} + 6e^{it}) dt = 0.$

- The 2π-periodicity of $(\sin(2\cdot))\cos$ is used to get the first equality in the last line.

$$\int_\gamma xy \, dz = \int_0^{2\pi} 2(\cos t)2(\sin t) 2i(\cos t + i \sin t) dt = 4i \int_0^{2\pi} (\sin(2t))(\cos t + i \sin t) dt$$
$$= 4i \int_{-\pi}^{\pi} \underbrace{(\sin(2t)) \cos t}_{\text{odd function}} \, dt - 2 \int_0^{2\pi} (\cos t - \cos(3t)) dt = 0.$$

Solution to Exercise 3.5.

- $\gamma(t) = (1+i)t$, $t \in [0,1]$. So $\int_\gamma \mathrm{Re}\, z\, dz = \int_0^1 t(1+i)dt = \frac{1+i}{2}$.
- $\gamma(t) = i + \exp(it)$, $t \in [-\frac{\pi}{2}, 0]$. So

$$\int_\gamma \mathrm{Re}\, z\, dz = \int_{-\frac{\pi}{2}}^0 (\cos t) i \exp(it) dt = \int_{-\frac{\pi}{2}}^0 \left(i(\cos t)^2 - (\cos t)(\sin t) \right) dt$$
$$= \int_{-\frac{\pi}{2}}^0 \left(i\frac{\cos(2t)+1}{2} - \frac{\sin(2t)}{2} \right) dt = 0 + i\frac{1}{2}\frac{\pi}{2} + \frac{1}{2} = \frac{1}{2} + i\frac{\pi}{4}.$$

- $\gamma(t) = t + it^2$, $t \in [0,1]$. So $\int_\gamma \mathrm{Re}\, z\, dz = \int_0^1 t(1 + 2it)\, dt = \frac{1}{2} + 2i\frac{1}{3} = \frac{1}{2} + \frac{2}{3}i$.

Solution to Exercise 3.6. By the Binomial Theorem, $(1+z)^n = \sum_{\ell=0}^n \binom{n}{\ell} z^\ell$.

For $0 \leqslant k \leqslant n$, and $z \in \mathrm{ran}\, C$, we have $\frac{(1+z)^n}{z^{k+1}} = \sum_{\ell=0}^n \binom{n}{\ell} z^{\ell-k-1}$. Consequently,

$$\frac{1}{2\pi i} \int_C \frac{(1+z)^n}{z^{k+1}} dz = \frac{1}{2\pi i} \int_C \sum_{\ell=0}^n \binom{n}{\ell} z^{\ell-k-1} dz = \sum_{\ell=0}^n \binom{n}{\ell} \frac{1}{2\pi i} \int_C z^{\ell-k-1} dz = \binom{n}{k}.$$

Solution to Exercise 3.7.

(1) Let $u_f, v_f, u_g, v_g : [a,b] \to \mathbb{R}$ be such that $f(\gamma(t))\gamma'(t) = u_f(t) + iv_f(t)$ and $g(\gamma(t))\gamma'(t) = u_g(t) + iv_g(t)$ for all $t \in [a,b]$. Then

$$\int_\gamma (f+g)(z)\, dz = \int_a^b (f+g)(\gamma(t))\gamma'(t)\, dt = \int_a^b \left(f(\gamma(t))\gamma'(t) + g(\gamma(t))\gamma'(t) \right) dt$$
$$= \int_a^b (u_f(t) + u_g(t))\, dt + i\int_a^b (v_f(t) + v_g(t))\, dt$$
$$= \int_a^b u_f(t)dt + \int_a^b u_g(t)dt + i\int_a^b v_f(t)dt + i\int_a^b v_g(t)dt$$
$$= \int_\gamma f(z)\, dz + \int_\gamma g(z)\, dz.$$

(2) Let $\alpha = p + iq$ where $p, q \in \mathbb{R}$, and let $u, v : [a,b] \to \mathbb{R}$ be such that for all $t \in [a,b]$, $f(\gamma(t))\gamma'(t) = u(t) + iv(t)$. Then

$$\int_\gamma (\alpha \cdot f)(z)\, dz = \int_a^b (p + iq)(u(t) + iv(t))\, dt$$
$$= \int_a^b (pu(t) - qv(t))\, dt + i\int_a^b (pv(t) + qu(t))\, dt$$
$$= p\left(\int_a^b u(t)\, dt + i\int_a^b v(t)\, dt \right) + iq\left(\int_a^b u(t)\, dt + i\int_a^b v(t)\, dt \right)$$
$$= (p + iq)\left(\int_a^b u(t)\, dt + i\int_a^b v(t)\, dt \right) = \alpha \int_\gamma f(z)\, dz.$$

Solution to Exercise 3.8. $-(-\gamma) : [a,b] \to \mathbb{C}$ is given by

$$(-(-\gamma))(t) = (-\gamma)(a + b - t) = \gamma(a + b - (a + b - t)) = \gamma(t), \quad t \in [a,b],$$

and so $-(-\gamma) = \gamma$. (This is also obvious visually.)

Solution to Exercise 3.9. We do have that $\gamma(b) = (-\gamma)(a)$, and so the two paths γ and $-\gamma$ can be concatenated, and we have

$$\int_{\gamma+(-\gamma)} f(z)\, dz = \int_\gamma f(z)\, dz + \int_{-\gamma} f(z)\, dz = \int_\gamma f(z)\, dz - \int_\gamma f(z)\, dz = 0.$$

Solution to Exercise 3.10. Let $z = \alpha + i\beta$, $\alpha, \beta \in \mathbb{R}$, and $\varphi(t) = x(t) + iy(t)$, where $x, y : [a, b] \to \mathbb{R}$. Then

$$
\begin{aligned}
\int_a^b z\varphi(t)\,dt &= \int_a^b (\alpha + i\beta)(x(t) + iy(t))\,dt \\
&= \int_a^b (\alpha x(t) - \beta y(t) + i(\alpha y(t) + \beta x(t)))\,dt \\
&= \int_a^b (\alpha x(t) - \beta y(t))\,dt + i\int_a^b (\alpha y(t) + \beta x(t))\,dt \\
&= \alpha \int_a^b x(t)\,dt - \beta \int_a^b y(t)\,dt + i\left(\alpha \int_a^b y(t)\,dt + \beta \int_a^b x(t)\,dt\right) \\
&= (\alpha + i\beta)\left(\int_a^b x(t)\,dt + i\int_a^b y(t)\,dt\right) = z\int_a^b \varphi(t)\,dt.
\end{aligned}
$$

Solution to Exercise 3.11. Define $\gamma : [0, 1] \to \mathbb{C}$ by $\gamma(t) = (1 + i)t$ for $t \in [0, 1]$. The length of γ is $\int_0^1 \sqrt{1^2 + 1^2}\,dt = \sqrt{2}$. (Also evident by Pythagoras's Theorem).

Also, $|(\gamma(t))^2| = |t + it|^2 = 2t^2$, and so $\max\limits_{t \in [0,1]} |(\gamma(t))^2| = 2(1^2) = 2$. Hence

$$
\left|\int_\gamma z^2\,dz\right| \le \left(\max_{t \in [0,1]} |(\gamma(t))^2|\right) (\text{length of } \gamma) = 2\sqrt{2}.
$$

As $\int_\gamma z^2\,dz = \int_0^1 (t + it)^2(1+i)\,dt = \int_0^1 (1 + i)^3 t^2\,dt = \frac{(1+i)^3}{3}$, $\left|\int_\gamma z^2\,dz\right| = \frac{(\sqrt{2})^3}{3} = \frac{2\sqrt{2}}{3}$.

Solution to Exercise 3.12. We have

$$
\binom{2n}{n} = \left|\binom{2n}{n}\right| = \left|\frac{1}{2\pi i}\int_C \frac{(1+z)^{2n}}{z^{n+1}}\,dz\right| \le \frac{1}{2\pi}\left(\max_{|z|=1} \left|\frac{(1+z)^{2n}}{z^{n+1}}\right|\right) 2\pi 1 = \max_{|z|=1} \frac{|1+z|^{2n}}{1}
$$
$$
\le (1 + 1)^{2n} = 2^{2n} = 4^n.
$$

Solution to Exercise 3.13. Set $\gamma(t) = (1-t)i + t$, $t \in [0, 1]$. By the ML inequality,

$$
\left|\int_\gamma \frac{1}{z^4}\,dz\right| \le (\text{length of } \gamma) \max_{t \in [0,1]} \left|\frac{1}{(\gamma(t))^4}\right| = \sqrt{2} \max_{t \in [0,1]} \frac{1}{((1-t)^2 + t^2)^2}.
$$

But for $a, b \in \mathbb{R}$, $a^2 + b^2 \ge \frac{(a+b)^2}{2}$ (since $(a - b)^2 \ge 0$). So with $a = (1 - t)$, $b = t$, we obtain $(1 - t)^2 + t^2 \ge \frac{(1-t+t)^2}{2} = \frac{1}{2}$. So $\left|\int_\gamma \frac{1}{z^4}\,dz\right| \le \sqrt{2} \max_{t \in [0,1]} \frac{1}{((1-t)^2 + t^2)^2} \le 4\sqrt{2}$.

Solution to Exercise 3.14. Write $z = x + iy$, where $x, y \in \mathbb{R}$. Let $f = u + iv$ be a primitive of the map $z \mapsto \bar{z}$ in $\mathbb{C}$, where $u := \operatorname{Re} f$ and $v := \operatorname{Im} f$. Then $\frac{\partial u}{\partial x} + i\frac{\partial v}{\partial x} = \frac{\partial v}{\partial y} - i\frac{\partial u}{\partial y} = f' = \bar{z} = x - iy$ in $\mathbb{R}^2$. Fix $x_0 \in \mathbb{R}$. Then for $(x, y) \in \mathbb{R}^2$, $v(x, y) - v(x_0, y) = \int_{x_0}^x \frac{\partial v}{\partial x}(\xi, y)\,d\xi = \int_{x_0}^x -y\,d\xi = -xy + x_0 y$. So $v(x, y) = -xy + \varphi(y)$, where $\varphi(y) := v(x_0, y) + x_0 y$. Hence $x = \frac{\partial v}{\partial y} = -x + \varphi'(y)$, that is, $\varphi'(y) = 2x$ for all $x \in \mathbb{R}$, which is impossible (otherwise, e.g., $2 = 2(1) = \varphi'(y) = 2(0) = 0$).

Solution to Exercise 3.15. The continuous map $D \ni \zeta \mapsto f(\zeta)g'(\zeta) + f'(\zeta)g(\zeta)$ has a primitive $fg \in \mathcal{O}(D)$, since $(fg)' = fg' + f'g$. By the Fundamental Theorem of Contour Integration,

$$\int_\gamma \left(f(\zeta)g'(\zeta) + f'(\zeta)g(\zeta) \right) d\zeta = \int_\gamma (fg)'(\zeta) \, d\zeta = f(z)g(z) - f(w)g(w),$$

and a rearrangement proves the claim.

For $r > |a|$, integrating by parts, $\int_{S_r} \frac{ze^{iz}}{z^2+a^2} \, dz = \frac{z}{z^2+a^2} \frac{e^{iz}}{i} \big|_r^{-r} - \int_{S_r} \frac{a^2-z^2}{(z^2+a^2)^2} \frac{e^{iz}}{i} \, dz$. From Exercise 1.37, there exists an $r > 0$ and constants C_1, C_2 such that for all $z \in \mathbb{C}$ such that $|z| \geqslant r$, $\left| \frac{z}{z^2+a^2} \right| \leqslant \frac{C_1}{|z|}$ and $\left| \frac{a^2-z^2}{(z^2+a^2)^2} \right| \leqslant \frac{C_2}{|z|^2}$. For $z \in \operatorname{ran} S_r$, $\operatorname{Im} z \geqslant 0$, and so $|e^{iz}| = e^{-\operatorname{Im} z} \leqslant e^0 = 1$. It follows from these estimates that $\lim\limits_{r\to\infty} \left| \frac{z}{z^2+a^2} \frac{e^{iz}}{i} \big|_r^{-r} \right| = 0$. Using the ML-inequality $\left| \int_{S_r} \frac{a^2-z^2}{(z^2+a^2)^2} \frac{e^{iz}}{i} \, dz \right| \leqslant \frac{C_2}{r^2} \pi r$, and so $\lim\limits_{r\to\infty} \left| \int_{S_r} \frac{a^2-z^2}{(z^2+a^2)^2} \frac{e^{iz}}{i} \, dz \right| = 0$. Consequently, $\lim\limits_{r\to\infty} \left| \int_{S_r} \frac{ze^{iz}}{z^2+a^2} \, dz \right| = 0$.

Solution to Exercise 3.16. $\cos$ is entire, and in particular continuous. Also $\cos$ has a primitive in $\mathbb{C}$: $\sin' z = \cos z$. Thus by the Fundamental Theorem of Contour Integration $\int_\gamma \cos z \, dz = \sin i - \sin(-i) = 2\frac{\exp(ii)-\exp(-ii)}{2i} = \frac{e^{-1}-e^1}{i} = (e - \frac{1}{e})i$.

Solution to Exercise 3.17. By the Fundamental Theorem of Contour Integration, if γ is the straight line path joining $z, w \in \mathbb{D}$, then $f(z) - f(w) = \int_\gamma f'(\zeta) \, d\zeta$. By the ML-inequality, $|f(z) - f(w)| = \left| \int_\gamma f'(\zeta) \, d\zeta \right| \leqslant \sup\limits_{\zeta \in \operatorname{ran} \gamma} |f'(\zeta)| (\text{length of } \gamma) \leqslant 1|z - w|$.

Solution to Exercise 3.18. Let $D := \mathbb{C} \backslash ((-\infty, 0] \times \{0\})$ and $\gamma : [0, 1] \to D$ be the straight line path joining 1 to $1 - i$, given by $\gamma(t) = (1-t)1 + t(1-i) = 1 - it$ for all $t \in [0, 1]$. As $\operatorname{Log} \in \mathcal{O}(D)$ and $\operatorname{Log}' z = \frac{1}{z}$ (which is continuous on D), it follows from the Fundamental Theorem of Contour Integration that

$$\int_0^1 \frac{1}{1-it} \, dt = \frac{1}{-i} \int_\gamma \frac{1}{z} \, dz = i \int_\gamma \operatorname{Log}' z \, dz = i(\operatorname{Log}(1-i) - \operatorname{Log} 1)$$
$$= i((\log \sqrt{2}) - i\frac{\pi}{4} - 0) = \frac{\pi}{4} + i\frac{1}{2}\log 2.$$

Solution to Exercise 3.19. Since $\exp' z = \exp z$ in $\mathbb{C}$, we have

$$\int_\gamma \exp z \, dz = \exp(a+ib) - \exp 0 = e^a(\cos b + i \sin b) - 1 = e^a \cos b - 1 + ie^a \sin b,$$

for a path γ joining 0 to $a + ib$. If $\gamma(x) = (a + ib)x$ for all $x \in [0, 1]$, then

$$\int_\gamma \exp z \, dz = \int_0^1 \exp((a + ib)x)(a + ib) \, dx = \int_0^1 e^{ax}(\cos(bx) + i \sin(bx))(a + ib) \, dx.$$

Hence $(a - ib) \int_\gamma \exp z \, dz = \int_0^1 e^{ax}(\cos(bx) + i \sin(bx))(a^2 + b^2) \, dx$. Thus

$$(a^2 + b^2) \int_0^1 e^{ax} \cos(bx) \, dx = \operatorname{Re}\left((a - ib) \int_\gamma \exp z \, dz \right)$$
$$= \operatorname{Re}\left((a - ib)(e^a \cos b - 1 + ie^a \sin b) \right)$$
$$= a(e^a \cos b - 1) + be^a \sin b.$$

Hence $\int_0^1 e^{ax} \cos(bx) \, dx = \frac{a(e^a \cos b - 1) + be^a \sin b}{a^2 + b^2}$.

Solution to Exercise 3.20. Let C be the circular path with centre at 0 and radius $r > 0$ traversed in the anticlockwise direction, i.e., $C(\theta) = re^{i\theta}$ for all $\theta \in [0, 2\pi]$. By the Fundamental Theorem of Contour Integration,

$$0 = \int_C \exp z\, dz = \int_0^{2\pi} e^{r\cos\theta + ir\sin\theta} rie^{i\theta}\, d\theta = \int_0^{2\pi} e^{r\cos\theta} rie^{i(r\sin\theta + \theta)}\, d\theta.$$

Equating real and imaginary parts, in particular, $\int_0^{2\pi} e^{r\cos\theta} \cos(r\sin\theta + \theta)\, d\theta = 0$.

Solution to Exercise 3.21. Suppose $f \in \mathcal{O}(\mathbb{C}\backslash\{0\})$ is such that $f' = \frac{1}{z}$ in $\mathbb{C}\backslash\{0\}$. Let C be a circular path with a positive radius centred at 0 traversed in the anticlockwise direction. By the Fundamental Theorem of Contour Integration, since C is closed, $\int_C f'(z)\, dz = 0$. But $\int_C f'(z)\, dz = \int_C \frac{1}{z}\, dz = 2\pi i$, a contradiction.

Solution to Exercise 3.22. E, C are both closed, since $E(0) = a = E(1)$ and $C(0) = r = C(1)$. Define $H : [0,1]\times[0,1] \to \mathbb{C}\backslash\{0\}$ by $H(t,\tau) = (1-\tau)C(t)+\tau E(t)$, for all $t, \tau \in [0,1]$. To see that $|H(t,\tau)| \neq 0$, setting $m = \min\{a, b, r\} > 0$,

$$\begin{aligned}
|H(t,\tau)|^2 &= ((1-\tau)r + \tau a)^2(\cos(2\pi t))^2 + ((1-\tau)r + \tau b)^2(\sin(2\pi t))^2 \\
&\geq ((1-\tau)m + \tau m)^2(\cos(2\pi t))^2 + ((1-\tau)m + \tau m)^2(\sin(2\pi t))^2 = m^2 > 0.
\end{aligned}$$

Then C is $\mathbb{C}\backslash\{0\}$-homotopic to E, since H is continuous, and moreover:

$$\begin{aligned}
H(t,0) &= C(t) \text{ for all } t \in [0,1], \\
H(t,1) &= E(t) \text{ for all } t \in [0,1], \\
H(0,\tau) &= (1-\tau)C(0) + \tau E(0) = (1-\tau)C(1) + \tau E(1) = H(1,\tau) \text{ for all } \tau \in [0,1].
\end{aligned}$$

Solution to Exercise 3.23.

(ER1) Let $\gamma : [0,1] \to D$ be a closed path. Define $H : [0,1] \times [0,1] \to D$ by $H(t,\tau) = \gamma(t)$, $t, \tau \in [0,1]$. Then H is continuous, and

$$\begin{aligned}
H(t,0) &= \gamma(t) \text{ for all } t \in [0,1], \\
H(t,1) &= \gamma(t) \text{ for all } t \in [0,1], \\
H(0,\tau) &= \gamma(0) = \gamma(1) = H(1,\tau) \text{ for all } \tau \in [0,1].
\end{aligned}$$

Hence γ is D-homotopic to itself. So the relation is reflexive.

(ER2) Let $\gamma_0, \gamma_1 : [0,1] \to D$ be closed paths such that γ_0 is D-homotopic to γ_1. Then there exists a continuous $H : [0,1] \times [0,1] \to D$ such that

$$\begin{aligned}
H(t,0) &= \gamma_0(t) \text{ for all } t \in [0,1], \\
H(t,1) &= \gamma_1(t) \text{ for all } t \in [0,1], \\
H(0,\tau) &= H(1,\tau) \text{ for all } \tau \in [0,1].
\end{aligned}$$

Let $\widetilde{H} : [0,1] \times [0,1] \to D$ be defined by $\widetilde{H}(t,\tau) = H(t, 1-\tau)$ for all $t, \tau \in [0,1]$. Then $\widetilde{H}$ is continuous and

$$\begin{aligned}
\widetilde{H}(t,0) &= H(t,1) = \gamma_1(t) \text{ for all } t \in [0,1], \\
\widetilde{H}(t,1) &= H(t,0) = \gamma_0(t) \text{ for all } t \in [0,1], \\
\widetilde{H}(0,\tau) &= H(0, 1-\tau) = H(1, 1-\tau) = \widetilde{H}(1,\tau) \text{ for all } \tau \in [0,1].
\end{aligned}$$

Thus γ_1 is D-homotopic to γ_0. Hence the relation is symmetric.

(ER3) Let $\gamma_0, \gamma_1, \gamma_2$ be closed paths such that γ_0 is D-homotopic to γ_1, and γ_1 is D-homotopic to γ_2. There are continuous $H, K : [0,1] \times [0,1] \to D$ such that

$$H(t,0) = \gamma_0(t) \text{ for all } t \in [0,1], \qquad K(t,0) = \gamma_1(t) \text{ for all } t \in [0,1],$$
$$H(t,1) = \gamma_1(t) \text{ for all } t \in [0,1], \qquad K(t,1) = \gamma_2(t) \text{ for all } t \in [0,1],$$
$$H(0,\tau) = H(1,\tau) \text{ for all } \tau \in [0,1], \qquad K(0,\tau) = K(1,\tau) \text{ for all } \tau \in [0,1].$$

Define $L : [0,1] \times [0,1] \to D$ by $L(t,\tau) = \begin{cases} H(t,2\tau) & \text{if } \tau \in [0, \frac{1}{2}] \\ K(t, 2(\tau - \frac{1}{2})) & \text{if } \tau \in (\frac{1}{2}, 1] \end{cases}$.

Then $L(t,0) = H(t,0) = \gamma_0(t)$ and $L(t,1) = K(t,1) = \gamma_2(t)$ for all $t \in [0,1]$. Also, for $0 \leqslant \tau \leqslant \frac{1}{2}$, $L(0,\tau) = H(0,2\tau) = H(1,2\tau) = L(1,\tau)$, and for $\frac{1}{2} < \tau \leqslant 1$, $L(0,\tau) = K(0, 2(\tau - \frac{1}{2})) = K(1, 2(\tau - \frac{1}{2})) = L(1,\tau)$. For the continuity of L, it is enough to show 'left continuity in the τ-variable' at $\tau = \frac{1}{2}$, as H, K are continuous. If $((t_n, \tau_n))_{n \in \mathbb{N}}$ is a sequence in $[0,1] \times (\frac{1}{2}, 1]$ that converges to $(t_0, \frac{1}{2})$, then

$$\lim_{n \to \infty} L(t_n, \tau_n) = \lim_{n \to \infty} K(t_n, 2(\tau_n - \frac{1}{2})) = K(t_0, 0) = \gamma_1(t_0) = H(t_0, 1) = L(t_0, \tfrac{1}{2}) = L(\lim_{n \to \infty}(t_n, \tau_n)).$$

So L is continuous, and γ_0 is D-homotopic to γ_2. Thus the relation is transitive.

As D-homotopy is reflexive, symmetric and transitive, it is an equivalence relation.

Solution to Exercise 3.24. The 'point' path $\gamma : [0,1] \to \mathbb{C}\backslash\{0\}$, $\gamma(t) := -2$, $t \in [0,1]$, is closed. Also, $\gamma, \widetilde{C}$ are $\mathbb{C}\backslash\{0\}$-homotopic: If $H : [0,1] \times [0,1] \to \mathbb{C}\backslash\{0\}$ is defined by $H(t,\tau) = (1-\tau)\widetilde{C}(t) + \tau(-2) = -2 + (1-\tau)e^{2\pi i t}$, $\tau, t \in [0,1]$, then $|H(t,\tau) + 2| \leqslant 1$, so that $|H(t,\tau)| \geqslant 1 > 0$, and for all $t, \tau \in [0,1]$, $H(t,0) = \widetilde{C}(t)$, $H(t,1) = -2 = \gamma(t)$ and $H(0,\tau) = H(1,\tau)$. Also, H is continuous.

If $C, \widetilde{C}$ are $\mathbb{C}\backslash\{0\}$-homotopic, then by the transitivity of $\mathbb{C}\backslash\{0\}$-homotopy, C, γ are $\mathbb{C}\backslash\{0\}$-homotopic, and the Cauchy Integral Theorem with $f := \frac{1}{z} \in \mathcal{O}(\mathbb{C}\backslash\{0\})$ implies that $2\pi i = \int_C \frac{1}{z}\, dz = \int_\gamma \frac{1}{z}\, dz = \int_0^1 \frac{1}{-2} 0\, dt = 0$, a contradiction.

Solution to Exercise 3.25. For a circular path $C : [0,1] \to \mathbb{C}\backslash\{0\}$ with centre 0 and radius $r > 0$, traversed once in the anticlockwise direction, $\int_C \frac{1}{z}\, dz = 2\pi i$. Define $\widetilde{E} : [0,1] \to \mathbb{C}$ by $\widetilde{E}(t) = a\cos(2\pi t) + ib\sin(2\pi t)$ for all $t \in [0,1]$. The elliptic path $\widetilde{E}$ is $\mathbb{C}\backslash\{0\}$-homotopic to C, as shown in Exercise 3.22, and $\widetilde{E}$ is a reparamterisation of E. So by the Cauchy Integral Theorem, $\int_{\widetilde{E}} \frac{1}{z}\, dz = \int_C \frac{1}{z}\, dz = 2\pi i$, and so

$$\begin{aligned}
2\pi i &= \int_{\widetilde{E}} \tfrac{1}{z}\, dz = \int_E \tfrac{1}{z}\, dz = \int_0^{2\pi} \frac{1}{a\cos\theta + ib\sin\theta}(-a\sin\theta + ib\cos\theta)\, d\theta \\
&= \int_0^{2\pi} \frac{(-a\sin\theta + ib\cos\theta)(a\cos\theta - ib\sin\theta)}{a^2(\cos\theta)^2 + b^2(\sin\theta)^2}\, d\theta \\
&= \int_0^{2\pi} \frac{(b^2 - a^2)(\cos\theta)(\sin\theta) + iab((\cos\theta)^2 + (\sin\theta)^2)}{a^2(\cos\theta)^2 + b^2(\sin\theta)^2}\, d\theta \\
&= \int_0^{2\pi} \frac{(b^2 - a^2)(\cos\theta)(\sin\theta) + iab(1)}{a^2(\cos\theta)^2 + b^2(\sin\theta)^2}\, d\theta.
\end{aligned}$$

Equating the imaginary parts, $\int_0^{2\pi} \frac{1}{a^2(\cos\theta)^2 + b^2(\sin\theta)^2}\, d\theta = \frac{2\pi}{ab}$.

Solution to Exercise 3.26. C is $\mathbb{C}\backslash\{0\}$-homotopic to S.

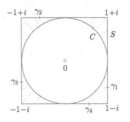

For $t \in [0,1]$, define

$$\gamma_1(t) := (1-t)(1-i) + t(1+i) = 1 + i(2t-1),$$
$$\gamma_2(t) := (1-t)(1+i) + t(-1+i) = (1-2t) + i,$$
$$\gamma_3(t) := (1-t)(-1+i) + t(-1-i) = -1 + i(1-2t),$$
$$\gamma_4(t) := (1-t)(-1-i) + t(1-i) = 2t-1-i.$$

Then $\int_S \frac{1}{z}\,dz = \int_{\gamma_1} \frac{1}{z}\,dz + \int_{\gamma_2} \frac{1}{z}\,dz + \int_{\gamma_3} \frac{1}{z}\,dz + \int_{\gamma_4} \frac{1}{z}\,dz$. We have

$$\int_{\gamma_1} \frac{1}{z}\,dz = \int_0^1 \frac{2i}{1+i(2t-1)}\,dt = \int_0^1 \frac{2i(1-i(2t-1))}{1+(2t-1)^2}\,dt = 2i\int_0^1 \frac{1}{1+(2t-1)^2}\,dt + 2\int_0^1 \frac{2t-1}{1+(2t-1)^2}\,dt$$

$$\overset{u=2t-1}{=} i\int_{-1}^1 \frac{1}{1+u^2}\,du + \int_{-1}^1 \frac{u}{1+u^2}\,du = i(\tan^{-1}1 - \tan^{-1}(-1)) + 0 = i(\tfrac{\pi}{4} - (-\tfrac{\pi}{4})) = i\tfrac{\pi}{2}.$$

Noting that $\int_0^1 \frac{2}{1+(2t-1)^2}\,dt = \frac{\pi}{2}$ and $\int_0^1 \frac{2t-1}{1+(2t-1)^2}\,dt = 0$, we obtain

$$\int_{\gamma_2} \frac{1}{z}\,dz = \int_0^1 \frac{-2}{1-2t+i}\,dt = \int_0^1 \frac{-2(-(2t-1)-i)}{1+(2t-1)^2}\,dt = 0 + (-1)(-i)\tfrac{\pi}{2} = i\tfrac{\pi}{2},$$

$$\int_{\gamma_3} \frac{1}{z}\,dz = \int_0^1 \frac{-2i}{-1+i(1-2t)}\,dt = \int_0^1 \frac{-2i(-1+i(2t-1))}{1+(2t-1)^2}\,dt = -i(-1)\tfrac{\pi}{2} + 0 = i\tfrac{\pi}{2},$$

$$\int_{\gamma_4} \frac{1}{z}\,dz = \int_0^1 \frac{2}{2t-1-i}\,dt = \int_0^1 \frac{2((2t-1)+i)}{1+(2t-1)^2}\,dt = 0 + i\tfrac{\pi}{2} = i\tfrac{\pi}{2}.$$

Thus $\int_S \frac{1}{z}\,dz = 4i\tfrac{\pi}{2} = 2\pi i$, as expected.

Solution to Exercise 3.27.

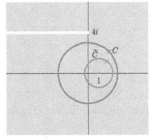

(1) $\operatorname{Log}(\cdot - 4i) \in \mathcal{O}(\mathbb{C}\backslash\{r+4i : r \leqslant 0\})$. So $\int_C \operatorname{Log}(z-4i)\,dz = 0$.

(2) If $\widetilde{C}$ is the circle with centre 1 and any radius $r > 0$, then $\int_{\widetilde{C}} \frac{1}{z-1}\,dz = 2\pi i$.

   As $\frac{1}{z-1} \in \mathcal{O}(\mathbb{C}\backslash\{1\})$, and since $C, \widetilde{C}$ are $\mathbb{C}\backslash\{1\}$-homotopic, it follows from the Cauchy Integral Theorem that $\int_C \frac{1}{z-1}\,dz = \int_{\widetilde{C}} \frac{1}{z-1}\,dz = 2\pi i$.

(3) $i^{z-3} = \exp((z-3)\operatorname{Log} i) = \exp((z-3)(\log 1 + i\frac{\pi}{2})) = \exp((z-3)(i\frac{\pi}{2})) = \exp(i\frac{\pi}{2}(z-3))$, and so $z \mapsto i^{z-3}$ is entire. By the Cauchy Integral Theorem, $\int_C i^{z-3}\,dz = 0$.

Solution to Exercise 3.28. We have

$$\int_{\gamma_0} \tfrac{1}{z}\,dz = \int_0^1 \tfrac{1}{e^{2\pi it}} 2\pi i e^{2\pi it}\,dt = \int_0^1 2\pi i\,dt = 2\pi i,$$

$$\int_{\gamma_1} \tfrac{1}{z}\,dz = \int_0^1 \tfrac{1}{e^{-2\pi it}}(-2\pi i e^{-2\pi it})\,dt = -\int_0^1 2\pi i\,dt = -2\pi i.$$

If γ_0 is $\mathbb{C}\backslash\{0\}$-homotopic to γ_1, then as $\tfrac{1}{z} \in \mathcal{O}(\mathbb{C}\backslash\{0\})$, it follows by the Cauchy Integral Theorem that $2\pi i = \int_{\gamma_0}\tfrac{1}{z}\,dz = \int_{\gamma_1}\tfrac{1}{z}\,dz = -2\pi i$, a contradiction. Thus γ_0 is not $\mathbb{C}\backslash\{0\}$-homotopic to γ_1.

Solution to Exercise 3.29.

(1) $\varphi'(t) = (\exp \int_0^t \tfrac{\gamma'(s)}{\gamma(s)}\,ds)\frac{d}{dt}\int_0^t \tfrac{\gamma'(s)}{\gamma(s)}\,ds = \varphi(t)\frac{\gamma'(t)}{\gamma(t)}$. Hence $\varphi'\gamma - \varphi\gamma' = 0$, and[1] $\frac{d}{dt}\frac{\varphi}{\gamma} = \frac{\varphi'\gamma - \varphi\gamma'}{\gamma^2} = \frac{0}{\gamma^2} = 0$. Thus $\frac{\varphi(0)}{\gamma(0)} = \frac{\varphi(1)}{\gamma(1)}$. But $\gamma(0) = \gamma(1)$ (since γ is closed). So $\varphi(1) = \varphi(0) = \exp\int_0^0 \tfrac{\gamma'(s)}{\gamma(s)}\,ds = \exp 0 = 1$. Hence $w(\gamma) \in \mathbb{Z}$.

(2) We have $w(\Gamma_1) = \frac{1}{2\pi i}\int_0^1 \frac{\Gamma_1'(t)}{\Gamma_1(t)}\,dt = \frac{1}{2\pi i}\int_0^1 \frac{2\pi i\exp(2\pi it)}{\exp(2\pi it)}\,dt = \frac{1}{2\pi i}2\pi i = 1$.

(3) If for all $t \in [0,1]$, $\gamma_1(t) \neq 0$ and $\gamma_2(t) \neq 0$, then also $(\gamma_1\cdot\gamma_2)(t) \neq 0$ for all $t \in [0,1]$, and so $\gamma_1\cdot\gamma_2$ does not pass through the origin. Also, $\gamma_1\cdot\gamma_2$ is closed since $(\gamma_1\cdot\gamma_2)(0) = \gamma_1(0)\gamma_2(0) = \gamma_1(1)\gamma_2(1) = (\gamma_1\cdot\gamma_2)(1)$. We have $(\gamma_1\cdot\gamma_2)'(t) = \gamma_1'(t)\gamma_2(t) + \gamma_1(t)\gamma_2'(t)$, $t \in [0,1]$. So

$$w(\gamma_1\cdot\gamma_2) = \frac{1}{2\pi i}\int_0^1 \frac{(\gamma_1\gamma_2)'(t)}{(\gamma_1\gamma_2)(t)}\,dt = \frac{1}{2\pi i}\int_0^1 \frac{\gamma_1'(t)\gamma_2(t) + \gamma_1(t)\gamma_2'(t)}{\gamma_1(t)\gamma_2(t)}\,dt$$

$$= \frac{1}{2\pi i}\int_0^1 \frac{\gamma_1'(t)\gamma_2(t)}{\gamma_1(t)\gamma_2(t)}\,dt + \frac{1}{2\pi i}\int_0^1 \frac{\gamma_1(t)\gamma_2'(t)}{\gamma_1(t)\gamma_2(t)}\,dt = w(\gamma_1) + w(\gamma_2).$$

(4) $\Gamma_m = \Gamma_1 \cdots \Gamma_1$ (m times). So $w(\Gamma_m) = w(\Gamma_1) + \cdots + w(\Gamma_1) = m\,w(\Gamma_1) = m\,1 = m$.

(5) Define the map $\varphi : [0,1] \to \mathbb{R}$ by $\varphi(t) = |\gamma_0(t)|$, $t \in [0,1]$, giving the distance of $\gamma_0(t)$ from 0. Being a real-valued continuous function on a compact interval, it has a minimum value d_0, and $d_0 > 0$ since γ_0 does not pass through 0. Take $\delta = d_0/2 > 0$. Suppose that γ is a smooth closed path such that we have $\|\gamma - \gamma_0\|_\infty := \max_{t\in[0,1]}|\gamma(t) - \gamma_0(t)| < \delta$. We will show γ is $\mathbb{C}\backslash\{0\}$-homotopic to γ_0. Define $H : [0,1] \times [0,1] \to \mathbb{C}\backslash\{0\}$ by $H(t,\tau) = (1-\tau)\gamma_0(t) + \tau\gamma(t)$, $t,\tau \in [0,1]$. Then H is continuous. Also, $H(t,0) = \gamma_0(t)$ and $H(t,1) = \gamma(t)$ for all $t \in [0,1]$, and $H(0,\tau) = (1-\tau)\gamma_0(0) + \tau\gamma(0) = (1-\tau)\gamma_0(1) + \tau\gamma(1) = H(1,\tau)$ for all $\tau \in [0,1]$. Finally, $H(t,\tau)$ is never 0, because being a convex combination

---

[1]Here we use the product and quotient rules for differentiation of *complex-valued* maps of a real variable. These follow easily by mimicking the analogous proofs in the context of real-valued functions. Firstly, if $\varphi : I \to \mathbb{C}$ is differentiable on an interval I, then $\varphi'(t) := (\operatorname{Re}\varphi)'(t) + i(\operatorname{Im}\varphi)'(t) = \lim_{h\to 0}\frac{\varphi(t+h)-\varphi(t)}{h}$. If $\varphi(t) \neq 0$ ($t \in I$), then $\frac{d}{dt}\frac{1}{\varphi} = \lim_{h\to 0}\frac{1}{h}(\frac{1}{\varphi(t+h)} - \frac{1}{\varphi(t)}) = -\frac{\varphi'(t)}{(\varphi(t))^2}$. Similarly, the product rule can be shown.

of $\gamma_0(t)$ and $\gamma(t)$, if $(1-\tau)\gamma_0(t)+\tau\gamma(t) = 0$ for some t,τ, then we arrive at a contradiction: $1\frac{d_0}{2} > \tau|\gamma_0(t)-\gamma(t)| = |\gamma_0(t)-((1-\tau)\gamma_0(t)+\tau\gamma(t))| = |\gamma_0(t)| \geqslant d_0$.

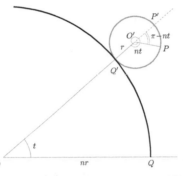

Using the Cauchy Integral Theorem, we get

$$w(\gamma) = \tfrac{1}{2\pi i}\int_0^1 \tfrac{\gamma'(t)}{\gamma(t)}\,dt = \tfrac{1}{2\pi i}\int_\gamma \tfrac{1}{z}\,dz = \tfrac{1}{2\pi i}\int_{\gamma_0}\tfrac{1}{z}\,dz = \tfrac{1}{2\pi i}\int_0^1\tfrac{\gamma_0'(t)}{\gamma_0(t)}\,dt = w(\gamma_0).$$

Solution to Exercise 3.30. The length of the arc on the big circle subtended by $\angle Q'OQ$ is $t(nr)$. As the small coin rolls without slipping, the angle made by $O'P$ with OO' is $(tnr)/r = nt$.

As $O' \equiv (nr+r)e^{it}$, and since $O'P$ is obtained from $O'P' = re^{it}$ by a clockwise rotation through $\pi - nt$, $P \equiv (nr+r)e^{it} + e^{-i(\pi-nt)}re^{it} = (n+1)re^{it}+(-1)\,r\,e^{(n+1)it}$. The area enclosed by the epicycloid γ is

$$\tfrac{1}{2i}\int_\gamma \bar{z}\,dz = \tfrac{1}{2i}\int_0^{2\pi}\overline{r((n+1)e^{it}-e^{(n+1)it})}r((n+1)ie^{it}-(n+1)ie^{(n+1)it})\,dt$$
$$= \tfrac{1}{2i}\int_0^{2\pi} r^2((n+1)e^{-it}-e^{-(n+1)it})(n+1)i(e^{it}-e^{(n+1)it})\,dt$$
$$= \tfrac{(n+1)r^2}{2}\int_0^{2\pi}((n+1)-(n+1)e^{int}-e^{-int}+1)\,dt$$
$$= \tfrac{(n+1)r^2}{2}((n+1)2\pi+0+0+2\pi) = \pi r^2(n+1)(n+2).$$

Solution to Exercise 3.31. For example, let $f(z) = \tfrac{1}{z}$ for $z \in D := \mathbb{C}\backslash\{0\}$. Then f does not have a primitive in D. (See Example 3.7 and Exercise 3.21.)

Solution to Exercise 3.32. Let $C(t) = e^{it}$ for all $t \in [0, 2\pi]$. Then

$$\int_C \tfrac{i}{(z-a)(az-1)}\,dz = \int_0^{2\pi}\tfrac{i}{(e^{it}-a)(ae^{it}-1)}ie^{it}\,dt = \int_0^{2\pi}\tfrac{-e^{it}}{(e^{it}-a)(a-e^{-it})e^{it}}\,dt$$
$$= \int_0^{2\pi}\tfrac{1}{(e^{it}-a)(e^{-it}-a)}\,dt = \int_0^{2\pi}\tfrac{1}{|e^{it}-a|^2}\,dt$$
$$= \int_0^{2\pi}\tfrac{1}{((\cos t)-a)^2+(\sin t)^2}\,dt = \int_0^{2\pi}\tfrac{1}{1-2a\cos t+a^2}\,dt.$$

Since $z \mapsto \frac{i}{az-1}$ is holomorphic in a disc containing the unit circle C (as $0 < a < 1$),
by the Cauchy Integral Formula, $\frac{1}{2\pi i} \int_\gamma \frac{\frac{i}{az-1}}{z-a} dz = \frac{i}{az-1}|_{z=a} = \frac{i}{a^2-1}$.

So $\int_0^{2\pi} \frac{1}{1-2a\cos t+a^2} dt = \int_\gamma \frac{\frac{i}{az-1}}{z-a} dz = 2\pi i \frac{i}{a^2-1} = \frac{2\pi}{1-a^2}$.

Solution to Exercise 3.33. Suppose F is a primitive. Consider the closed circular path γ given by $|z - 0| = \frac{1}{2}$ traversed once anticlockwise. By the Fundamental Theorem of Contour Integration, $\int_\gamma \frac{1}{z(1-z^2)} dz = \int_\gamma F'(z) dz = 0$, since γ is closed. On the other hand, by the Cauchy Integral Formula, as $\frac{1}{1-z^2} \in \mathcal{O}(D(0,1))$, we get

$$\int_\gamma \frac{1}{z(1-z^2)} dz = \int_\gamma \frac{\frac{1}{1-z^2}}{z-0} dz = 2\pi i \frac{1}{1-z^2}|_{z=0} = 2\pi i,$$

a contradiction. So $\frac{1}{z(1-z^2)}$ does not have a primitive in $\{z \in \mathbb{C} : 0 < |z| < 1\}$.

Solution to Exercise 3.34. As $|z| < 1$, and $\zeta \mapsto f(\zeta)(1 - \zeta\bar{z})$ is entire, by the Cauchy integral formula, we obtain

$$(1-|z|^2)f(z) = (1 - z\bar{z})f(z) = f(\zeta)(1-\zeta\bar{z})|_{\zeta=z} = \frac{1}{2\pi i} \int_C f(\zeta)\frac{1-\zeta\bar{z}}{\zeta-z} d\zeta.$$

Since $|z| < 1$,

$$(1-|z|^2)|f(z)| = |(1-|z|^2)f(z)| = |\frac{1}{2\pi i} \int_C f(\zeta)\frac{1-\zeta\bar{z}}{\zeta-z} d\zeta| = |\int_0^{2\pi} f(e^{i\theta})\frac{1-e^{i\theta}\bar{z}}{e^{i\theta}-z} ie^{i\theta} d\theta|$$

$$\leqslant \frac{1}{2\pi} \int_0^{2\pi} |f(e^{i\theta})\frac{1-e^{i\theta}\bar{z}}{e^{i\theta}-z} ie^{i\theta}| d\theta.$$

The result now follows by observing that for $\theta \in [0, 2\pi]$,

$$|f(e^{i\theta})\frac{1-e^{i\theta}\bar{z}}{e^{i\theta}-z} ie^{i\theta}| = |f(e^{i\theta})|\frac{|e^{i\theta}(e^{-i\theta}-\bar{z})|}{|e^{i\theta}-z|}|ie^{i\theta}| = |f(e^{i\theta})|\frac{1|e^{i\theta}-z|}{|e^{i\theta}-z|}1 = |f(e^{i\theta})|.$$

Solution to Exercise 3.35.

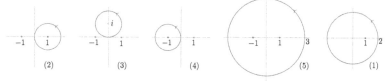

(2)   (3)   (4)   (5)   (1)

(1) $\int_\gamma \frac{\exp z}{z-1} dz = 2\pi i \exp z|_{z=1} = 2\pi i \exp 1 = 2\pi i e$.

(2) $\int_\gamma \frac{z^2+1}{z^2-1} dz = \int_\gamma \frac{\frac{z^2+1}{z+1}}{z-1} dz = 2\pi i \frac{z^2+1}{z+1}|_{z=1} = 2\pi i \frac{1^2+1}{1+1} = 2\pi i$.

(3) $\int_\gamma \frac{z^2+1}{z^2-1} dz = 0$, since e.g. $\frac{z^2+1}{z^2-1} \in \mathcal{O}(D(i, \sqrt{2}))$, $D(i, \sqrt{2})$ is simply connected, and γ is a closed path in $D(i, \sqrt{2})$.

(4) $\int_\gamma \frac{z^2+1}{z^2-1} dz = \int_\gamma \frac{\frac{z^2+1}{z-1}}{z-(-1)} dz = 2\pi i \frac{z^2+1}{z-1}|_{z=-1} = 2\pi i \frac{(-1)^2+1}{-1-1} = -2\pi i$.

(5) $\int_\gamma \frac{z^2+1}{z^2-1} dz = \int_\gamma \frac{z^2+1}{2}(\frac{1}{z-1} - \frac{1}{z+1}) dz = \int_\gamma \frac{\frac{z^2+1}{2}}{z-1} dz - \int_\gamma \frac{\frac{z^2+1}{2}}{z-(-1)} dz$

$\qquad = 2\pi i \frac{z^2+1}{2}|_{z=1} - 2\pi i \frac{z^2+1}{2}|_{z=-1} = 2\pi i(1) - 2\pi i(1) = 0$.

Solution to Exercise 3.36.

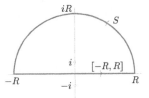

(1) By the Cauchy Integral Formula, we have

$$\int_\sigma F(z)dz = \int_\sigma \frac{\exp(iz)}{z^2+1}dz = \int_\sigma \frac{\frac{\exp(iz)}{z+i}}{z-i}dz = 2\pi i \frac{\exp(iz)}{z+i}\Big|_{z=i} = 2\pi i \frac{\exp(ii)}{i+i} = 2\pi i \frac{e^{-1}}{2i} = \frac{\pi}{e}.$$

(2) Let $z = x+iy$, with $x,y \in \mathbb{R}$, $y \geqslant 0$. Then $|\exp(iz)| = |\exp(-y+ix)| = e^{-y} \leqslant 1$.
Thus $|F(z)| = \frac{|\exp(iz)|}{|z^2+1|} \leqslant \frac{1}{|z^2+1|}$. But $|z^2| - |-1| \leqslant |z^2 - (-1)| = |z^2+1|$, and
so if $|z| > 1$, then $|F(z)| \leqslant \frac{1}{|z^2+1|} \leqslant \frac{1}{|z|^2-1}$. Finally, for $|z| > 1$, we have that
$\frac{1}{|z|^2-1} \leqslant \frac{2}{|z|^2}$ if $|z|^2 \leqslant 2|z|^2 - 2$, i.e., $|z^2| \geqslant 2$, i.e., $|z| \geqslant \sqrt{2}$.

(3) For $R \geqslant \sqrt{2}$, $|\int_S F(z)dz| \leqslant 2\pi R \max_{z \in S}|F(z)| \leqslant 2\pi R \frac{2}{R^2}$. So $\lim\limits_{R\to\infty} \int_S F(z)dz = 0$.
As $\int_{[-R,R]} F(z)dz = \int_\sigma F(z)dz - \int_S F(z)dz = \frac{\pi}{e} - \int_S F(z)dz$, we obtain that
$\lim\limits_{R\to\infty} \int_{[-R,R]} F(z)dz = \frac{\pi}{e} - \lim\limits_{R\to\infty} \int_S F(z)dz = \frac{\pi}{e} - 0 = \frac{\pi}{e}.$

(4) Using the parameterisation $x \mapsto x$ for the straight line path $[-R,R]$, we have

$$\int_{[-R,R]} F(z)dz = \int_{-R}^R \frac{\exp(ix)}{x^2+1} 1\,dx = \int_{-R}^R \frac{\cos x}{x^2+1}\,dx + i\int_{-R}^R \frac{\sin x}{x^2+1}\,dx = \int_{-R}^R \frac{\cos x}{x^2+1}\,dx + 0,$$

where we have used the fact that $\frac{\sin x}{x^2+1}$ is an odd function to get the last equality.
Hence $\lim\limits_{R\to\infty} \int_{-R}^R \frac{\cos x}{x^2+1}\,dx = \lim\limits_{R\to\infty} \int_S F(z)dz = \frac{\pi}{e}.$

Solution to Exercise 3.37. Let $C(\theta) := \exp(i\theta)$ for all $\theta \in [0, 2\pi]$. As $\exp \in \mathcal{O}(\mathbb{C})$,
by the Cauchy Integral Formula, $\frac{1}{2\pi i} \int_C \frac{\exp z}{z-0}\,dz = \exp z|_{z=0} = \exp 0 = 1$. But

$$\begin{aligned}
\int_C \frac{\exp z}{z-0}\,dz &= \int_0^{2\pi} \frac{\exp(\exp(i\theta))}{\exp(i\theta)} i \exp(i\theta)\,d\theta = \int_0^{2\pi} i\exp(\exp(i\theta))\,d\theta \\
&= \int_0^{2\pi} i\exp(\cos\theta + i\sin\theta)\,d\theta = \int_0^{2\pi} ie^{\cos\theta}(\cos(\sin\theta) + i\sin(\sin\theta))\,d\theta \\
&= -\int_0^{2\pi} e^{\cos\theta}\sin(\sin\theta)\,d\theta + i\int_0^{2\pi} e^{\cos\theta}\cos(\sin\theta)\,d\theta
\end{aligned}$$

Hence $\int_0^{2\pi} e^{\cos\theta}\cos(\sin\theta)\,d\theta = 2\pi$.

Solution to Exercise 3.38. If f is holomorphic, then $f^{(n)}$ is holomorphic too, and
so is its derivative $f^{(n+1)}$. But $f^{(n+1)}$, being complex differentiable, is in particular
continuous. So $f^{(n)}$ has a continuous complex derivative.

Solution to Exercise 3.39. For $n \in \mathbb{N}$, let $C^n(U)$ denote the set of all real-valued functions on U possessing continuous partial derivatives of order $\leqslant n$ on U. We use induction to show that $u \in C^n(U)$ for all $n \in \mathbb{N}$, hence proving the claim. Consider the following sequence of statements S_n, $n \in \mathbb{N}$:

S_n: For every holomorphic function on U, its real part belongs to $C^n(U)$.

We show that S_1 is true. Let $f \in \mathcal{O}(U)$. We want to show that $u := \operatorname{Re} f \in C^1(U)$. As $f = u + iv \in \mathcal{O}(U)$, $f' \in \mathcal{O}(U)$. Also, $f' = \frac{\partial u}{\partial x} + i\frac{\partial v}{\partial x} = \frac{\partial u}{\partial x} - i\frac{\partial u}{\partial y}$. It follows from Theorem 2.4 that for the holomorphic function f', its real and imaginary parts are real differentiable, and in particular, continuous, everywhere in U. So $\frac{\partial u}{\partial x}, \frac{\partial u}{\partial y}$ are continuous on U. Thus $u \in C^1(U)$. This shows S_1 is true.

Suppose S_n is true. Let $f \in \mathcal{O}(U)$. We want to show that $u := \operatorname{Re} f \in C^{n+1}(U)$ in order to complete the induction step. As $f' \in \mathcal{O}(U)$ and $f' = \frac{\partial u}{\partial x} - i\frac{\partial u}{\partial y}$, it follows from the validity of S_n that $\frac{\partial u}{\partial x} = \operatorname{Re}(f') \in C^n(U)$, and $\frac{\partial u}{\partial y} = \operatorname{Re}(if') \in C^n(U)$. We already know that $u = \operatorname{Re} f \in C^n(U)$. To show that $u \in C^{n+1}(U)$, we need to show that the partial derivatives of order $n+1$ of u are continuous. But any such partial derivative will be an nth order partial derivative of $\frac{\partial u}{\partial x}$ or of $\frac{\partial u}{\partial y}$, which we have established is continuous. Thus $u \in C^{n+1}(U)$. Hence S_{n+1} is true, completing the induction step. By the induction principle, S_n is true for all $n \in \mathbb{N}$.

So if $f \in \mathcal{O}(U)$, then $u := \operatorname{Re} f \in C^\infty(U)$. Also, $v := \operatorname{Im} f = \operatorname{Re}(-if) \in C^\infty(U)$.

Solution to Exercise 3.40. Since for all $z \in \mathbb{C}$, $|f(z)| \geqslant \delta > 0$, in particular $f(z) \neq 0$ for all $z \in \mathbb{C}$, and so $\frac{1}{f}$ is entire. For all $z \in \mathbb{C}$, $|\frac{1}{f(z)}| \leqslant \frac{1}{\delta}$.
By Liouville's Theorem, $\frac{1}{f}$ is constant, say C ($\neq 0$, as f is pointwise nonzero). Thus $f \equiv \frac{1}{C}$. Hence f is a constant.

Solution to Exercise 3.41. Let g be defined by $g(z) := f(z) - w_0$ for $z \in \mathbb{C}$. Then g is entire and $|g(z)| \geqslant r > 0$ for all $z \in \mathbb{C}$ (Why?). By Exercise 3.40, g is constant. So $f = g + w_0$ must be constant too.

Solution to Exercise 3.42. Let $K := \{(x, y) \in \mathbb{R}^2 : 0 \leqslant x \leqslant T_1, \ 0 \leqslant y \leqslant T_2\}$. Then K is compact. (Why?) By the Weierstrass Theorem, the continuous function $K \ni (x, y) \mapsto |f(x + iy)| \in \mathbb{R}$ assumes a maximum value M on K. Given any $x, y \in \mathbb{R} = \cup_{n \in \mathbb{Z}}[nT_1, (n+1)T_1) = \cup_{m \in \mathbb{Z}}[mT_1, (m+1)T_1)$, there exist integers n, m such that $x + iy = x_0 + nT_1 + i(y_0 + mT_2)$ for some $x_0 \in [0, T_1)$ and $y_0 \in [0, T_2)$. By the periodicity of f, $f(x + iy) = f(x_0 + nT_1 + i(y_0 + mT_2)) = f(x_0 + iy_0) \in f(K)$. So for all $x, y \in \mathbb{R}$, $|f(x + iy)| \leqslant M$. Hence f is bounded on $\mathbb{C}$, and by Liouville's Theorem, must be constant.

The exponential function is 2π-periodic the imaginary direction (since we have $\exp(z + 2\pi i) = (\exp z)(\exp(2\pi i)) = (\exp z)(1) = \exp z$, but exp is not constant. By Exercise 1.56, $\sin(z + 2\pi) = (\sin z)(\cos(2\pi)) + (\sin(2\pi))(\cos z) = \sin z$, and so sin is 2π-periodic the real direction, but not constant.

Solution to Exercise 3.43.

(1) Define g by $g(z) = (\exp(-z))f(z)$ for all $z \in \mathbb{C}$. Then g is entire. Moreover, since $|f(z)| \leqslant |\exp z|$, by rearranging, we get $|g(z)| = |(\exp(-z))f(z)| \leqslant 1$ for all $z \in \mathbb{C}$. So by Liouville's Theorem, g is constant with value say, c, i.e., $g(z) = (\exp(-z))f(z) = c$, giving $f(z) = c \exp z$ for all $z \in \mathbb{C}$. Also, since $|g(z)| \leqslant 1$, we obtain $|c| \leqslant 1$.

(2) By Exercise 1.37, if p is a polynomial of degree $d \geqslant 1$, then there exist $m, R > 0$ such that $|p(z)| \geqslant m|z|^d$ for all $|z| > R$. Thus for real $z = x < -R < 0$, we have $|z| = -x > R$, and so $m|x|^d \leqslant |p(x)| \leqslant |e^x| = e^x \leqslant 1$ (because $x < 0$). Hence $|x|^d \leqslant 1/m$ for all $x < -R$, a contradiction. So p is a constant, say equal to c_0. But then $|p(z)| \leqslant |\exp z|$ again gives with $z = x < 0$ that $|c_0| \leqslant |e^x| = e^x$ for all $x < 0$, and so $|c_0| = 0$, i.e., $c_0 = 0$. Consequently, $p = c_0 = 0$.

Solution to Exercise 3.44. For all $z \in \mathbb{C}$, $|f(z)| > |f'(z)| \geqslant 0$, and so f is never 0. Thus $g := \frac{f'}{f}$ is well-defined, and is entire. Also from the given inequality, we see that $|g| < 1$ everywhere in $\mathbb{C}$, that is, g is bounded. By Liouville's theorem, g must be a constant, say C. Hence $f' = Cf$ in $\mathbb{C}$.

We have $(e^{-Cz}f)' = -Ce^{-Cz}f + e^{-Cz}f' = e^{-Cz}(f' - Cf) = e^{-Cz}0 = 0$. As $\mathbb{C}$ is domain and $e^{-Cz}f$ is entire, it follows that it must be a constant there, say $k \in \mathbb{C}$. This k must be nonzero, since we know that f is nowhere 0 in $\mathbb{C}$. So $f(z) = ke^{Cz}$ with $k \neq 0$. But $|f'| < |f|$ gives $|kCe^{Cz}| < |ke^{Cz}|$, showing that $|C| < 1$. So f must have the form $f(z) = ke^{Cz}$, with $k \in \mathbb{C}\backslash\{0\}$ and $C \in \mathbb{D} := \{z \in \mathbb{C} : |z| < 1\}$. Vice versa, if f has this form, then f is entire, and moreover, as $|f| = |ke^{Cz}| > 0$, we have for all $z \in \mathbb{C}$ that $|f'(z)| = |kCe^{Cz}| = |C||ke^{Cz}| < 1|f(z)| = |f(z)|$.

Solution to Exercise 3.45. The function e^f is entire, being the composition of the entire functions $\exp$ and f. We have $|e^{f(z)}| = e^{\operatorname{Re} f(z)} \leqslant e^M$. By Liouville's Theorem, e^f is a constant function. Thus $(e^f)' \equiv 0$. By the Chain Rule, we now obtain $0 = (e^f)' = e^f f'$. As $e^{f(z)} \neq 0$ for all $z \in \mathbb{C}$, it follows from the above that $f'(z) = 0$. As $f' \equiv 0$, we conclude (e.g. by C-R equations or the Fundamental Theorem of Contour Integration) that f is a constant function.

Solution to Exercise 3.46. We have that $if = i(u + iv) = -v + iu$, and so $g := f + if = (u-v) + i(u+v)$. But g is entire too, and so is e^g (being the composition of the entire functions $\exp$ and g). We have $|e^{g(z)}| = e^{\operatorname{Re} g(z)} = e^{u-v} \leqslant e^0 = 1$. By Liouville's Theorem, e^g is a constant function. Thus $(e^g)' \equiv 0$. By the Chain Rule, we now obtain $0 = (e^g)' = e^g g'$. As $e^{g(z)} \neq 0$ for all $z \in \mathbb{C}$, it follows from the above that $g'(z) = 0$. But $g' = (1 + i)f'$. So $f' \equiv 0$.

Solution to Exercise 3.47.

(1) For $z \in C$, $|z - a_1| \geqslant |z| - |a_1| = R - |a_1|$ and $|z - a_2| \geqslant |z| - |a_2| = R - |a_2|$. Thus by the ML-inequality, with $M := \max_{z \in C} |f(z)|$,

$$\left|\int_C \frac{f(z)}{(z-a_1)(z-a_2)} dz\right| \leqslant \max_{z \in C} \frac{|f(z)|}{|z-a_1||z-a_2|} 2\pi R \leqslant \frac{M}{(R-|a_1|)(R-|a_2|)} 2\pi R.$$

(2) For $z \in C$, $\frac{1}{z-a_1} - \frac{1}{z-a_2} = \frac{z-a_2-(z-a_1)}{(z-a_1)(z-a_2)} = \frac{a_1-a_2}{(z-a_1)(z-a_2)}$. As $a_1 \neq a_2$, we obtain $\frac{1}{(z-a_1)(z-a_2)} = \frac{1}{a_1-a_2}(\frac{1}{z-a_1} - \frac{1}{z-a_2})$. Hence $\alpha := -\beta := \frac{1}{a_1-a_2}$.

(3) $\int_C \frac{f(z)}{(z-a_1)(z-a_2)}dz = \int_C \frac{1}{a_1-a_2}(\frac{f(z)}{z-a_1} - \frac{f(z)}{z-a_2})dz = \frac{1}{a_1-a_2}(\int_C \frac{f(z)}{z-a_1}dz - \int_C \frac{f(z)}{z-a_2}dz)$.

Consider a small disc Δ_1 with centre a_1 and radius $r_1 > 0$ with boundary C_1. Then C and C_1 are $\mathbb{C}\backslash\{a_1\}$-homotopic, and $g := \frac{f(\cdot)}{\cdot-a_1} \in \mathcal{O}(\mathbb{C}\backslash\{a_1\})$. Thus by the Cauchy Integral Theorem, $\int_C \frac{f(z)}{z-a_1}dz = \int_{C_1} \frac{f(z)}{z-a_1}dz$. By the Cauchy Integral Formula, $\frac{1}{2\pi i}\int_{C_1} \frac{f(z)}{z-a_1}dz = f(a_1)$. Thus $\int_C \frac{f(z)}{z-a_1}dz = 2\pi i f(a_1)$. Similarly, we have $\int_C \frac{f(z)}{z-a_2}dz = 2\pi i f(a_2)$. Consequently, $\int_C \frac{f(z)}{(z-a_1)(z-a_2)}dz = \frac{2\pi i(f(a_1)-f(a_2))}{a_1-a_2}$.

(4) Let f be entire and bounded. Let $M > 0$ be such that $|f(z)| \leqslant M$ for all $z \in \mathbb{C}$. Let a_1, a_2 be any two distinct points in $\mathbb{C}$. Let $C(t) = Re^{it}$, $t \in [0, 2\pi]$, with $R > 0$ large enough so that a_1, a_2 lie in the interior of C. By the above,

$$|f(a_1) - f(a_2)| = \frac{|a_1-a_2|}{2\pi}|\frac{2\pi i(f(a_1)-f(a_2))}{a_1-a_2}| = \frac{|a_1-a_2|}{2\pi}|\int_C \frac{f(z)}{(z-a_1)(z-a_2)}dz|$$
$$\leqslant \frac{|a_1-a_2|}{2\pi}\frac{M}{(R-|a_1|)(R-|a_2|)}2\pi R.$$

As R can be made as large as we please and since $\frac{2\pi RM}{(R-|a_1|)(R-|a_2|)} \to 0$ as $R \to \infty$, it follows that $|f(a_1) - f(a_2)| = 0$, and so $f(a_1) = f(a_2)$. Hence f is constant.

Solution to Exercise 3.48. ('If part') Let $p(\alpha) = 0$ and $p'(\alpha) \neq 0$. So α is a zero of p. Suppose that α is not a simple zero of p. Then $p(z) = (z - \alpha)^m q(z)$, where the integer $m > 1$, and q is a polynomial. Thus $p' = m(z - \alpha)^{m-1}q + (z - \alpha)^m q'$, and so $p'(\alpha) = 0$, a contradiction. So α is a simple zero of p.

('Only if part') Let α be a simple zero of p. Then $p(\alpha) = 0$ and $p(z) = (z-\alpha)q(z)$, where q is a polynomial such that $q(\alpha) \neq 0$. Thus $p' = q + (z - \alpha)q'$, and so $p'(\alpha) = q(\alpha) + 0 = q(\alpha) \neq 0$.

If the degree of p is 1, then $p(z) = az + b$, where $a \neq 0$, and so taking $w = 0$, we have that $\tilde{p} = p + w = p + 0 = p$ has the only zero at $-\frac{b}{a}$, which is a simple zero.

Next suppose that the degree of p is > 1. Then p' has degree $d - 1 \geqslant 1$. By the Fundamental Theorem of Algebra and the Division Algorithm, p' has $d - 1$ zeroes (possibly with repetition), say $\beta_1, \cdots, \beta_{d-1}$. Let $w \in \mathbb{C}\backslash\{-p(\beta_1), \cdots, -p(\beta_{d-1})\}$. We claim $\tilde{p}+w$ has only simple zeroes. Indeed, if α is a zero of $\tilde{p}$, then $\tilde{p}(\alpha) = 0$, i.e., $p(\alpha) + w = 0$, i.e., $p(\alpha) = -w \notin \{p(\beta_1), \cdots, p(\beta_{d-1})\}$. Hence $\alpha \notin \{\beta_1, \cdots, \beta_{d-1}\}$, and so $\tilde{p}'(\alpha) = p'(\alpha) \neq 0$. By the previous part of the exercise, we conclude that α is a simple zero of $\tilde{p}$. Consequently, all the zeroes of $\tilde{p}$ are simple.

Solution to Exercise 3.49. As $g := f - e^z \sin z$ is entire and bounded, by Liouville's Theorem, g is a constant, say c. In particular, $c = f(0) - e^0 \sin 0 = 0 - 0 = 0$. Thus $g \equiv c = 0$, and so $f(z) = e^z \sin z$ for all $z \in \mathbb{C}$.

Solution to Exercise 3.50. By Morera's Theorem, it is enough to show that for every rectangular path R, $\int_R f(z)\,dz = 0$. If the rectangular path R lies entirely within $D := \mathbb{C}\backslash(\mathbb{R} \times \{0\})$, then $\int_R f(z)\,dz = 0$ since $f \in \mathcal{O}(D)$. We will show the result for rectangles with a side on the real axis. From here the result follows also for rectangles going across the real axis (since it can be decomposed into two rectangles, each of which has a common side lying on the real axis). Consider a rectangle in the closed upper half-plane with the bottom side s as $[a, b] \subset \mathbb{R}$, and $s(t) := t$, $t \in [a, b]$. (The lower half-plane case is dealt analogously.) Let h be the height of the rectangle R. Displacing s upwards through a distance $\frac{h}{n}$, we get the straight line path s_n, $s_n(t) = t + \frac{ih}{n}$, $t \in [a, b]$. Call the rectangle obtained from R by this side displacement as R_n. Thus $R_n \subset D$, and so $\int_{R_n} f(z)\,dz = 0$. As f is continuous, it is uniformly continuous on compact sets. Thus

$$\int_{s_n} f(z)\,dz = \int_a^b f(t + \tfrac{ih}{n})\,1\,dt \xrightarrow{n\to\infty} \int_a^b f(t)\,1\,dt = \int_s f(z)\,dz.$$

Also the contributions of the small vertical segments $v_{a,n}$, $v_{b,n}$ at a, b, each of height $\frac{h}{n}$ converge to 0 as $n \to \infty$: For e.g.,

$$\int_{v_{a,n}} f(z)\,dz = \int_0^1 f(a + \tfrac{iht}{n})\tfrac{ih}{n}\,dt = \tfrac{1}{n}\int_0^1 f(a + \tfrac{iht}{n})ih\,dt \xrightarrow{n\to\infty} 0\int_0^1 f(a)ih\,dt = 0.$$

Hence

$$\int_R f(z)\,dz = \int_s - \int_{s_n} + \int_{v_{b,n}} - \int_{v_{a,n}} + \int_{R_n} f(z)\,dz$$

$$= \int_s - \int_{s_n} + \int_{v_{b,n}} - \int_{v_{a,n}} f(z)\,dz + 0$$

$$= \lim_{n\to\infty} \left(\int_s - \int_{s_n} + \int_{v_{b,n}} - \int_{v_{a,n}} f(z)\,dz \right) = 0.$$

Solutions to the exercises from Chapter 4

Solution to Exercise 4.1. As $\sum\limits_{n=1}^{\infty} a_n$ converges, so do the real series $\sum\limits_{n=1}^{\infty} \mathrm{Re}\, a_n$ and $\sum\limits_{n=1}^{\infty} \mathrm{Im}\, a_n$. Hence $\lim\limits_{n\to\infty} \mathrm{Re}\, a_n = 0$ and $\lim\limits_{n\to\infty} \mathrm{Im}\, a_n = 0$. Thus $\lim\limits_{n\to\infty} a_n = 0$ too.

Solution to Exercise 4.2. As $\sum\limits_{n=1}^{\infty} |a_n|$ converges, and for all $n \in \mathbb{N}$, $|\mathrm{Re}\, a_n| \leqslant |a_n|$ and $|\mathrm{Im}\, a_n| \leqslant |a_n|$, it follows that $\sum\limits_{n=1}^{\infty} |\mathrm{Re}\, a_n|$ and $\sum\limits_{n=1}^{\infty} |\mathrm{Im}\, a_n|$ converge by the Comparison Test. Thus the two real series $\sum\limits_{n=1}^{\infty} \mathrm{Re}\, a_n$ and $\sum\limits_{n=1}^{\infty} \mathrm{Im}\, a_n$ are absolutely convergent and hence convergent (using the real analysis result that a real series which is absolutely convergent is convergent). So $\sum\limits_{n=1}^{\infty} a_n$ converges.

Solution to Exercise 4.3. Let $s_n := a_1 + \cdots + a_n$, $n \in \mathbb{N}$, and $L := \sum\limits_{n=1}^{\infty} a_n$. Then $(s_n)_{n\in\mathbb{N}}$ converges to L. Let $N \in \mathbb{N}$ be such that $|s_N - L| < \epsilon$. The subsequence $(s_{N+k})_{k\in\mathbb{N}}$ of $(s_n)_{n\in\mathbb{N}}$ converges to L too. Thus $(s_{N+k} - s_N)_{k\in\mathbb{N}}$ converges to $L - s_N$, i.e., $(a_{N+1} + \cdots + a_{N+k})_{k\in\mathbb{N}}$ converges to $L - s_N$. In other words, $\sum\limits_{n=N+1}^{\infty} a_n = L - s_N$. Moreover, $\left| \sum\limits_{n=N+1}^{\infty} a_n \right| = |L - s_N| = |s_N - L| < \epsilon$.

Solution to Exercise 4.4. The nth partial sum is $s_n := 1 + z + \cdots + z^{n-1}$. Thus $z s_n = z + z^2 + \cdots + z^{n-1} + z^n$, and so $(1-z)s_n = 1 - z^n$. Since $|z| < 1$, $z \neq 1$ and so $1 - z \neq 0$. So $s_n = 1 + z + \cdots + z^{n-1} = \frac{1-z^n}{1-z}$.

As $|z| < 1$, $\lim\limits_{n\to\infty} z^n = 0$ (Example 1.1, p. 17). Thus $\lim\limits_{n\to\infty} s_n = \lim\limits_{n\to\infty} \frac{1-z^n}{1-z} = \frac{1-0}{1-z}$, and so $\sum\limits_{n=0}^{\infty} z^n$ converges, with the sum $\sum\limits_{n=0}^{\infty} z^n = \lim\limits_{n\to\infty} s_n = \frac{1}{1-z}$.

Solution to Exercise 4.5. Let $\sigma_n := 1 + 2z + 3z^2 + \cdots + (n-1)z^{n-2} + nz^{n-1}$, $n \in \mathbb{N}$. Then $z\sigma_n = z + 2z^2 + \cdots + (n-1)z^{n-1} + nz^n$. Subtracting, we get $(1-z)\sigma_n = \sigma_n - z\sigma_n = 1 + z + z^2 + \cdots + z^{n-1} - nz^n = \frac{1-z^n}{1-z} - nz^n$. Thus $\sigma_n = \frac{1-z^n}{(1-z)^2} - \frac{nz^n}{1-z}$. (We can also obtain this by differentiating $1 + z + \cdots + z^n = \frac{1-z^{n+1}}{1-z}$ with respect to z.) If we set $r := |z|$, then $0 \leqslant r < 1$ and so $r = \frac{1}{1+h}$ where $h := \frac{1}{r} - 1 > 0$. We have $(1+h)^n = 1 + \binom{n}{1}h + \binom{n}{2}h^2 + \cdots + \binom{n}{n}h^n \geqslant \binom{n}{2}h^2 = \frac{n(n-1)}{2}h^2$. Hence $0 \leqslant nr^n = \frac{n}{(1+h)^n} \leqslant n\frac{2}{n(n-1)h^2} = \frac{2}{(n-1)h^2}$, and so by the Sandwich Theorem, $\lim\limits_{n\to\infty} nr^n = 0$. Consequently, $\lim\limits_{n\to\infty} \sigma_n = \lim\limits_{n\to\infty} \left(\frac{1-z^n}{(1-z)^2} - \frac{nz^n}{1-z} \right) = \frac{1-0}{(1-z)^2} - \frac{0}{1-z} = \frac{1}{(1-z)^2}$.

Solution to Exercise 4.6. We have

$$\left| \frac{1}{n^s} \right| = \left| \frac{1}{\exp(s\, \mathrm{Log}\, n)} \right| = \left| \frac{1}{\exp(s \log n)} \right| = \frac{1}{e^{\mathrm{Re}(s \log n)}} = \frac{1}{e^{(\log n)\mathrm{Re}\, s}} = \frac{1}{(e^{\log n})^{\mathrm{Re}\, s}} = \frac{1}{n^{\mathrm{Re}\, s}}.$$

Recall that $\sum\limits_{n=1}^{\infty} \frac{1}{n^p}$ converges if $p > 1$. Hence if $\mathrm{Re}\, s > 1$, then $\sum\limits_{n=1}^{\infty} \frac{1}{n^{\mathrm{Re}\, s}}$ converges. Thus $\sum\limits_{n=1}^{\infty} \frac{1}{n^s}$ converges absolutely for $\mathrm{Re}\, s > 1$, and so it converges for $\mathrm{Re}\, s > 1$.

Solution to Exercise 4.7. Suppose that $\sum_{n=0}^{\infty} c_n(z-z_0)^n$ converges for a $z_* \in \mathbb{C}$.
If $R < |z_* - z_0|$, then $\sum_{n=0}^{\infty} c_n(z-z_0)^n$ diverges at z_*, a contradiction. So $R \geqslant |z_* - z_0|$.
Suppose that $\sum_{n=0}^{\infty} c_n(z-z_0)^n$ diverges for a $z_* \in \mathbb{C}$. If $R > |z_* - z_0|$, then the series
$\sum_{n=0}^{\infty} c_n(z-z_0)^n$ converges at z_*, a contradiction. Thus $R \leqslant |z_* - z_0|$.

Solution to Exercise 4.8. Each term $|z^{n!}|$, $n = 1, \cdots, N$, appears in the list of
terms $|z^n|$, $n = 1, \cdots, N!$. As $|z^k| \geqslant 0$ for all $k \in \mathbb{N}$, it follows immediately that
$|z^1| + |z^{2!}| + 0 + 0 + 0 + |z^{3!}| + \cdots + |z^{N!}| \leqslant |z^1| + |z^2| + \cdots + |z^{N!}|$.
For $|z| < 1$, the geometric series converges: $\sum_{n=1}^{\infty} |z|^n = \frac{|z|}{1-|z|}$. So the increasing real
sequence $(\sum_{n=1}^{N} |z^{n!}|)_{N \in \mathbb{N}}$ is bounded: $\sum_{n=1}^{N} |z^{n!}| \leqslant \sum_{n=1}^{N!} |z^n| = \sum_{n=1}^{N!} |z|^n \leqslant \sum_{n=1}^{\infty} |z|^n = \frac{|z|}{1-|z|}$.
Thus $(\sum_{n=1}^{N} |z^{n!}|)_{N \in \mathbb{N}}$ converges, that is, $\sum_{n=1}^{\infty} z^{n!}$ converges absolutely. As absolutely
convergent complex series are convergent in $\mathbb{C}$, for $z \in \mathbb{D}$, $\sum_{n=1}^{\infty} z^{n!}$ converges in $\mathbb{C}$.
We show that the radius of convergence $R = 1$. From the above, it is enough to
show that $\sum_{n=1}^{\infty} z^{n!}$ diverges if $|z| > 1$. Suppose for a z with $|z| > 1$, $\sum_{n=1}^{\infty} z^{n!}$ converges.
If c_n denote the coefficients of the power series, then $(c_n z^n)_{n \in \mathbb{N}}$ converges to 0.
Hence the subsequence $(c_{n!} z^{n!})_{n \in \mathbb{N}} = (z^{n!})_{n \in \mathbb{N}}$ also converges to 0. So there exists
an N large enough so that for all $n > N$, $|z^{n!}| < 1$. But as $|z| > 1$, this leads to
the contradiction that for $n > N$, $1 < |z|^{n!} = |z^{n!}| < 1$.

Solution to Exercise 4.9.
$1°$ Let $L \neq 0$. For all z such that $|z| < \frac{1}{L}$, there exists a $q < 1$ and an N large
enough such that $\sqrt[n]{|c_n z^n|} = \sqrt[n]{|c_n|} \, |z| \leqslant q < 1$ for all $n > N$. This is because
$\sqrt[n]{|c_n|} \, |z| \xrightarrow{n \to \infty} L|z| < 1$. (E.g., take $q = \frac{L|z|+1}{2} < 1$.) So by the Root Test, the
power series is absolutely convergent for such z.
If $|z| > \frac{1}{L}$, then there exists an $N \in \mathbb{N}$ such that $\sqrt[n]{|c_n z^n|} = \sqrt[n]{|c_n|} \, |z| > 1$ for
all $n > N$. This is because $\sqrt[n]{|c_n|} \, |z| \xrightarrow{n \to \infty} L|z| > 1$. So again by the Root Test,
the power series diverges.
$2°$ Let $L = 0$. For $z \in \mathbb{C}$, there exists a $q < 1$ such that $\sqrt[n]{|c_n z^n|} = \sqrt[n]{|c_n|} \, |z| \leqslant q < 1$
for all $n > N$. This is because $\sqrt[n]{|c_n|} \, |z| \xrightarrow{n \to \infty} 0|z| = 0 < 1$. (E.g., $q = \frac{1}{2} < 1$.)
So again by the Root Test, the power series is absolutely convergent for such z.

Solution to Exercise 4.10. We have $\lim_{n \to \infty} \sqrt[n]{\frac{1}{n^n}} = \lim_{n \to \infty} \frac{1}{n} = 0$. So the radius of
convergence of $\sum_{n=1}^{\infty} \frac{z^n}{n^n}$ is infinite, and the power series converges for all $z \in \mathbb{C}$.

Solution to Exercise 4.11. For $z = 0$, the series converges with sum 0. For
$z \neq 0$, $|z| \neq 0$, and let $N \in \mathbb{N}$ be such that $N > \frac{1}{|z|}$. For $n > N$, $|nz| > N|z| > 1$,
giving $|n^n z^n - 0| = |nz|^n > 1^n = 1$, showing $\neg(\lim_{n \to \infty} n^n z^n = 0)$. So $\sum_{n=1}^{\infty} n^n z^n$ diverges.

Solution to Exercise 4.12.

As $\lim\limits_{n\to\infty}\left|\dfrac{\frac{(-1)^{n+1}}{n+1}}{\frac{(-1)^n}{n}}\right| = \lim\limits_{n\to\infty}\dfrac{n}{n+1} = 1$, the radius of convergence of $\sum\limits_{n=1}^{\infty}\dfrac{(-1)^n}{n}z^n$ is 1.

As $\lim\limits_{n\to\infty}\left|\dfrac{(n+1)^{2022}}{n^{2022}}\right| = \lim\limits_{n\to\infty}\left(1+\tfrac{1}{n}\right)^{2022} = 1$, the radius of convergence of $\sum\limits_{n=0}^{\infty}n^{2022}z^n$ is 1.

As $\lim\limits_{n\to\infty}\left|\dfrac{\frac{1}{(n+1)!}}{\frac{1}{n!}}\right| = \lim\limits_{n\to\infty}\dfrac{1}{n+1} = 0$, the radius of convergence of $\sum\limits_{n=0}^{\infty}\dfrac{1}{n!}z^n$ is infinite.

Solution to Exercise 4.13.

(1) Let $(\sqrt[n]{|c_n|})_{n\in\mathbb{N}}$ be unbounded. Suppose for some $z\in\mathbb{C}\backslash\{0\}$, $\sum\limits_{n=0}^{\infty}c_nz^n$ converges. Then $\lim\limits_{n\to\infty}c_nz^n = 0$, and so $(c_nz^n)_{n\in\mathbb{N}}$ is bounded. Let $M>0$ be such that for all $n\in\mathbb{N}$, $|c_nz^n|\leqslant M$. Thus $\sqrt[n]{|c_n|}\leqslant\dfrac{1}{|z|\sqrt[n]{M}}$, a contradiction. So for all $z\in\mathbb{C}\backslash\{0\}$, $\sum\limits_{n=0}^{\infty}c_nz^n$ diverges, i.e., the radius of convergence of $\sum\limits_{n=0}^{\infty}c_nz^n$ is 0.

(2) We have: $\begin{aligned}M_1 &= \sup\{|c_1|, \sqrt{|c_2|}, \sqrt[3]{|c_3|},\cdots\},\\ M_2 &= \sup\{\quad\;\; \sqrt{|c_2|}, \sqrt[3]{|c_3|},\cdots\},\\ M_3 &= \sup\{\qquad\qquad\; \sqrt[3]{|c_3|},\cdots\},\\ &\quad\cdots.\end{aligned}$

$(M_n)_{n\in\mathbb{N}}$ converges to $L := \inf\limits_{n\in\mathbb{N}}M_n$ (as it's decreasing and bounded below by 0).

If $z = 0$, then convergence of the power series is obvious.

Let $z\in\mathbb{C}\backslash\{0\}$ satisfy $\frac{1}{|z|} > L$. As there is a gap between $\frac{1}{|z|}$ and L, there exists an α such that $L = \inf\limits_{n\in\mathbb{N}}M_n < \alpha < \frac{1}{|z|}$. Then there exists an $N\in\mathbb{N}$ such that for all $n>N$, $M_n\leqslant M_N < \alpha$. Thus for all $n>N$, $\sqrt[n]{|c_n|} < \alpha < \frac{1}{|z|}$, i.e., $\sqrt[n]{|c_nz^n|} < \alpha|z| =: r < 1$. By the Root Test, $\sum\limits_{n=0}^{\infty}c_nz^n$ is absolutely convergent. Hence we have shown:

- If $L>0$, then $\sum\limits_{n=0}^{\infty}c_nz^n$ is absolutely convergent for all $z\in D(0,L)$.
- If $L=0$, then $\sum\limits_{n=0}^{\infty}c_nz^n$ is absolutely convergent for all $z\in\mathbb{C}$.

It remains to show that if $L>0$, then we have divergence for $|z|>\frac{1}{L}$. If $|z|>\frac{1}{L}$, then $\frac{1}{|z|} < L = \inf\limits_{n\in\mathbb{N}}M_n$. So for all $n\in\mathbb{N}$, $\frac{1}{|z|} < M_n = \sup\{\sqrt[n]{|c_n|}, \sqrt[n+1]{|c_{n+1}|},\cdots\}$, and in particular, there exists an $m_n > n$ such that $|c_{m_n}z^{m_n}| > 1$. So for all $n\in\mathbb{N}$, there exists an $m_n > n$ such that $|c_{m_n}z^{m_n}| > 1$. Thus it is *not* the case that $\lim\limits_{n\to\infty}c_nz^n = 0$, and hence $\sum\limits_{n=0}^{\infty}c_nz^n$ diverges.

Solution to Exercise 4.14. For $n\geqslant 0$, let A_n, B_n, C_n be $\sum\limits_{k=0}^{n}a_k, \sum\limits_{k=0}^{n}b_k, \sum\limits_{k=0}^{n}c_k$. Then:

$$C_{2n} - A_nB_n = \begin{aligned}&a_0(b_{n+1}+\cdots+b_{2n}) + a_1(b_{n+1}+\cdots+b_{2n-1})+\cdots+a_{n-1}(b_{n+1})\\ &+b_0(a_{n+1}+\cdots+a_{2n})+b_1(a_{n+1}+\cdots+a_{2n-1})+\cdots+b_{n-1}(a_{n+1}).\end{aligned}$$

See the following picture.

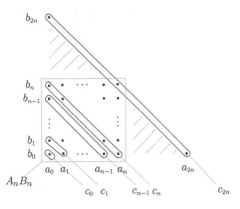

So
$$|C_{2n} - A_n B_n| \leqslant |a_0|(|b_{n+1}| + \ldots) + |a_1|(|b_{n+1}| + \ldots) + \ldots + |a_{n-1}|(|b_{n+1}| + \ldots)$$
$$+ |b_0|(|a_{n+1}| + \ldots) + |b_1|(|a_{n+1}| + \ldots) + \ldots + |b_{n-1}|(|a_{n+1}| + \ldots)$$
$$\leqslant \left(\sum_{k=0}^{\infty} |a_k|\right) \sum_{k=n+1}^{\infty} |b_k| + \left(\sum_{k=0}^{\infty} |b_k|\right) \sum_{k=n+1}^{\infty} |a_k| \xrightarrow{n \to \infty} 0.$$

Thus $\lim_{n \to \infty} C_{2n} = \lim_{n \to \infty} A_n B_n = \lim_{n \to \infty} A_n \lim_{n \to \infty} B_n = \left(\sum_{n=0}^{\infty} a_n\right)\left(\sum_{n=0}^{\infty} b_n\right)$. Similarly, one can

show that $|C_{2n+1} - A_n B_n| \xrightarrow{n \to \infty} 0$, and so $\lim_{n \to \infty} C_{2n+1} = \lim_{n \to \infty} A_n B_n = \left(\sum_{n=0}^{\infty} a_n\right)\left(\sum_{n=0}^{\infty} b_n\right)$.

Consequently, $\sum_{n=0}^{\infty} c_n = \lim_{n \to \infty} C_n = \left(\sum_{n=0}^{\infty} a_n\right)\left(\sum_{n=0}^{\infty} b_n\right)$.

Solution to Exercise 4.15. In one jump, the 'edge-distance' (i.e., smallest number of edges to E) changes by 1. As the initial edge-distance is 4, which is even, an odd number of jumps cannot reduce it to 0. Thus $a_{2n-1} = 0$ for all $n \in \mathbb{N}$.

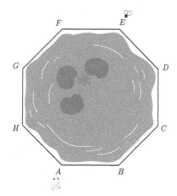

In two jumps, the frog can only stay at A, or reach C or G, and so $a_2 = 0$. In four jumps, the frog can reach E either following the path $ABCDE$ or $AHGFE$, and so $a_4 = 2$.

b_n is also the number of paths of length n from C to E. For a path of $2n$ jumps starting from A, the first two jumps can land the frog back at A in two ways (ABA or AHA), or at C in one way (ABC), or at G in one way (AHG). Thus $a_{2n} = 2a_{2n-2} + 2b_{2n-2}$ for all $n > 1$. Next, for a path of $2n > 2$ jumps starting from C, the first two jumps can land the frog back at C in two ways (CDC or CBC), or at A in one way (CBA). Thus $b_{2n} = 2b_{2n-2} + a_{2n-2}$ for all $n > 1$.

From $a_{2n} = 2a_{2n-2} + 2b_{2n-2}$, we obtain $b_{2n-2} = \frac{a_{2n} - 2a_{2n-2}}{2}$. So $b_{2n} = \frac{a_{2n+2} - 2a_{2n}}{2}$. Substituting these expressions for b_{2n-2} and b_{2n} in $b_{2n} = 2b_{2n-2} + a_{2n-2}$, we get $a_{2n+2} = 4a_{2n} - 2a_{2n-2}$ for all $n > 1$.

We have $A(z) = a_0 + a_2 z + a_4 z^2 + a_6 z^3 + \cdots = 2z^2 + a_6 z^3 + \cdots$. For all $n > 1$, $a_{2n+2} = 4a_{2n} - 2a_{2n-2}$, and so we get $a_{2n+4} = 4a_{2n+2} - 2a_{2n}$ for all $n \in \mathbb{N}$. Multiplying by z^{n+2} and adding, we obtain $A(z) - 2z^2 = 4zA(z) - 2z^2 A(z)$, and so $(1 - 4z + 2z^2)A(z) = 2z^2$, i.e., $(1 - \alpha z)(1 - \beta z)A(z) = 2z^2$, where $\alpha = 2 - \sqrt{2}$, $\beta = 2 + \sqrt{2}$. Hence for $|z|$ small enough, and with $c_n = \sum\limits_{k=0}^{n} \alpha^{n-k} \beta^k$,

$$A(z) = \frac{2z^2}{(1-\alpha z)(1-\beta z)} = 2z^2(1 + \alpha z + \alpha^2 z^2 + \cdots)(1 + \beta z + \beta^2 z^2 + \cdots) = 2z^2 \sum_{n=0}^{\infty} c_n z^n,$$

'Comparing coefficients' (justified by Corollary 4.5; see Remark 4.2), for $n \geqslant 2$,

$$a_{2n} = 2c_{n-2} = 2\sum_{k=0}^{n-2} \alpha^{n-2-k} \beta^k = 2(\alpha^{n-2} + \alpha^{n-3}\beta + \cdots + \alpha\beta^{n-3} + \beta^{n-2})\frac{\beta - \alpha}{\beta - \alpha}$$
$$= 2\frac{\beta^{n-1} - \alpha^{n-1}}{\beta - \alpha} = \frac{(2+\sqrt{2})^{n-1} - (2-\sqrt{2})^{n-1}}{\sqrt{2}}.$$

If $n = 1$, then the right-hand side is 0, which matches $a_2 = 0$. So the formula holds for all $n \in \mathbb{N}$.

As a_{2n} is the number of paths, a_{2n} is a nonnegative integer, and so the right-hand side is necessarily a nonnegative integer. But this also follows from the Binomial Theorem, since

$$(2 + \sqrt{2})^{n-1} - (2 - \sqrt{2})^{n-1} = \sum_{k=0}^{n-1} \binom{n-1}{k} \sqrt{2}^k 2^{n-1-k} - \sum_{k=0}^{n-1} \binom{n-1}{k}(-1)^k \sqrt{2}^k 2^{n-1-k}$$
$$= \sum_{k=0}^{n-1} \binom{n-1}{k}(1 - (-1)^k)\sqrt{2}^k 2^{n-1-k}$$

As $1 - (-1)^k$ is 0 for even k, the only terms that survive are the ones with an odd k. If $k = 2m + 1 \leqslant n - 1$, m a nonnegative integer, then the corresponding summand is $\binom{n-1}{2m+1}(2)2^m \sqrt{2}\, 2^{n-1-(2m+1)}$, which is equal to $\sqrt{2}$ times a nonnegative integer. So when the sum of such terms is divided by $\sqrt{2}$, we get a nonnegative integer.

Solution to Exercise 4.16. $f(z) := 1 + 2z + 3z^2 + 4z^3 + \cdots = \frac{1}{(1-z)^2}$ for $|z| < 1$. Multiplying by z, we get $zf(z) = g(z) := z + 2z^2 + 3z^3 + 4z^4 + \cdots = \frac{z}{(1-z)^2}$ for $|z| < 1$. As $g(z) := z + 2z^2 + 3z^3 + 4z^4 + \cdots$ converges for $|z| < 1$, g is holomorphic in the disc $D(0,1)$, with $g'(z) = 1 + 2^2 z + 3^2 z^2 + 4^2 z^3 + \cdots$ for $|z| < 1$. But $g'(z) = \frac{d}{dz}\frac{z}{(1-z)^2} = 1\frac{1}{(1-z)^2} + z\frac{2}{(1-z)^3} = \frac{1-z+2z}{(1-z)^3} = \frac{1+z}{(1-z)^3}$. Hence $1 + 2^2 z + 3^2 z^2 + 4^2 z^3 + \cdots = \frac{1+z}{(1-z)^3}$ for $|z| < 1$.

Solution to Exercise 4.17.

(1) False. For example, $\{z \in \mathbb{C} : \sum\limits_{n=1}^{\infty} \frac{z^n}{n^2} \text{ converges}\} = \{z \in \mathbb{C} : |z| \leqslant 1\}$.

(2) True.

(3) False. For example $\sum\limits_{n=1}^{\infty} \frac{(-1)^n}{n} z^n$ converges for $z = 1$, but diverges for $z = -1$.

(4) False. See the example in (3).

(5) True. Same example as in (3).

(6) True. For example, consider $\sum\limits_{n=1}^{\infty} \frac{z^n}{n^2}$.

(7) True. The radius of convergence R satisfies $R \leqslant |i| = 1$, and $|1+i| = \sqrt{2} > 1 = R$.

Solution to Exercise 4.18. As $a \in (0,1)$, none of the coefficients is 0. We have

$$\lim_{n\to\infty} \left| \frac{c_{n+1}}{c_n} \right| = \lim_{n\to\infty} \left| \frac{\frac{a(a-1)\cdots(a-n)}{(n+1)!}}{\frac{a(a-1)\cdots(a-(n-1))}{n!}} \right| = \lim_{n\to\infty} \left| \frac{a-n}{n+1} \right| = \lim_{n\to\infty} \left| \frac{\frac{a}{n}-1}{1+\frac{1}{n}} \right| = \left| \frac{0-1}{1+0} \right| = 1.$$

So the radius of convergence is $\frac{1}{\lim\limits_{n\to\infty} \left| \frac{c_{n+1}}{c_n} \right|} = \frac{1}{1} = 1$.

As $R = 1$, $D(0,R) = \{z \in \mathbb{C} : |z| < 1\}$. By termwise differentiation, for $|z| < 1$,

$$f'(z) = 0 + \frac{a}{1!} 1 + \frac{a(a-1)}{2!} 2z + \frac{a(a-1)(a-2)}{3!} 3z^2 + \cdots = a + \frac{a(a-1)}{1!} z + \frac{a(a-1)(a-2)}{2!} z^2 + \cdots .$$

Multiplying by z, $zf'(z) = \frac{a}{1!} z + \frac{a(a-1)}{2!} 2z^2 + \frac{a(a-1)(a-2)}{3!} 3z^3 + \cdots$. So for all $|z| < 1$,

$$(1+z)f'(z) = a + \frac{a(a-1)}{1!} z + \cdots + \frac{a(a-1)\cdots(a-n)}{n!} z^n + \cdots$$
$$+ \frac{a}{1!} z + \cdots + \frac{a(a-1)\cdots(a-(n-1))}{n!} nz^n + \cdots$$
$$= a + a^2 z + \cdots + \frac{a(a-1)\cdots(a-(n-1))}{n!} (a - n + n) z^n + \cdots$$
$$= a + a^2 z + \cdots + \frac{a^2(a-1)\cdots(a-(n-1))}{n!} z^n + \cdots$$
$$= a(1 + az + \cdots + \frac{a(a-1)\cdots(a-(n-1))}{n!} z^n + \cdots) = af(z).$$

For $z \neq -1$, we have $(1+z)^a = \exp(a\mathrm{Log}(1+z))$. As $\mathrm{Log} \in \mathbb{C} \backslash ((-\infty, 0] \times \{0\})$, we have $\mathrm{Log}(1 + \cdot) \in \mathcal{O}(\mathbb{C}\backslash((-\infty, -1] \times \{0\}))$. In particular, $\mathrm{Log}(1 + \cdot) \in \mathcal{O}(D(0,1))$. Thus $(1+z)^a \in \mathcal{O}(D(0,1))$. For $z \in D(0,1)$,

$$((1+\cdot)^{-a} f)'(z) = (\exp(-a\mathrm{Log}(1 + \cdot)))'(z) f(z) + \exp(-a\mathrm{Log}(1+z)) f'(z)$$
$$= \exp(-a\mathrm{Log}(1+z)) \frac{-a}{1+z} 1 f(z) + \exp(-a\mathrm{Log}(1+z)) \frac{af(z)}{1+z}$$
$$= \exp(-a\mathrm{Log}(1+z)) (\frac{-a}{1+z} f(z) + \frac{af(z)}{1+z}) = 0.$$

By Exercise 2.18, $(1 + \cdot)^{-a} f(\cdot)$ is constant in $D(0,1)$. Hence for all $z \in D(0,1)$, $(1+z)^{-a} f(z) = (1+0)^{-a} f(0) = 1$. Thus $f(z) = (1+z)^a$ for all $z \in D(0,1)$.

Solution to Exercise 4.19. Consider first $\sum\limits_{n=0}^{\infty} c_n w^n$, where $c_n = \frac{(-1)^n}{2^{2n}(n!)^2}$. We have

$$\left| \frac{c_{n+1}}{c_n} \right| = \left| \frac{\frac{(-1)^{n+1}}{2^{2(n+1)}((n+1)!)^2}}{\frac{(-1)^n}{2^{2n}(n!)^2}} \right| = \frac{2^{2n} \cdot n! \cdot n!}{2^{2n} \cdot 2^2 \cdot (n+1)! \cdot (n+1)!} = \frac{1}{4(n+1)^2} \xrightarrow{n\to\infty} 0. \text{ So } \sum_{n=0}^{\infty} \frac{(-1)^n}{2^{2n}(n!)^2} w^n$$

converges everywhere. For $z \in \mathbb{C}$, putting $w = z^2$, $J_0(z) = \sum\limits_{n=0}^{\infty} \frac{(-1)^n}{2^{2n}(n!)^2} z^{2n}$ converges.

Power series can be differentiated termwise inside the disc of their convergence. So for $z \in \mathbb{C}\backslash\{0\}$, $J_0'(z) = \sum\limits_{n=1}^{\infty} \frac{(-1)^n}{2^{2n}(n!)^2} 2n z^{2n-1}$, $J_0''(z) = \sum\limits_{n=1}^{\infty} \frac{(-1)^n}{2^{2n}(n!)^2} 2n(2n-1)z^{2n-2}$, and

$$J_0''(z) + \tfrac{1}{z} J_0'(z) = \sum_{n=1}^{\infty} \frac{(-1)^n}{2^{2n}(n!)^2} 2n((2n-1)+1)z^{2n-2} = (-1)\sum_{n=1}^{\infty}\frac{(-1)^{n-1}}{2^{2n}(n!)^2}2n(2n)z^{2(n-1)}$$

$$= (-1)\sum_{n=1}^{\infty}\frac{(-1)^{n-1}}{2^{2(n-1)}((n-1)!)^2}z^{2(n-1)} = (-1)\sum_{m=0}^{\infty}\frac{(-1)^m}{2^{2m}(m!)^2}z^{2m} = -J_0(z).$$

Solution to Exercise 4.20. As $\sum\limits_{n=0}^{\infty} c_n z^n$ is absolutely convergent for $z = 1$, it is convergent for $z = 1$, and the radius of convergence $R \geqslant 1$. If $R > 1$, then with $f(z) := \sum\limits_{n=0}^{\infty} c_n z^n$ for $z \in D(0,R)$, we have $f \in \mathcal{O}(D(0,R))$, and $f'(z) = \sum\limits_{n=1}^{\infty} n c_n z^{n-1}$ in $D(0,R)$. So $\sum\limits_{n=1}^{\infty} n c_n z^{n-1}$ is absolutely convergent for all $z \in D(0,R)$. In particular, as $R > 1$, with $z = 1$, we get $\sum\limits_{n=1}^{\infty} n|c_n|$ converges, a contradiction. Thus $R = 1$.

Solution to Exercise 4.21. As $\frac{d^{2n}}{dz^{2n}}\sin z = (-1)^n \sin z$, $\frac{d^{2n+1}}{dz^{2n+1}}\sin z = (-1)^n \cos z$, $\sin 0 = 0$ and $\cos 0 = 1$, we have $\sin z = \sum\limits_{n=0}^{\infty} \frac{1}{n!}(\frac{d^n}{dz^n}\sin z)|_{z=0} z^n = z - \frac{z^3}{3!} + \frac{z^5}{5!} - + \cdots$.
Similarly, $\cos z = 1 - \frac{z^2}{2!} + \frac{z^4}{4!} - + \cdots$.
Alternatives: As $\cos z = \frac{\exp(iz)+\exp(-iz)}{2} = \frac{1}{2}(\sum\limits_{n=0}^{\infty}\frac{i^n}{n!}z^n + \sum\limits_{n=0}^{\infty}\frac{(-1)^n i^n}{n!}z^n)$, and $i^{2n} = (-1)^n$,

$$\cos z = \tfrac{1}{2}\big(1 + iz - \tfrac{z^2}{2!} - \tfrac{iz^3}{3!} + \tfrac{z^4}{4!} + \tfrac{iz^5}{5!} - \tfrac{z^6}{6!} + \cdots$$
$$+ 1 - iz - \tfrac{z^2}{2!} + \tfrac{iz^3}{3!} + \tfrac{z^4}{4!} - \tfrac{iz^5}{5!} - \tfrac{z^6}{6!} + \cdots\big)$$
$$= 1 - \tfrac{z^2}{2!} + \tfrac{z^4}{4!} - \tfrac{z^6}{6!} + - \cdots.$$

Or termwise differentiate the power series for sin:
$$\cos z = \sin' z = \sum_{n=0}^{\infty}\frac{(-1)^n}{(2n+1)!}(2n+1)z^{2n} = \sum_{n=0}^{\infty}\frac{(-1)^n}{(2n)!}z^{2n} = 1 - \tfrac{z^2}{2!} + \tfrac{z^4}{4!} - \tfrac{z^6}{6!} + - \cdots.$$

Solution to Exercise 4.22. Let $p(z) = z^6 - z^4 + z^2 - 1$, $z \in \mathbb{C}$. Then
$$p'(z) = 6z^5 - 4z^3 + 2z \qquad p'''(z) = 120z^3 - 24z \qquad p^{(5)}(z) = 720z \qquad p^{(7)}(z) = 0$$
$$p''(z) = 30z^4 - 12z^2 + 2 \qquad p^{(4)}(z) = 360z^2 - 24 \qquad p^{(6)}(z) = 720 \qquad \cdots.$$
So $p(1) = 0$, $\frac{p'(1)}{1!} = 4$, $\frac{p''(1)}{2!} = 10$, $\frac{p'''(1)}{3!} = 16$, $\frac{p^{(4)}(1)}{4!} = 14$, $\frac{p^{(5)}(1)}{5!} = 6$, $\frac{p^{(6)}(1)}{6!} = 1$. So $\forall z \in \mathbb{C}$,
$$z^6 - z^4 + z^2 - 1 = p(1) + \frac{p'(1)}{1!}(z-1) + \cdots + \frac{p^{(6)}(1)}{6!}(z-1)^6 + 0$$
$$= 4(z-1) + 10(z-1)^2 + 16(z-1)^3 + 14(z-1)^4 + 6(z-1)^5 + (z-1)^6.$$
Or, use Binomial Theorem: $p(z) = ((z-1)+1)^6 - ((z-1)+1)^4 + (z-1+1)^2 - 1$.

Solution to Exercise 4.23.
(1) $z \mapsto \exp(z^2)$ has a primitive, say g, in the simply connected domain $\mathbb{C}$. Thus $f(z) = \int_{[0,z]}\exp(\zeta^2)d\zeta = \int_{[0,z]}g'(\zeta)d\zeta = g(z) - g(0)$. So $f \in \mathcal{O}(\mathbb{C})$ and $f'(z) = g'(z) = \exp(z^2) = \sum\limits_{n=0}^{\infty}\frac{1}{n!}z^{2n}$. Thus for all $n \geqslant 0$, $\frac{1}{(2n)!}\frac{d^{2n}}{dz^{2n}}f'(z)|_{z=0} = \frac{1}{n!}$ and $\frac{1}{(2n+1)!}\frac{d^{2n+1}}{dz^{2n+1}}f'(z)|_{z=0} = 0$, i.e., $f^{(2n+1)}(0) = \frac{(2n)!}{n!}$ and $f^{(2n+2)}(0) = 0$. Also, $f(0) = \int_{[0,0]}\exp(\zeta^2)d\zeta = 0$. So $f(z) = \sum\limits_{m=0}^{\infty}\frac{f^{(m)}(0)}{m!}z^m = \sum\limits_{n=0}^{\infty}\frac{f^{(2n+1)}(0)}{(2n+1)!}z^{2n+1} = \sum\limits_{n=0}^{\infty}\frac{z^{2n+1}}{(2n+1)(n!)}$.

(2) For $|z| < 1$, $\frac{1}{z+1} = 1 - z + z^2 - z^3 + z^4 - + \cdots$. As power series are holomorphic in the region of convergence with complex derivative obtained by termwise differentiation, $-\frac{1}{(z+1)^2} = \frac{d}{dz}\frac{1}{z+1} = -1 + 2z - 3z^2 + 4z^3 - + \cdots$ for $|z| < 1$. Multiplying by $-z^2$ gives $\frac{z^2}{(z+1)^2} = z^2 - 2z^3 + 3z^4 - + \cdots = \sum_{n=2}^{\infty} (-1)^n (n-1) z^n$ for $|z| < 1$. So we have $c_0 = c_1 = 0$ and $c_n = (-1)^n (n-1)$ for $n \geqslant 2$.

Solution to Exercise 4.24. Let $C_r(t) = r e^{2\pi i t}$, $t \in [0,1]$. Then for all integers $n \geqslant 0$, we have $c_n = \frac{1}{2\pi i} \int_{C_r} \frac{f(\zeta)}{\zeta^{n+1}} d\zeta$. Consequently, by the ML-inequality, we get $|c_n| \leqslant \frac{2\pi r}{2\pi} \max_{\zeta \in C_r} \frac{|f(\zeta)|}{r^{n+1}} \leqslant r \frac{1}{r^{n+1}} = \frac{1}{r^n}$. Passing to the limit as $r \nearrow 1$, $|c_n| \leqslant \lim_{r \to 1^-} \frac{1}{r^n} = 1$.

Solution to Exercise 4.25. Suppose $f \in \mathcal{O}(D(0,r))$ is such that $f^{(n)}(0) = (n!)^2$ for all $n \geqslant 0$. Then for $z \in D(0,r)$, $f(z) = \sum_{n=0}^{\infty} \frac{f^{(n)}(0)}{n!} z^n = \sum_{n=0}^{\infty} \frac{(n!)^2}{n!} z^n = \sum_{n=0}^{\infty} n! z^n$. Setting $z = \frac{r}{2} \in D(0,r)$, $\sum_{n=0}^{\infty} n! (\frac{r}{2})^n$ converges. If $N \in \mathbb{N}$ is such that $N\frac{r}{2} > 1$, then for all $n > N$, $n!(\frac{r}{2})^n = N!(N+1) \cdots (N+(n-N))(\frac{r}{2})^N (\frac{r}{2})^{n-N} > N!(\frac{r}{2})^N 1^{n-N}$, showing that $\neg(\lim_{n\to\infty} n!(\frac{r}{2})^n = 0)$, a contradiction. Hence, there is no $f \in \mathcal{O}(D(0,r))$ such that $f^{(n)}(0) = (n!)^2$ for all $n \geqslant 0$.

Solution to Exercise 4.26. Let the Taylor series of f be given by $f(z) = \sum_{n=0}^{\infty} c_n z^n$. The series is (absolutely) convergent for all $z \in \mathbb{C}$ since f is entire. As $f(z^2) = f(z)$, we have $c_0 + c_1 z^2 + c_2 z^4 + \cdots = c_0 + c_1 z + c_2 z^2 + \cdots$, that is, for all $z \in \mathbb{C}$, $c_1 z + (c_2 - c_1) z^2 + \cdots + c_{2n+1} z^{2n+1} + (c_{2n+2} - c_{n+1}) z^{2n+2} + \cdots = 0$. Since the right-hand side is the zero function, which has the Taylor expansion given by the left-hand side, it follows that the kth coefficient of the left-hand side power series is zero (being $\frac{0^{(k)}(0)}{k!}$) for all integers $k \geqslant 0$. So we get $0 = c_1 = c_3 = c_5 = \cdots$ and $c_2 - c_1 = 0$, $c_4 - c_2 = 0, \cdots$, $c_{2n+2} - c_{n+1} = 0, \cdots$, so that all the coefficients $c_n = 0$ for $n \in \mathbb{N}$. Thus $f(z) = c_0$ for all $z \in \mathbb{C}$, that is, f is a constant. Conversely, if f is constant, it satisfies the given relation and is entire.

Solution to Exercise 4.27. For $\delta > 0$, let $S_k := \{z \in \mathbb{C} : |z - a_k| + \mathrm{Re}(z - a_k) \leqslant \delta\}$. Then S_k is a 'parabolic sector': Writing $z - a_k = x + iy$, the above qualifying inequality becomes $x \leqslant \frac{\delta}{2} - \frac{y^2}{2\delta}$, and a picture of S_k is displayed below.

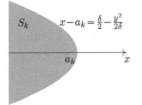

S_k $x - a_k = \frac{\delta}{2} - \frac{y^2}{2\delta}$

a_k x

The Taylor series centered at a_k of $f \in \mathcal{O}(\mathbb{C})$ is $f(z) = \sum_{n=0}^{\infty} \frac{f^{(n)}(a_k)}{n!}(z - a_k)^n$ $(z \in \mathbb{C})$.

In particular,
$$|f(z)| = |\sum_{n=0}^{\infty} \frac{f^{(n)}(a_k)}{n!}(z-a_k)^n| \leq \sum_{n=0}^{\infty} \frac{|f^{(n)}(a_k)|}{n!}|z-a_k|^n$$
$$\leq \sum_{n=0}^{\infty} \frac{e^{-a_k}}{n!}|z-a_k|^n = e^{-a_k}\sum_{n=0}^{\infty} \frac{1}{n!}|z-a_k|^n = e^{-a_k}e^{|z-a_k|}.$$

Thus for $z \in S_k$, $|\frac{f(z)}{e^{-z}}| \leq e^{\text{Re}\,z}e^{-a_k}e^{|z-a_k|} \leq e^{\delta}$.

As $\lim_{k\to\infty} a_k = \infty$, the sectors S_k (which are all congruent, being shifted-versions of each other) together cover $\mathbb{C}$, i.e., $\mathbb{C} = \bigcup_{k=0}^{\infty} S_k$. Hence

$$\text{for } z \in \mathbb{C} = \bigcup_{k=0}^{\infty} S_k, \quad |\frac{f(z)}{e^{-z}}| \leq e^{\delta}. \qquad (\star)$$

By Liouville's theorem, the bounded entire function $z \mapsto \frac{f(z)}{e^{-z}}$ is a constant. So there exists a $c \in \mathbb{C}$ such that $f(z) = ce^{-z}$ ($z \in \mathbb{C}$). Substituting this expression for f in $(\star)$ yields $|c| \leq e^{\delta}$. As $\delta > 0$ was arbitrary, we obtain $|c| \leq \inf_{\delta>0} e^{\delta} = e^0 = 1$.

Solution to Exercise 4.28. Let $z \in \mathbb{C}$. Let $R > |z|$. Then
$$|f^{(n+1)}(z)| \leq \frac{(n+1)!}{R^{n+1}}\max_{|z|\leq R}|f(z)| \leq \frac{(n+1)!}{R^{n+1}}\max_{|z|\leq R}M|z|^n = \frac{(n+1)!}{R^{n+1}}MR^n = \frac{(n+1)!M}{R}.$$
But the choice of $R > |z|$ was arbitrary, and so $f^{(n+1)}(z) = 0$. Since $z \in \mathbb{C}$ was arbitrary, we obtain that $f^{(n+1)} \equiv 0$ in $\mathbb{C}$. By Taylor's Theorem, for all $z \in \mathbb{C}$,
$$f(z) = \sum_{k=0}^{\infty} \frac{f^{(k)}(0)}{k!}(z-0)^k = \sum_{k=0}^{n} \frac{f^{(k)}(0)}{k!}z^k. \text{ So } f \text{ is a polynomial of degree at most } n.$$

If $n = 0$, then f is a bounded entire function, and our conclusion obtained above says that f is constant. So the special case when $n = 0$ is Liouville's Theorem.

Solution to Exercise 4.29. $\int_C \frac{\sin z}{z^{2023}}dz = 0$: By the Cauchy Integral Formula,
$$\frac{2022!}{2\pi i}\int_C \frac{\sin z}{(z-0)^{2022+1}}dz = \frac{d^{2022}}{dz^{2022}}\sin z|_{z=0} = (-1)^{\frac{2022}{2}}\sin z|_{z=0} = 0.$$

Solution to Exercise 4.30. We have $f(z_0) = 0^m g(z_0) = 0$, showing that z_0 is a zero of f. We claim that z_0 is an isolated zero of f. By the continuity of g at z_0, there exists a $r \in (0,R)$, such that $g(z) \neq 0$ for $z \in D(z_0,r)$. From the factorisation $f(z) = (z-z_0)^m g(z)$ for all $z \in D(z_0,R)$, we obtain $f(z) \neq 0$ for all $z \in D(z_0,r)\backslash\{z_0\}$. So z_0 is an isolated zero, i.e., we are in case 2° of the Theorem on Classification of Zeroes. Hence there exists an $\tilde{m} \in \mathbb{N}$ (which is the order of z_0 as a zero of f), and there exists a $\tilde{g} \in \mathcal{O}(D(z_0,R))$, such that $\tilde{g}(z_0) \neq 0$ and $f(z) = (z-z_0)^{\tilde{m}}\tilde{g}(z)$ for all $z \in D(z_0,R)$. Then for $z \in D(z_0,R)$, we have $(z-z_0)^{\tilde{m}}\tilde{g}(z) = (z-z_0)^m g(z)$. We show that this implies $\tilde{m} = m$. For if $\tilde{m} > m$, then for all $z \in D(z_0,R)\backslash\{z_0\}$, $(z-z_0)^{\tilde{m}-m}\tilde{g}(z) = g(z)$, and we get the contradiction
$$0 \neq g(z_0) = \lim_{z\to z_0} g(z) = \lim_{z\to z_0}(z-z_0)^{\tilde{m}-m}\tilde{g}(z) = 0\cdot\tilde{g}(z_0) = 0.$$

Similarly, $m > \tilde{m}$ also yields a contradiction. Consequently, $m = \tilde{m}$, and so z_0 is a zero of order $m = \tilde{m}$. (Also, then $g \equiv \tilde{g}$ on $D(z_0,R)\backslash\{z_0\}$, and since they are continuous, also on $D(z_0,R)$.)

Solution to Exercise 4.31.

(1) $f(z) = (1 + z^2)^4 = ((z - i)(z + i))^4 = (z - i)^4 (z + i)^4$ for all $z \in \mathbb{C}$. So with $g(z) := (z + i)^4$, g is entire, $g(i) = (2i)^4 = 16 \neq 0$ and $f(z) = (z - i)^4 g(z)$. Thus i is a zero of f of order 4.

(2) For $n \in \mathbb{Z}$, $f(2n\pi i) = 1 - 1 = 0$, and $f'(2n\pi i) = \exp z|_{z = 2n\pi i} = 1 \neq 0$. So $2n\pi i$ is a zero of f of order 1.

(3) $f(0) = \cos 0 - 1 + \frac{1}{2}(\sin 0)^2 = 1 - 1 + \frac{1}{2}(0)^2 = 0$, and we have

$$f(z) = \cos z - 1 + \frac{1}{2}(\sin z)^2 = \cos z - 1 + \frac{1}{2}\frac{(1 - \cos(2z))}{2}$$
$$= \cos z - \frac{3}{4} - \frac{1}{4}\cos(2z)$$
$$= (1 - \frac{z^2}{2!} + \frac{z^4}{4!} - \frac{z^6}{6!} + - \cdots) - \frac{3}{4} - \frac{1}{4}(1 - \frac{4z^2}{2!} + \frac{16z^4}{4!} - \frac{2^6 z^6}{6!} + - \cdots)$$
$$= (1 - \frac{3}{4} - \frac{1}{4}) + (-\frac{1}{2!} + \frac{1}{4}\frac{4}{2!})z^2 + (\frac{1}{4!} - \frac{1}{4}\frac{16}{4!})z^4 + \cdots$$
$$= 0 + 0z^2 - \frac{3}{4!}z^4 + \cdots,$$

and so z_0 is a zero of order 4.

Solution to Exercise 4.32. In the disc, there is only one zero of f, which is at z_0, and so for z in the disc such that $z \neq z_0$, we have $f(z) \neq 0$. By the result on the classification of zeroes, there exists a function g that is holomorphic in the disc, $g(z_0) \neq 0$, and $f(z) = (z - z_0)g(z)$ for all z in the disc. It follows that g is nonzero at each point of C. Thus

$$\frac{1}{2\pi i}\int_C \frac{zf'(z)}{f(z)}dz = \frac{1}{2\pi i}\int_C \frac{z(1g(z) + (z - z_0)g'(z))}{(z - z_0)g(z)}dz = \frac{1}{2\pi i}\int_C \frac{\frac{z(g(z) + (z - z_0)g'(z))}{g(z)}}{z - z_0}dz$$
$$= \frac{z(g(z) + (z - z_0)g'(z))}{g(z)}\Big|_{z = z_0} \quad \text{(Cauchy Integral Formula)}$$
$$= \frac{z_0(g(z_0) + 0g'(z_0))}{g(z_0)} = z_0.$$

Solution to Exercise 4.33. By the result on the classification of zeroes, there exists a $g \in \mathcal{O}(D)$ such that $f(z) = (z - z_0)^m g(z)$, and $g(z_0) \neq 0$. Thus with $G(z) := (g(z))^2$, we have $(f(z))^2 = (z - z_0)^{2m}G(z)$. Clearly $(f(z_0))^2 = 0$, and $G \in \mathcal{O}(D)$, with $G(z_0) = (g(z_0))^2 \neq 0$. So z_0 is a zero of $z \mapsto (f(z))^2$ of order $2m$. We have $f'(z) = m(z - z_0)^{m-1}g(z) + (z - z_0)^m g'(z) = (z - z_0)^{m-1}g_1(z)$, where $g_1(z) := mg(z) + (z - z_0)g'(z)$. As $m > 1$, $f'(z_0) = 0g_1(z_0) = 0$. Clearly, $g_1 \in \mathcal{O}(D)$ and $g_1(z_0) = mg(z_0) + 0g'(z_0) = mg(z_0) \neq 0$. Thus z_0 is a zero of f' of order $m - 1$.

Solution to Exercise 4.34. We have $g(z_0) = \cos(f(z_0)) - 1 = \cos 0 - 1 = 0$. By the classification of zeroes theorem, there exists an entire function h such that $f(z) = (z - z_0)h(z)$ for all $z \in \mathbb{C}$, and $h(z_0) \neq 0$. Moreover,

$$g'(z) = (-\sin((z - z_0)h))(h(z) + (z - z_0)h'(z)),$$
$$g''(z) = -(\cos((z - z_0)h))(h(z) + (z - z_0)h'(z))^2$$
$$+ (-\sin((z - z_0)h))(h'(z) + h'(z) + (z - z_0)h''(z)).$$

Thus $g'(z_0) = 0$, $g''(z_0) = -1(h(z_0))^2 + 0 \neq 0$. So z_0 is a zero of g of order 2.

Solution to Exercise 4.35. We have $|f(\frac{1}{\log n})| \leq \frac{1}{n}$ for $n = 2, 3, \cdots$. As f is continuous at 0, by passing to the limit as $n \to \infty$ in the above, we get $|f(0)| \leq 0$. Thus $f(0) = 0$. Suppose f is not identically 0 in $D(0,1)$. By the theorem on the classification of zeroes, 0 is a zero of f of some order $m \in \mathbb{N}$, and there exists a $g \in \mathcal{O}(D(0,1))$ such that $f = z^m g$ in $\mathbb{D}$, and $g(0) \neq 0$. Setting $z = \frac{1}{\log n}$, for integers $n \geq 2$, we get $\frac{1}{(\log n)^m}|g(\frac{1}{\log n})| = |f(\frac{1}{\log n})| \leq \frac{1}{n}$, which gives upon rearranging that $|g(\frac{1}{\log n})| \leq \frac{(\log n)^m}{n}$ for all integers $n \geq 2$. Passing to the limit as $n \to \infty$, we obtain, thanks to the continuity of g at 0, that $|g(0)| \leq \lim\limits_{n\to\infty} \frac{(\log n)^m}{n} = 0$, and so $g(0) = 0$, a contradiction. Hence $f \equiv 0$ in $D(0,1)$.

Proof that $\lim\limits_{n\to\infty} \frac{(\log n)^m}{n} = 0$: It is enough to show $\lim\limits_{n\to\infty} \frac{\log n^{1/m}}{n^{1/m}} = 0$, since then

$$\lim_{n\to\infty} \frac{(\log n)^m}{n} = \lim_{n\to\infty} (m\frac{\log n^{1/m}}{n^{1/m}})^m = (\lim_{n\to\infty} m\frac{\log n^{1/m}}{n^{1/m}})^m = (m \cdot 0)^m = 0.$$

We have $0 \leq \frac{\log n^{1/m}}{n^{1/m}} = \frac{1}{n^{1/m}} \int_1^{n^{1/m}} \frac{1}{t} \, dt \leq \frac{1}{n^{1/m}} \int_1^{n^{1/m}} \frac{1}{\sqrt{t}} \, dt = \frac{1}{n^{1/m}} 2(\sqrt{n^{1/m}} - 1)$, and so by the Sandwich Theorem, $\lim\limits_{n\to\infty} \frac{\log n^{1/m}}{n^{1/m}} = 0$.

Solution to Exercise 4.36. We know that for all $x, y \in \mathbb{R}$,

$$\cos(x + y) = (\cos x)(\cos y) - (\sin x)(\sin y). \qquad (*)$$

Let $y \in \mathbb{R}$. Define $f_y \in \mathcal{O}(\mathbb{C})$ by $f_y(z) := \cos(z+y) - ((\cos z)(\cos y) - (\sin z)(\sin y))$ for all $z \in \mathbb{C}$. We have $f_y(x) = 0$ for all $x \in \mathbb{R}$, thanks to $(*)$, and so by the Identity Theorem, $f_y(z) = 0$ for all $z \in \mathbb{C}$, that is,

$$\cos(z + y) = (\cos z)(\cos y) - (\sin z)(\sin y) \text{ for all } z \in \mathbb{C}. \qquad (**)$$

But the choice of $y \in \mathbb{R}$ was arbitrary, and so $(**)$ holds for *all* $y \in \mathbb{R}$. Fix $z \in \mathbb{C}$. Define $g_z \in \mathcal{O}(\mathbb{C})$ by $g_z(w) := \cos(z + w) - ((\cos z)(\cos w) - (\sin z)(\sin w))$ for all $w \in \mathbb{C}$. Then $g_z(y) = 0$ for all $y \in \mathbb{R}$ by $(**)$. Another application of the Identity Theorem yields $g_z(w) = 0$ for all $w \in \mathbb{C}$. Hence

$$\cos(z+w) = (\cos z)(\cos w) - (\sin z)(\sin w) \text{ for all } w \in \mathbb{C}. \qquad (***)$$

As $z \in \mathbb{C}$ was arbitrary, $(***)$ holds for all $z, w \in \mathbb{C}$.

Solution to Exercise 4.37. Let $f, g \in \mathcal{O}(D)$ be such that $(f \cdot g)(z) = f(z)g(z) = 0$ for all $z \in D$. Suppose that there exists a $z_0 \in D$ such that $f(z_0) \neq 0$. By the continuity of f, there exists a $\delta > 0$ such that $f(z) \neq 0$ whenever $|z - z_0| < \delta$. The equation $f(z)g(z) = 0$ $(z \in D)$ then implies $g(z) = 0$ for $|z - z_0| < \delta$. By the Identity Theorem, $g \equiv 0$ in D. So $\mathcal{O}(D)$ has no zero divisors.

We now show that $C(D)$ is not an integral domain. Let $z_0 \in D$ and let $\delta > 0$ be such that the disc $D(z_0, \delta) \subset D$. Define the continuous function $\varphi : \mathbb{R} \to \mathbb{R}$ by

$$\varphi(t) = \begin{cases} 0 & \text{if } t \leq 0, \\ t & \text{if } t > 0. \end{cases}$$

Define f, g by $f(z) = \varphi(\text{Re}(z - z_0))$ and $g(z) = \varphi(-\text{Re}(z - z_0))$ for all $z \in D$.

Being the composition of continuous functions, $f, g \in C(D)$. Also $f(z) > 0$ for all z in the right half of Δ, and so $f \neq 0$ in $C(D)$. Similarly $g(z) > 0$ for all z belonging to the left half of Δ, and so $g \neq 0$ in $C(D)$. Nevertheless, $f \cdot g = 0$.

Solution to Exercise 4.38. Let $f \in \mathcal{O}(\mathbb{C})$ be such that $Z(f) = \{z \in \mathbb{C} : f(z) = 0\}$ is uncountable. For $n \in \mathbb{N}$, let $S_n := Z(f) \cap D(0, n)$. Then we have $\bigcup_{n \in \mathbb{N}} S_n = Z(f)$. If each S_n is finite, then their countable union, $Z(f)$, which we know is infinite, must be countable, which is not true. Thus there exists an $N \in \mathbb{N}$ such that S_N is an infinite set. Take any sequence $(z_n)_{n \in \mathbb{N}}$ of distinct elements from S_N. Then for all $n \in \mathbb{N}$, $|z_n| \leq N$, that is, the sequence $(z_n)_{n \in \mathbb{N}}$ is bounded. Using the Bolzano-Weierstrass Theorem, there exists a subsequence $(z_{n_k})_{k \in \mathbb{N}}$ which is convergent, with limit say $z_* \in \mathbb{C}$. As $f(z_{n_k}) = 0$ for all $k \in \mathbb{N}$, it follows by the Identity Theorem that $f \equiv 0$. So there is no nonzero entire function with an uncountable zero set.

Solution to Exercise 4.39. Suppose that $z_0 \in \mathbb{D} \backslash \{0\}$ is a zero of f. Then we have $|z_0^2| = |z_0||z_0| < 1 |z_0|$. So $z_0^2 \neq z_0$. Also $0 < |z_0|^2 < 1$, i.e., $z_0^2 \in \mathbb{D} \backslash \{0\}$. Finally, $f(z_0^2) = f(z_0) f(-z_0) = 0 f(-z_0) = 0$.

It is given that f has some zero $z_0 \in \mathbb{D} \backslash \{0\}$. By the above, each term in the sequence $z_0, z_0^2, z_0^4, z_0^8, \cdots$ is a zero of f. Moreover, these terms are distinct: Using $z_0^{2^n} \in \mathbb{D} \backslash \{0\}$, for $N > n$, $|z_0^{2^N}| = |z_0^{2^n}||z_0^{2^N - 2^n}| < |z_0^{2^n}| \cdot 1$, so that $z_0^{2^N} \neq z_0^{2^n}$. Also, since $|z_0| < 1$, the geometric sequence $z_0, z_0^2, z_0^3, \cdots$ converges to 0, so that its subsequence $(z_0^{2^n})_{n \in \mathbb{N}}$ also converges to $0 \in \mathbb{D}$. By the identity theorem, since f is holomorphic in the domain $\mathbb{D}$, we conclude that $f \equiv 0$ in $\mathbb{D}$.

Vice versa, $f \equiv 0$ is holomorphic, satisfies $f(z)f(-z) = f(z^2)$ in $\mathbb{D}$, and is not zero-free in $\mathbb{D} \backslash \{0\}$.

Solution to Exercise 4.40.
(1) No. Take $D = \mathbb{C}$, $f = \exp$, $g = 1$. Then $f(2\pi i n) = \exp(2\pi i n) = 1 = g(2\pi i n)$ for all $n \in \mathbb{N}$, but $f \neq g$ (e.g., because $f(i\pi) = -1 \neq 1 = g(i\pi)$).
(2) Yes.
(3) Yes. Let $\gamma(t) = x(t) + iy(t)$ for all $t \in [a, b]$, where $x, y : [a, b] \to \mathbb{R}$. Take $t_0 \in (a, b)$ so that $x'(t_0)$ or $y'(t_0)$ is nonzero. (If they are both always 0, then $z = w$, a contradiction.) Let $x'(t_0) > 0$ (the other cases are handled similarly). Then $x'(t) > 0$ in a neighbourhood of t_0. So x is strictly increasing there. Take $t_n = t_0 + \frac{1}{n}$, $n \geq N$, with N large enough so that $t_0 + \frac{1}{N} \in [a, b]$. Set $z_n = \gamma(t_n)$. Then $(z_n)_{n \geq N}$ is a sequence of *distinct* points (as the real parts are distinct), which converges to $\gamma(t_0)$. By the Identity Theorem, $f = g$ in D.
(4) Yes. By the Taylor expansion in a small disc $D(w, \delta) \subset D$ around w, $f = g$ in $D(w, \delta)$, and so by the Identity Theorem, $f = g$ in D.

Solution to Exercise 4.41. $\sin(\log x) = 0$ if and only if $\log x \in \{n\pi : n \in \mathbb{Z}\}$, i.e., $x \in \{e^{n\pi} : n \in \mathbb{Z}\}$. Let F be an entire extension of f. Then $F(e^{-n\pi}) = f(e^{-n\pi}) = 0$ for every $n \in \mathbb{N}$. As e^x is strictly increasing, the points $e^{-n\pi}$, $n \in \mathbb{N}$, are distinct. Also, $e^{-n\pi} \to 0 \in \mathbb{C}$ as $n \to \infty$. By the Identity Theorem, $F \equiv 0$. But this is absurd since, e.g., $F(e^{\frac{\pi}{2}}) = f(e^{\frac{\pi}{2}}) = \sin(\log e^{\frac{\pi}{2}}) = \sin \frac{\pi}{2} = 1 \neq 0$.

Solution to Exercise 4.42. For $z \in D(0, r)$, $g(z) = \sum\limits_{n=0}^{\infty} \frac{g^{(n)}(0)}{n!} z^n$. As $\sum\limits_{n=0}^{\infty} g^{(n)}(0)$ converges, $g^{(n)}(0) \to 0$ as $n \to \infty$. Thus $\lim\limits_{n \to \infty} \sqrt[n]{|\frac{g^{(n)}(0)}{n!}|} = 0$. So $\sum\limits_{n=0}^{\infty} \frac{g^{(n)}(0)}{n!} z^n$ has an infinite radius of convergence, and defines an entire function that coincides with g on $D(0, r)$ (as both have the same power series there). Also, such an entire function has to be unique by the Identity Theorem (as any two such coincide with g on $D(0, r)$).

Solution to Exercise 4.43. Let $K = \{z \in \mathbb{C} : |z| \leqslant 1\}$. For each $z \in K$, there exists a smallest integer $n(z) \geqslant 0$ such that for all $w \in \mathbb{C}$, $f(w) = \sum\limits_{n=0}^{\infty} c_n(z)(w - z)^n$ and $c_{n(z)}(z) = 0$. Hence $\frac{f^{(n(z))}(z)}{(n(z))!} = 0$, and so $f^{(n(z))}(z) = 0$. Let $\varphi : K \to \mathbb{N} \cup \{0\}$ be defined by $\varphi(z) = n(z)$. Since K is uncountable, while $\mathbb{N} \cup \{0\}$ is countable, there exists an N such that $\varphi^{-1}(N)$ is infinite. Let $(z_n)_{n \in \mathbb{N}}$ be a sequence of distinct points in $\varphi^{-1}(N) \subset K$. In particular, $|z_n| \leqslant 1$ for all $n \in \mathbb{N}$, and so by the Bolzano-Weierstrass Theorem, it follows that $(z_n)_{n \in \mathbb{N}}$ has a convergent subsequence $(z_{n_k})_{k \in \mathbb{N}}$ with limit, say $z_* \in K$. As $f^{(N)}(z_{n_k}) = 0$ for all $k \in \mathbb{N}$, by the Identity Theorem (applied to $f^{(N)} \in \mathcal{O}(\mathbb{C})$), we have $f^{(N)} = 0$ in $\mathbb{C}$. By Taylor's Theorem, $f(z) = \sum\limits_{n=0}^{\infty} \frac{f^{(n)}(0)}{n!} z^n = \sum\limits_{n=0}^{N-1} \frac{f^{(n)}(0)}{n!} z^n$, for all $z \in \mathbb{C}$, and so f is a polynomial.

Solution to Exercise 4.44. We have $(\frac{f}{g})' = \frac{f'g - fg'}{g^2}$, and so for all integers $n \geqslant 2$, $(\frac{f}{g})'(\frac{1}{n}) = \frac{f'(\frac{1}{n})g(\frac{1}{n}) - f(\frac{1}{n})g'(\frac{1}{n})}{(g(\frac{1}{n}))^2} = \frac{0}{(g(\frac{1}{n}))^2} = 0$. By the Identity Theorem applied to the function $(\frac{f}{g})' \in \mathcal{O}(\mathbb{D})$, we obtain $(\frac{f}{g})' \equiv 0$ in $\mathbb{D}$. By the Fundamental Theorem of Contour Integration, if $\lambda := \frac{f(0)}{g(0)} \in \mathbb{C} \backslash \{0\}$, and $[0, z]$ is the line segment from 0 to $z \in \mathbb{D}$, then $\frac{f(z)}{g(z)} - \lambda = \int_{[0,z]} (\frac{f}{g})'(\zeta) \, d\zeta = \int_{[0,z]} 0 \, d\zeta = 0$. So $f(z) = \lambda g(z)$, $z \in \mathbb{D}$.

Solution to Exercise 4.45. g given by $g(z) = f(z) - z^2$ ($z \in \mathbb{C}$) is entire, and $g(\frac{1}{n}) = f(\frac{1}{n}) - \frac{1}{n^2} = 0$. As $(\frac{1}{n})_{n \in \mathbb{N}}$ is a sequence of distinct zeroes of g in the domain $\mathbb{C}$, such that $\frac{1}{n} \to 0 \in \mathbb{C}$, the Identity Theorem implies $g \equiv 0$. So $f(z) = z^2$ for all $z \in \mathbb{C}$. Vice versa, if $f(z) = z^2$ ($z \in \mathbb{C}$), then $f \in \mathcal{O}(\mathbb{C})$ and $f(\frac{1}{n}) = \frac{1}{n^2}$ for all $n \in \mathbb{N}$. Define h by $h(z) = f(z^2) - z$ for all $z \in \mathbb{C}$. The composition $f \circ (z \mapsto z^2)$ is entire. So h is entire. We have $h(\frac{1}{n}) = f(\frac{1}{n^2}) - \frac{1}{n} = 0$. As $(\frac{1}{n})_{n \in \mathbb{N}}$ is a sequence of distinct zeroes of h in the domain $\mathbb{C}$, such that $\frac{1}{n} \to 0 \in \mathbb{C}$, $h \equiv 0$ by the Identity Theorem. Thus $f(z^2) = z$, $z \in \mathbb{C}$. Taking $z = 1$, $f(1) = f(1^2) = 1$. Taking $z = -1$, we obtain $f(1) = f((-1)^2) = -1$, a contradiction to $f(1) = 1$ obtained above.

Solution to Exercise 4.46. As $K := \{z \in \mathbb{C} : |z - z_0| \leqslant R\}$ is compact, the Identity Theorem implies that K has only finitely many distinct zeroes of f. (Otherwise, there exists a sequence of distinct zeroes of f in K, which will possess a convergent subsequence in the compact set K, and so $f \equiv 0$ in D, a contradiction to f being nonzero on the range of $C \subset D$.) So $Z(f)$ is a finite set.

If $Z(f) = \varnothing$, then by definition the empty sum $\sum\limits_{z \in Z(f)} m(z) = 0$. Furthermore, as $\frac{f'}{f}$ is holomorphic in a neighbourhood of K, $\int_C \frac{f'}{f} dz = 0$, as wanted.

Next, let $\varnothing \neq Z(f) = \{\zeta_1, \cdots, \zeta_k\}$ for some $k \in \mathbb{N}$. By a repeated application of the theorem on the classification of zeroes, $f = (z - \zeta_1)^{m(\zeta_1)} \cdots (z - \zeta_k)^{m(\zeta_k)} g$, for some $g \in \mathcal{O}(D)$ such that g is never zero in K. For all $z \in \operatorname{ran} C$,

$$f'(z) = \sum_{\ell=1}^{k} \frac{m(\zeta_\ell) f}{z - \zeta_\ell} + \frac{f}{g} g'.$$

Hence $\int_C \frac{f'}{f} dz = \sum\limits_{\ell=1}^{k} m(\zeta_\ell) \int_C \frac{1}{z - \zeta_\ell} dz + \int_C \frac{g'}{g} dz \overset{(*)}{=} 2\pi i \sum\limits_{\ell=1}^{k} m(\zeta_\ell) + 0.$

Justification of (*): Since $\frac{g'}{g}$ is holomorphic in a neighbourhood of K, $\int_C \frac{g'}{g} dz = 0$, while $\int_C \frac{1}{z - \zeta_\ell} dz = 2\pi i$ (as $\frac{1}{\cdot - \zeta_\ell} \in \mathcal{O}(\mathbb{C} \backslash \{\zeta_\ell\})$), and since C is $\mathbb{C} \backslash \{\zeta_\ell\}$-homotopic to a little circular path centred at ζ_ℓ with a positive radius traversed once in the counterclockwise direction).

Solution to Exercise 4.47. Let $z_0 \in D$ be such that $|f(z_0)| \geqslant |f(z)|$ for all $z \in D$. By the Maximum Modulus Theorem, f is constant in D, a contradiction.

Solution to Exercise 4.48. Let $f(z_0) \neq 0$. Then $|f(z_0)| > 0$ and so for all $z \in D$, $|f(z)| \geqslant |f(z_0)| > 0$, implying that for all $z \in D$, $f(z) \neq 0$. Define $g = \frac{1}{f} \in \mathcal{O}(D)$. Then $|g(z_0)| = \frac{1}{|f(z_0)|} \geqslant \frac{1}{|f(z)|} = |g(z)|$ for all $z \in D$, and so by the Maximum Modulus Theorem, g is constant. But then f is constant too.

Solution to Exercise 4.49. Let z_0 be a maximiser, which exists since $|f|$ is continuous and $K := \{z \in \mathbb{C} : |z| \leqslant 1\}$ is compact. But z_0 cannot be in the interior of K: Indeed, if $|z_0| < 1$, then by the Maximum Modulus Theorem (applied to f on $\mathbb{D} := \{z \in \mathbb{C} : |z| < 1\}$), f would be constant in $\mathbb{D}$, which is not true (e.g. $f(0) = 0 \neq \frac{1}{4} - 2 = f(\frac{1}{2})$). Hence $z_0 \in \mathbb{T} := \{z \in \mathbb{C} : |z| = 1\}$. Referring to the picture below, we see that $\max\limits_{z \in K} |f(z)| = \max\limits_{|z|=1} |f(z)| = \max\limits_{t \in [0, 2\pi)} |e^{2it} - 2| = |-1 - 2| = 3.$

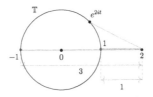

Similarly, if z_1 is a minimiser, z_1 cannot be in the interior of K: Indeed, $z_1^2 - 2 \neq 0$, and so by the Minimum Modulus Theorem, f would be a constant, a contradiction. So $z_1 \in \mathbb{T}$. Hence $\min\limits_{z \in K} |f(z)| = \min\limits_{|z|=1} |f(z)| = \min\limits_{t \in [0, 2\pi)} |e^{2it} - 2| = |1 - 2| = 1.$

Solution to Exercise 4.50. For $z \in D(0,1)$, $|f(z)| = e^{|z|} > 0$, and so f is never
0 on $D(0,1)$. Consider $g := \frac{1}{f}$. Then as $f \in \mathcal{O}(D(0,1))$, $g \in \mathcal{O}(D(0,1))$ too. For
$z \in D(0,1)$, $|z| \geqslant 0$, and so $|g(z)| = \frac{1}{|f(z)|} = e^{-|z|} \leqslant e^0 = e^{-|0|} = \frac{1}{|f(0)|} = |g(0)|$,
and so, by the Maximum Modulus Theorem, g must be a constant. Then f is also
a constant. But $|f(0)| = e^{|0|} = 1 \neq e^{\frac{1}{2}} = |f(\frac{1}{2})|$. So we arrive at a contradiction.
Thus there is no $f \in \mathcal{O}(D(0,1))$ such that $|f(z)| = e^{|z|}$ for all $z \in D(0,1)$.

Solution to Exercise 4.51. For $z \in \mathbb{A}_1 := \{z \in \mathbb{C} : 0 < |z-1| < 1\}$, we have
$$\tfrac{1}{z(z-1)} = \tfrac{1}{(z-1+1)(z-1)} = \tfrac{1}{z-1}(1 - (z-1) + (z-1)^2 - (z-1)^3 + - \cdots)$$
$$= \tfrac{1}{z-1} - 1 + (z-1) - (z-1)^2 + (z-1)^3 - + \cdots.$$
On the other hand, for $z \in \widetilde{\mathbb{A}}_1 := \{z \in \mathbb{C} : 1 < |z-1|\}$,
$$\tfrac{1}{z(z-1)} = \tfrac{1}{(z-1+1)(z-1)} = \tfrac{1}{(z-1)^2(1+\frac{1}{z-1})} = \tfrac{1}{(z-1)^2}(1 - \tfrac{1}{z-1} + \tfrac{1}{(z-1)^2} - + \cdots)$$
$$= \tfrac{1}{(z-1)^2} - \tfrac{1}{(z-1)^3} + \tfrac{1}{(z-1)^4} - \tfrac{1}{(z-1)^5} + - \cdots.$$

Solution to Exercise 4.52. $f_{-w}(-z) = \exp(\frac{-w}{2}(-z-\frac{1}{-z})) = \exp(\frac{w}{2}(z-\frac{1}{z})) = f_w(z)$.
For all $z \in \mathbb{C}\backslash\{0\}$, $\sum\limits_{n=-\infty}^{\infty} J_n(-w)(-z)^n = f_{-w}(-z) = f_w(z) = \sum\limits_{n=-\infty}^{\infty} J_n(w)z^n$. By the
uniqueness of coefficients, $J_n(-w)(-1)^n = J_n(w)$, i.e., $J_n(-w) = (-1)^n J_n(w)$.
Let C be circular path $C(\theta) = e^{i\theta}$, $\theta \in [0, 2\pi]$. For all $n \in \mathbb{Z}$,
$$J_n(w) = \tfrac{1}{2\pi i}\int_C \tfrac{f_w(z)}{(z-0)^{n+1}}dz = \tfrac{1}{2\pi i}\int_0^{2\pi} \tfrac{\exp(\frac{w}{2}(e^{i\theta}-e^{-i\theta}))}{(e^{i\theta})^{n+1}}ie^{i\theta}d\theta$$
$$= \tfrac{1}{2\pi}\int_0^{2\pi} \tfrac{\exp(wi\sin\theta)}{\exp(in\theta)}d\theta = \tfrac{1}{2\pi}\int_0^{2\pi} \exp(iw\sin\theta - in\theta)d\theta.$$

Solution to Exercise 4.53. $e^{-\frac{1}{z^2}} \in \mathcal{O}(\mathbb{C}\backslash\{0\})$, and has the Laurent series
$$e^{-\frac{1}{z^2}} = \sum\limits_{n=0}^{\infty} \tfrac{(-1)^n}{n!}\tfrac{1}{z^{2n}} = 1 - \tfrac{1}{z^2} + \tfrac{1}{2!}\tfrac{1}{z^4} - \tfrac{1}{3!}\tfrac{1}{z^6} + - \cdots.$$
The coefficient of z^{-6} is $-\frac{1}{3!} = -\frac{1}{6}$. So $-\frac{1}{6} = \frac{1}{2\pi i}\int_C \frac{e^{-\frac{1}{z^2}}}{(z-0)^{-6+1}}dz = \frac{1}{2\pi i}\int_C z^5 e^{-\frac{1}{z^2}}dz$,
giving $\int_C z^5 e^{-\frac{1}{z^2}}dz = -\frac{\pi i}{3}$.

Solution to Exercise 4.54. $f \in \mathcal{O}(\mathbb{C}\backslash\{0\})$, and so 0 is an isolated singularity.
We will first show that 0 is not a removable singularity of f by showing it is not
the case that $\lim\limits_{z\to 0} f(z)$ exists. If $z_n := \frac{1}{\log(n\pi)}$, $n \in \mathbb{N}$, then $z_n \to 0$ as $n \to \infty$. But
$$f(z_n) = \cos(e^{1/z_n}) = \cos(e^{\log(n\pi)}) = \cos(n\pi) = (-1)^n.$$
So it is not the case that $\lim\limits_{z\to 0} f(z)$ exists. So 0 is not a removable singularity of f.
We will show 0 is not a pole by showing it is not the case that $\lim\limits_{z\to 0}|f(z)| = +\infty$:
For all $n \in \mathbb{N}$, $|f(z_n)| = 1$, showing that it cannot be the case that $\lim\limits_{z\to 0}|f(z)| = \infty$.
So the isolated singularity 0 is essential, as it is neither removable nor a pole.

Solution to Exercise 4.55. If $D := \mathbb{C}\backslash((-\infty,0] \times \{0\})$, then Log $\in \mathcal{O}(D)$, and
Log$'z = \frac{1}{z}$. Hence $\lim\limits_{z\to 0} \frac{\text{Log}(1+z)}{z} = \lim\limits_{z\to 0}\frac{\text{Log}(1+z)-\text{Log}1}{(1+z)-1} = \text{Log}'1 = \frac{1}{1} = 1$. Thus
$\lim\limits_{z\to 0} z\frac{\text{Log}(1+z)}{z^2} = 1 \neq 0$. Hence $z = 0$ is a pole of order 1 of $\frac{\text{Log}(1+z)}{z^2}$.

Solution to Exercise 4.56. Let $z \in Z$. Write $z = x + iy$, where $x, y \in \mathbb{R}$. Then $e^z - 1 = 0$ implies $e^x(\cos y + i \sin y) = 1$. Taking complex absolute values, $e^x = 1$, and so $x = 0$. Thus $1(\cos y + i \sin y) = 1$, giving $\cos y = 1$ and $\sin y = 0$. So $y \in 2\pi\mathbb{Z}$. Vice versa, if $z = 0 + i2\pi n$ for some $n \in \mathbb{Z}$, then $e^z = e^{0+i2\pi n} = 1(1 + i0) = 1$, so that $z \in Z$. Consequently, $Z = 2\pi i\mathbb{Z}$.

For $z = 2\pi in \in Z$, $n \in \mathbb{Z}$, $(e^z - 1)'|_{z=2\pi in} = e^z|_{z=2\pi in} = 1 \neq 0$, and so each zero in Z has order 1 as a zero of $e^z - 1$.

The numerator z of $\frac{z}{e^z-1}$ is entire. Since the denominator $e^z - 1$ is nonzero in the punctured disc $D(0, 2\pi)\backslash\{0\}$ of radius 2π, and is holomorphic there, 0 is an isolated singularity of f. We have $\lim_{z \to 0} f(z) = \lim_{z \to 0} \frac{z}{e^z - 1} = \lim_{z \to 0} \frac{1}{\frac{e^z - e^0}{z - 0}} = \frac{1}{(e^z)'|_{z=0}} = \frac{1}{e^0} = 1$. Thus $\lim_{z \to 0}(z - 0)^1 f(z) = 0 \cdot 1 = 0$. By Theorem 4.16, f has a removable singularity at 0.

Let F be the holomorphic extension of f to the disc $D(0, 2\pi)$. By Taylor's theorem, F has a power series expansion in $D(0, 2\pi)$, with coefficients, say, c_n, $n \geqslant 0$. In particular, for $0 < |z| < 2\pi =: r$, $f(z) = F(z) = \sum_{n=0}^{\infty} c_n z^n$.

We define $B_n := n! c_n$. Hence $B_0 = 1c_0 = F(0) = \lim_{z \to 0} F(z) = \lim_{z \to 0} f(z) = 1$.

Using the identity $\frac{z}{e^z - 1} - \frac{(-z)}{e^{-z} - 1} = -z$, and the series expansion for $0 < |z| < r$,

$$-z = \sum_{n=0}^{\infty} \frac{B_n}{n!} z^n - \sum_{n=0}^{\infty} \frac{B_n}{n!}(-z)^n = 0 + 2\frac{B_1}{1!}z + 0 + 2\frac{B_3}{3!}z^3 + 0 + \cdots.$$

By the uniqueness of coefficients in the Laurent series expansion, it follows that $B_1 = -\frac{1}{2}$ and the rest are zeroes, that is, $0 = B_3 = B_5 = \cdots$.

Solution to Exercise 4.57. For all $z \in \mathbb{C}$, we have

$$\cos z - 1 = (1 - \tfrac{z^2}{2!} + \tfrac{z^4}{4!} - +\cdots) - 1 = z^2(-\tfrac{1}{2!} + \tfrac{1}{4!}z^2 - +\cdots).$$

If $h := -\frac{1}{2!} + \frac{1}{4!}z^2 - +\cdots$, then as h is defined by a power series with an infinite radius of convergence, h is entire. Also $h(0) = -\frac{1}{2} \neq 0$. Thus h is nonzero in a disc centred at 0. So 0 is an isolated singularity of $\frac{1}{\cos z - 1} = \frac{1}{z^2 h(z)}$. Moreover $\lim_{z \to 0} z^2 \frac{1}{\cos z - 1} = \lim_{z \to 0} \frac{1}{h(z)} = -2 \neq 0$, and $\lim_{z \to 0} z^3 \frac{1}{\cos z - 1} = 0(-2) = 0$. By the theorem on classification of singularities via limits, 0 is a pole of f, and its order is 2.

Solution to Exercise 4.58. By Theorem 4.11 (classification of zeroes), we have $f(z) = (z - z_0)^m g(z)$, $z \in D$, where $g \in \mathcal{O}(D)$ and $g(z_0) \neq 0$. Hence g is nonzero in a disc $D(z_0, r) \subset D$ for some $r > 0$. Then $f(z) = (z - z_0)^m g(z) \neq 0$ for all $z \in D(z_0, r)\backslash\{z_0\}$, and so $\frac{1}{f}$ is well-defined in $D(z_0, r)\backslash\{z_0\}$. As $\frac{1}{g} \in \mathcal{O}(D(z_0, r))$, it has a Taylor expansion in $D(z_0, r)$: $\frac{1}{g(z)} = \sum_{n=0}^{\infty} c_n(z - z_0)^n$, and $c_0 = \frac{1}{g(z_0)} \neq 0$. Thus

$$\frac{1}{f(z)} = \frac{1}{(z-z_0)^m g(z)} = \frac{1}{(z-z_0)^m} \sum_{n=0}^{\infty} c_n(z - z_0)^n = \frac{c_0}{(z-z_0)^m} + \cdots + \frac{c_{m-1}}{z - z_0} + \sum_{n=0}^{\infty} c_{m+n}(z - z_0)^n$$

for $0 < |z - z_0| < r$. So $\frac{1}{f}$ has a pole of order m at z_0. (Alternatively, use the classification of singularities via limits, noting that $\lim_{z \to z_0}(z - z_0)^m \frac{1}{f(z)} = \frac{1}{g(z_0)} \neq 0$.)

Solution to Exercise 4.59. The function f has a Laurent series expansion in $D(z_0, r)\backslash\{z_0\}$: $f(z) = \frac{c_{-m}}{(z-z_0)^m} + \cdots + \frac{c_{-1}}{z-z_0} + c_0 + c_1(z-z_0) + \cdots$, where $c_{-m} \neq 0$. Hence $(z-z_0)^m f = c_{-m} + c_{-m+1}(z-z_0) + \cdots$ for all $z \in D(z_0, r)\backslash\{z_0\}$. So $h := c_{-m} + c_{-m+1}(z-z_0) + \cdots \in \mathcal{O}(D(z_0, r))$, and $h(z_0) = c_{-m} \neq 0$. Since for all $z \in D(z_0, r)\backslash\{z_0\}$, $h(z) = (z-z_0)^m f(z)$, and as $f(z) \neq 0$, also $h(z) \neq 0$. Thus $h(z) \neq 0$ for $z \in D(z_0, r)$. So $\frac{1}{h} \in \mathcal{O}(D(z_0, r))$. Define $g := (z-z_0)^m \frac{1}{h} \in \mathcal{O}(D(z_0, r))$. Then $g|_{D(z_0, r)\backslash\{z_0\}} = \frac{1}{f}$, and by Exercise 4.30, z_0 is a zero of g of order m .

Solution to Exercise 4.60. By the classification of singularities via Laurent series coefficients, $c_n = 0$ for all $n < -m$. So for all $z \in D(z_0, r)\backslash\{z_0\}$,

$$f(z) = \frac{c_{-m}}{(z-z_0)^m} + \cdots + \frac{c_{-1}}{z-z_0} + \sum_{n=0}^{\infty} c_n(z-z_0)^n.$$

Thus $(z-z_0)^m f(z) = c_{-m} + c_{-m+1}(z-z_0) + \cdots$ for all $z \in D(z_0, r)\backslash\{z_0\}$. Let $g \in \mathcal{O}(D(z_0, r))$ be defined by $g(z) := c_{-m} + c_{-m+1}(z-z_0) + \cdots + c_{-1}(z-z_0)^{m-1} + \cdots$ for all $z \in D(z_0, r)$. By Taylor's Theorem, $c_{-1} = \frac{1}{(m-1)!}\frac{d^{m-1}g}{dz^{m-1}}(z_0)$. But the function $g^{(m-1)} \in \mathcal{O}(D(z_0, r))$ and in particular, it is continuous at z_0. So

$$g^{(m-1)}(z_0) = \lim_{z \to z_0} g^{(m-1)}(z).$$

For $z \in D(z_0, r)\backslash\{z_0\}$, $g(z) = (z-z_0)^m f(z)$. So $g^{(m-1)}(z) = \frac{d^{m-1}}{dz^{m-1}}((z-z_0)^m f(z))$. Hence $c_{-1} = \frac{g^{(m-1)}(z_0)}{(m-1)!} = \frac{1}{(m-1)!} \lim_{z \to z_0} g^{(m-1)}(z) = \frac{1}{(m-1)!} \lim_{z \to z_0} \frac{d^{m-1}}{dz^{m-1}}((z-z_0)^m f(z))$.

Solution to Exercise 4.61.
(1) True, since $c_{-1} = 1 \neq 0$, and $c_{-2} = c_{-3} = \cdots = 0$.
(2) True.
(3) True.
(4) True.

Solution to Exercise 4.62.
(1) $\sin z$ does not have a singularity at 0, and for $z \in \mathbb{C}$, $\sin z = z - \frac{z^3}{3!} + \frac{z^5}{5!} - + \cdots$.
(2) $\sin \frac{1}{z}$ has an essential singularity at 0: For $z \neq 0$, $\sin \frac{1}{z} = \cdots - + \frac{1}{5!z^5} - \frac{1}{3!z^3} + \frac{1}{z}$.
(3) As $\frac{\sin z}{z} = 1 - \frac{1}{3!}z^2 + \frac{1}{5!}z^4 - + \cdots$ for $z \neq 0$, 0 is a removable singularity.
(4) $\frac{\sin z}{z^2}$ has a pole of order 1 at 0, since for $z \neq 0$, $\frac{\sin z}{z^2} = \frac{1}{z} - \frac{z}{3!} + \frac{z^3}{5!} - \frac{z^5}{7!} + - \cdots$.
(5) $\frac{1}{\sin \frac{1}{z}}$ does not have an isolated singularity at 0 (see Example 4.15).
(6) As $z \sin \frac{1}{z} = \cdots - + \frac{1}{5!z^4} - \frac{1}{3!z^2} + 1$ for $z \neq 0$, 0 is an essential singularity.

Solution to Exercise 4.63. Let $R > 0$ be such that $f \in \mathcal{O}(D(z_0, R)\backslash\{z_0\})$. By the theorem on classification of singularities via Laurent series coefficients, f has a Laurent series $f(z) = \frac{c_{-m}}{(z-z_0)^m} + \cdots + \frac{c_{-1}}{z-z_0} + \sum_{n=0}^{\infty} c_n(z-z_0)^n$ in $D(z_0, R)\backslash\{z_0\}$, where $c_{-m} \neq 0$. Thus, $(z-z_0)^m f(z) = c_{-m} + c_{-m+1}(z-z_0) + \cdots$ for all $z \in D(z_0, R)\backslash\{z_0\}$. Define $g \in \mathcal{O}(D(z_0, R))$ by $g(z) = c_{-m} + c_{-m+1}(z-z_0) + \cdots + c_{-1}(z-z_0)^{m-1} + \cdots$ for all $z \in D(z_0, r)$. Then $\lim_{z \to z_0} (z-z_0)^m f(z) = \lim_{z \to z_0} g(z) = g(z_0) = c_{-m} \neq 0$.

Solution to Exercise 4.64.

(1) False. $\lim\limits_{x\to 0-} |e^{\frac{1}{x}}| = \lim\limits_{x\to 0-} e^{\frac{1}{x}} = 0$, and so $\neg(\lim\limits_{z\to 0} |\exp\frac{1}{z}| = +\infty)$.

(2) True. There exists an $R > 0$ such that for $0 < |z - z_0| < R$

$$f(z) = \frac{c_{-m}}{(z-z_0)^m} + \frac{c_{-m+1}}{(z-z_0)^{m-1}} + \cdots + \frac{c_{-1}}{z-z_0} + \sum_{n=0}^{\infty} c_n(z-z_0)^n,$$

and so with $p := c_{-m} + c_{-m+1}(z - z_0) + \cdots + c_{-1}(z - z_0)^{m-1}$, we have for $0 < |z - z_0| < R$ that $f(z) - \frac{p(z)}{(z-z_0)^m} = \sum_{n=0}^{\infty} c_n(z-z_0)^n$.

(3) True. Let the order of 0 as a zero of f be $m \in \mathbb{N}$. Set $m = 0$ if $f(0) \neq 0$. Then there exists a holomorphic function g such that $f(z) = z^m g(z)$ and $g(0) \neq 0$. For $n > m$, and $z \neq 0$, $\frac{f(z)}{z^n} = \frac{z^m g(z)}{z^n} = \frac{g(z)}{z^{n-m}}$. Using $g(0) \neq 0$ and $n > m$,

$$\lim_{z\to 0} \left|\frac{f(z)}{z^n}\right| = \lim_{z\to 0} \frac{|g(z)|}{|z|^{n-m}} = |g(0)| \lim_{z\to 0} \frac{1}{|z|^{n-m}} = +\infty.$$

(4) True. In some punctured disc $D(z_0, r)\setminus\{z_0\}$, $r > 0$, f, g are nonzero, and by Exercise 4.59, there exist $h_f, h_g \in \mathcal{O}(D(z_0, r))$ such that $h_f(z_0) \neq 0$, $h_g(z_0) \neq 0$, and for all $z \in D(z_0, r)\setminus\{z_0\}$, $\frac{1}{f(z)} = (z - z_0)^{m_f} h_f(z)$, $\frac{1}{g(z)} = (z - z_0)^{m_g} h_g(z)$. So $h_f(z_0)h_g(z_0) \neq 0$, and $\forall z \in D(z_0, r)\setminus\{z_0\}$, $\frac{1}{f(z)g(z)} = (z-z_0)^{m_f+m_g} h_f(z)h_g(z)$. Thus fg has a pole of order $m_f + m_g$ at z_0, by Exercise 4.58. (Alternatively, by Exercise 4.63, $\lim\limits_{z\to z_0} (z - z_0)^{m_f} f(z) =: \alpha \neq 0$ and $\lim\limits_{z\to z_0} (z - z_0)^{m_g} g(z) =: \beta \neq 0$, giving $\lim\limits_{z\to z_0} (z - z_0)^{m_f+m_g} f(z)g(z) = \alpha\beta \neq 0$, and so by Corollary 4.17, z_0 is a pole of fg of order $m_f + m_g$.)

Solution to Exercise 4.65. Define f by $f(z) = (\exp\frac{1}{z}) + \exp\frac{1}{z-1}$, $z \in \mathbb{C}\setminus\{0,1\}$. Then $f \in \mathcal{O}(\mathbb{C}\setminus\{0,1\})$. The function $\exp\frac{1}{z-1}$ is holomorphic in the neighbourhood $D(0,1)$ of $z = 0$, while the function $\exp\frac{1}{z}$ has an essential singularity at 0. Thus, considering the Laurent series expansion of their sum f in the punctured disc $D(0,1)\setminus\{0\}$, f has an essential singularity at 0 (because the negatively indexed coefficients in $f = \sum_{n\in\mathbb{Z}} c_n z^n$ are given by $c_{-n} = \frac{1}{n!} \neq 0$, $n \in \mathbb{N}$).

Similarly, as $\exp\frac{1}{z}$ is holomorphic in the neighbourhood $D(1,1)$ of 1, and as $\exp\frac{1}{z-1}$ has an essential singularity at 1, f has an essential singularity at $z = 1$ (as the negatively indexed coefficients in $f = \sum_{n\in\mathbb{Z}} c_n(z-1)^n$ are $c_{-n} = \frac{1}{n!} \neq 0$, $n \in \mathbb{N}$).

Solution to Exercise 4.66. If z_0 is an isolated singularity of a function g with the Laurent series expansion $g(z) = \sum_{n\in\mathbb{Z}} c_n(z - z_0)^n$ for $0 < |z - z_0| < R$, where $R > 0$, and there are infinitely many indices $n < 0$ such that $c_n \neq 0$, then z_0 is an essential singularity of g.

However, for the given f, the annulus for the Laurent expansion $z^{-1} + z^{-2} + z^{-3} + \cdots$ is given by $|z| > 1$. The correct annulus to consider for deciding the nature of the singularity at $z = 0$ is of the form $0 < |z| < R$ for some $R > 0$. In fact, for $|z| < 1$ we have $f(z) = -\frac{1}{1-z} = -(1 + z + z^2 + z^3 + \cdots)$, showing that f is holomorphic for $|z| < 1$, and f does not have a singularity at $z = 0$.

Solution to Exercise 4.67. z_0 is an isolated singularity of fg: As f, g have an isolated singularity at z_0, $f \in \mathcal{O}(D(z_0, r_f) \backslash \{z_0\})$ for an $r_f > 0$, and $g \in \mathcal{O}(D(z_0, r_g) \backslash \{z_0\})$ for an $r_g > 0$, and so $fg \in \mathcal{O}(D(z_0, r) \backslash \{z_0\})$ where $r := \min\{r_f, r_g\}$.

If fg has a removable singularity or a pole at z_0, then there exists an $m \in \mathbb{N}$ such that $\lim\limits_{z \to z_0} (z - z_0)^m f(z) g(z) = 0$. Since f has a pole at z_0, say of order $\ell \in \mathbb{N}$, f is nonzero near z_0. Also, $f(z) = \frac{c_{-\ell}}{(z-z_0)^\ell} + \frac{c_{-\ell+1}}{(z-z_0)^{\ell-1}} + \cdots + \frac{c_{-1}}{z-z_0} + \sum\limits_{n=0}^{\infty} c_n (z - z_0)^n$, for $0 < |z - z_0| < r_f$, and $c_{-\ell} \neq 0$. So for $z \neq z_0$, but near z_0, we have

$$(z - z_0)^m g(z) = \underbrace{\frac{1}{(z-z_0)^\ell f(z)}}_{} \underbrace{(z - z_0)^\ell}_{\to 0} \underbrace{(z - z_0)^m f(z) g(z)}_{\to 0} \xrightarrow{z \to z_0} \frac{1}{c_{-\ell}} 0 \cdot 0 = 0.$$

So g has a pole at z_0 or a removable singularity at z_0, a contradiction. Consequently, fg has an essential singularity at z_0.

Solution to Exercise 4.68. For $n \in \mathbb{N}$, set $\epsilon := \frac{1}{n} =: \delta$ (> 0). By the Casorati-Weierstrass Theorem, there exists a z_n in the punctured disc $D(z_0, \delta) \backslash \{z_0\}$, such that $|f(z_n) - w| < \epsilon$. So for all $n \in \mathbb{N}$, $|z_n - z_0| < \delta = \frac{1}{n}$ and $|f(z_n) - w| < \epsilon = \frac{1}{n}$. So $(z_n)_{n \in \mathbb{N}}$ converges to z_0, and $(f(z_n))_{n \in \mathbb{N}}$ converges to w.

Solution to Exercise 4.69. Let the Taylor series of f be $f(z) = \sum\limits_{n=0}^{\infty} c_n z^n$, $z \in \mathbb{C}$. As f is not a polynomial, infinitely many c_n are nonzero. Define $g \in \mathcal{O}(\mathbb{C} \backslash \{0\})$ by $g(\zeta) = f(\frac{1}{\zeta})$ for all $\zeta \in \mathbb{C} \backslash \{0\}$. Then $g(\zeta) = \sum\limits_{n=0}^{\infty} c_n \zeta^{-n}$ for all $\zeta \in \mathbb{C} \backslash \{0\}$, and so g has an essential singularity at 0. By Exercise 4.68, there exists a sequence $(\zeta_n)_{n \in \mathbb{N}}$ in $\mathbb{C} \backslash \{0\}$ which converges to 0, while $(g(\zeta_n))_{n \in \mathbb{N}}$ converges to w. If $z_n := \frac{1}{\zeta_n}$, $n \in \mathbb{N}$, then $(z_n)_{n \in \mathbb{N}}$ converges to ∞, and $(f(z_n))_{n \in \mathbb{N}} = (g(\zeta_n))_{n \in \mathbb{N}}$ converges to w.

Solution to Exercise 4.70. $1 + \exp z = 0 \Leftrightarrow z \in \{\pi i + 2\pi n i : n \in \mathbb{Z}\} =: Z$. Each zero $z \in Z$ of $1 + \exp$ has order 1, since $(1 + \exp)'(\pi i + 2\pi n i) = \exp(\pi i) = -1 \neq 0$. The function f defined by $f(z) := \frac{\mathrm{Log}\, z}{1 + \exp z}$ is holomorphic in $(\mathbb{C} \backslash ((-\infty, 0] \times \{0\})) \backslash Z$, and has singularities only at the points of Z, each of which is an isolated singularity, and in fact each is a pole of order 1 (easily checked using the classification of singularities via limits). Of these poles, exactly two lie inside the given path γ: $-\pi i$ and $3\pi i$.

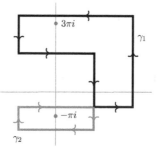

We have $\int_\gamma f(z)\,dz = \int_{\gamma_1} f(z)\,dz + \int_{\gamma_2} f(z)\,dz = 2\pi i(\text{Res}(f,3\pi i) - \text{Res}(f,-\pi i))$.

Writing $\frac{\text{Log}\,z}{1+\exp z} = \frac{c_{-1,3\pi i}}{z-3\pi i} + h_{3\pi i}$, where $h_{3\pi i} \in \mathcal{O}(D(3\pi i, r))$ for some $r > 0$, we have

$$c_{-1,3\pi i} = \lim_{z\to 3\pi i} \frac{(z-3\pi i)\text{Log}\,z}{1+\exp z} = \lim_{z\to 3\pi i} \frac{z-3\pi i}{\exp z - \exp(3\pi i)}\text{Log}\,z = \frac{1}{\exp'(3\pi i)}\lim_{z\to 3\pi i}\text{Log}\,z$$

$$= \frac{1}{\exp(3\pi i)}\text{Log}(3\pi i) = -(\log|3\pi i| + i\tfrac{\pi}{2}) = -\log 3 - \log\pi - i\tfrac{\pi}{2}.$$

Writing $\frac{\text{Log}\,z}{1+\exp z} = \frac{c_{-1,-\pi i}}{z-(-\pi i)} + h_{-\pi i}$, where $h_{-\pi i} \in \mathcal{O}(D(-\pi i,\tilde r))$ for some $\tilde r > 0$,

$$c_{-1,-\pi i} = \lim_{z\to -\pi i} \frac{(z-(-\pi i))\text{Log}\,z}{1+\exp z} = \lim_{z\to -\pi i} \frac{z-(-\pi i)}{\exp z - \exp(-\pi i)}\text{Log}\,z = \frac{1}{\exp'(-\pi i)}\lim_{z\to -\pi i}\text{Log}\,z$$

$$= \frac{1}{\exp(-\pi i)}\text{Log}(-\pi i) = -(\log|-\pi i| + i(-\tfrac{\pi}{2})) = -\log\pi + i\tfrac{\pi}{2}.$$

So $\int_\gamma \frac{\text{Log}\,z}{1+\exp z}\,dz = 2\pi i(-\log 3 - \log\pi - i\tfrac{\pi}{2} + \log\pi - i\tfrac{\pi}{2}) = 2\pi^2 - (2\pi\log 3)i$.

Solution to Exercise 4.71. Let $C(t) = e^{it}$, $t \in [-\pi,\pi]$. Then

$$\int_{-\pi}^\pi \frac{1}{2+\cos\theta}\,d\theta = \int_{-\pi}^\pi \frac{1}{2+\frac{e^{i\theta}+e^{-i\theta}}{2}}\frac{ie^{i\theta}}{ie^{i\theta}}\,d\theta = \int_C \frac{1}{2+\frac{z+1/z}{2}}\frac{1}{iz}\,dz = \int_C \frac{1}{2+\frac{z^2+1}{2z}}\frac{1}{iz}\,dz$$

$$= \int_C \frac{2z}{4z+z^2+1}\frac{1}{iz}\,dz = \frac{2}{i}\int_C \frac{1}{z^2+4z+1}\,dz = \frac{2}{i}\int_C \frac{1}{(z-p_1)(z-p_2)}\,dz,$$

where $p_1 := -2+\sqrt 3$, $p_2 := -2-\sqrt 3$. As $|p_1| < 1 < |p_2|$, by the Residue Theorem,

$$\int_{-\pi}^\pi \frac{1}{2+\cos\theta}\,d\theta = \frac{2}{i}2\pi i\,\text{Res}\Big(\frac{1}{(z-p_1)(z-p_2)}, p_1\Big) = 4\pi\lim_{z\to p_1}(z-p_1)\frac{1}{(z-p_1)(z-p_2)} = \frac{4\pi}{p_1-p_2} = \frac{2\pi}{\sqrt 3}.$$

Solution to Exercise 4.72. Let $\gamma(\theta) := e^{i\theta}$ for $\theta \in [0,2\pi)$. Then

$$\int_0^{2\pi} \frac{\cos\theta}{5+4\cos\theta}\,d\theta = \int_\gamma \frac{\frac{z+\frac12}{2}}{5+4\frac{z+\frac12}{2}}\frac{1}{iz}\,dz = \int_\gamma \frac{z^2+1}{2iz(2z^2+5z+2)}\,dz = \int_\gamma \frac{z^2+1}{2iz(2z+1)(z+2)}\,dz.$$

Let $f(z) := \frac{z^2+1}{2iz(2z+1)(z+2)}$. Then f has three poles, at 0, $-\frac12$, -2, and each is of order 1 (which can be seen using the classification of singularities via limits). Of these, the poles at 0 and $-\frac12$ lie inside γ. So by the Residue Theorem,

$$\int_0^{2\pi} \frac{\cos\theta}{5+4\cos\theta}\,d\theta = 2\pi i(\text{Res}(f,0) + \text{Res}(f,-\tfrac12))$$

$$= 2\pi i\Big(\lim_{z\to 0}\frac{z(z^2+1)}{2iz(2z+1)(z+2)} + \lim_{z\to -\frac12}\frac{(z+\frac12)(z^2+1)}{2iz(2z+1)(z+2)}\Big)$$

$$= 2\pi i\Big(\frac{1}{2i\cdot 1\cdot 2} + \frac{1\cdot\frac54}{2i\cdot(-\frac12)\cdot 2\cdot\frac32}\Big) = 2\pi i(\tfrac{1}{4i} - \tfrac{5}{12i}) = -\tfrac{\pi}{3}.$$

Solution to Exercise 4.73.

For $z \neq 0$, $\frac{\exp z}{z^{n+1}} = \frac{1}{z^{n+1}}(1 + \frac{z}{1!} + \frac{z^2}{2!} + \cdots + \frac{z^n}{n!} + \cdots) = \frac{1}{z^{n+1}} + \cdots + \frac{1}{n!}\frac{1}{z} + \cdots$.

Thus $\text{Res}(\frac{\exp z}{z^{n+1}},0) = \frac{1}{n!}$. Also, 0 is a pole of order $n+1$ of $\frac{\exp z}{z^{n+1}}$. By the Residue Theorem, $\int_C \frac{\exp z}{z^{n+1}}\,dz = 2\pi i\,\text{Res}(\frac{\exp z}{z^{n+1}},0) = \frac{2\pi i}{n!}$. Hence

$$\frac{2\pi i}{n!} = \int_0^{2\pi} \frac{e^{\cos\theta+i\sin\theta}}{(e^{i\theta})^{n+1}}ie^{i\theta}\,d\theta = i\int_0^{2\pi} e^{\cos\theta}e^{i\sin\theta}e^{-in\theta}\,d\theta$$

$$= i\int_0^{2\pi} e^{\cos\theta}(\cos(n\theta-\sin\theta) - i\sin(n\theta-\sin\theta))\,d\theta.$$

Equating the imaginary parts, we obtain $\int_0^{2\pi} e^{\cos\theta}\cos(n\theta-\sin\theta)\,d\theta = \frac{2\pi}{n!}$.

Solution to Exercise 4.74. For z in a punctured disc $D(z_0,r)\backslash\{z_0\}$, $r > 0$, $f(z) \neq 0$ and $f(z) = (z - z_0)h(z)$ for some $h \in \mathcal{O}(D(z_0,r))$ such that $h(z_0) \neq 0$. From $f = (z - z_0)h$, we have $f'(z) = h(z) + (z - z_0)h'(z)$, and so $f'(z_0) = h(z_0)$. As $\frac{1}{h} \in \mathcal{O}(D(z_0,r))$, $\frac{1}{h(z)} = d_0 + d_1(z - z_0) + \cdots$ for $z \in D(z_0,r)$ with $d_0 = \frac{1}{h(z_0)} = \frac{1}{f'(z_0)}$. So $\frac{1}{f(z)} = \frac{1}{z-z_0}(d_0 + d_1(z - z_0) + \cdots) = \frac{d_0}{z-z_0} + d_1 + \cdots$ for $z \in D(z_0,r)\backslash\{z_0\}$. Hence $\mathrm{Res}(\frac{1}{f}, z_0) = d_0 = \frac{1}{f'(z_0)}$.

Solution to Exercise 4.75. The set of zeroes of $\sin$ is $\{k\pi : k \in \mathbb{Z}\}$, and each zero has order 1. By Exercise 4.74, $\mathrm{Res}(\frac{1}{\sin z}, k\pi) = \frac{1}{\sin' z|_{z=k\pi}} = \frac{1}{\cos(k\pi)} = \frac{1}{(-1)^k} = (-1)^k$.

Solution to Exercise 4.76. Let C be given by $C(t) = e^{it}$ for all $t \in [0, 2\pi]$. Then $\int_0^{2\pi}(\sin t)^{2n}dt = \int_0^{2\pi}(\frac{e^{it}-e^{-it}}{2i})^{2n}\frac{1}{ie^{it}}ie^{it}dt = \int_C(\frac{z-z^{-1}}{2i})^{2n}\frac{1}{iz}dz = \frac{1}{2^{2n}(-1)^ni}\int_C(z - \frac{1}{z})^{2n}\frac{1}{z}dz$. Define the function $f \in \mathcal{O}(\mathbb{C}\backslash\{0\})$ by $f(z) := (z - \frac{1}{z})^{2n}\frac{1}{z}$, $z \in \mathbb{C}\backslash\{0\}$. By the Binomial Theorem, $f(z) = (z^{2n} + \cdots + (-1)^n\binom{2n}{n} + \cdots + \frac{1}{z^{2n}})\frac{1}{z}$, $z \in \mathbb{C}\backslash\{0\}$. So f has a pole at $z = 0$. Also, $\mathrm{Res}(f, 0) = (-1)^n\binom{2n}{n}$. The Residue Theorem gives $\int_C f(z)dz = 2\pi i\,\mathrm{Res}(f, 0)$. So $\int_0^{2\pi}(\sin t)^{2n}dt = \frac{1}{2^{2n}(-1)^ni}2\pi i(-1)^n\binom{2n}{n} = \frac{\pi}{2^{2n-1}}\binom{2n}{n}$.

Solution to Exercise 4.77.

(1) For $z \in \mathbb{C}\backslash\{0\}$, we have
$$g(z)(1+\tfrac{1}{z^2}) = (g(0) + g'(0)z + \tfrac{g''(0)}{2!}z^2 + \cdots) + (\tfrac{g(0)}{z^2} + \tfrac{g'(0)}{z} + \tfrac{g''(0)}{2!} + \cdots),$$
and so $\mathrm{Res}(g(z)(1+\tfrac{1}{z^2}), 0) = $ coefficient of $z^{-1} = g'(0)$.

(2) Let $C(t) = e^{it}$, $t \in [0, 2\pi]$. Then by the Residue Theorem,
$$\begin{aligned}\int_0^{2\pi} g(e^{it})\cos t\,dt &= \int_0^{2\pi} g(e^{it})\tfrac{e^{it}+e^{-it}}{2}\tfrac{1}{ie^{it}}ie^{it}dt = \int_C g(z)\tfrac{z+\frac{1}{z}}{2}\tfrac{1}{iz}dz \\ &= \tfrac{1}{2i}\int_C g(z)(1+\tfrac{1}{z^2})\,dz = \tfrac{1}{2i}2\pi i\,\mathrm{Res}(g(z)(1+\tfrac{1}{z^2}), 0) \\ &= \tfrac{1}{2i}2\pi i\,g'(0) = \pi g'(0).\end{aligned}$$

Alternatively, one could use the Cauchy Integral Formula:
$$\begin{aligned}\int_0^{2\pi} g(e^{it})\cos t\,dt &= \tfrac{1}{2i}\int_C g(z)(1+\tfrac{1}{z^2})\,dz = \tfrac{1}{2i}(\int_C g(z)\,dz + \int_C \tfrac{g(z)}{z^2}\,dz) \\ &= \tfrac{1}{2i}(0 + 2\pi i\,g'(0)) = \pi g'(0).\end{aligned}$$

Solution to Exercise 4.78.

(1) $f_0 = 1 \leqslant 2^0 = 1$, $f_1 = 1 \leqslant 2^1 = 2$. If for some $n \geqslant 1$, $f_m \leqslant 2^m$ for all $m \leqslant n$, then $f_{n+1} = f_n + f_{n-1} \leqslant 2^n + 2^{n-1} = 2^{n-1}3 < 2^{n-1}4 = 2^{n+1}$.

(2) For $|z| < \frac{1}{2}$, $\sqrt[n]{|f_n z^n|} = \sqrt[n]{|f_n|}\,|z| \leqslant \sqrt[n]{2^n}|z| = 2|z| < 1$ for all $n \in \mathbb{N}$. By the Root Test, $\sum_{n=0}^{\infty}|f_n z^n|$ converges if $|z| < \frac{1}{2}$. So the radius of convergence of F is $\geqslant \frac{1}{2}$.

(3) For $|z| < \frac{1}{2}$, $zF(z) = f_0 z + f_1 z^2 + f_2 z^3 + \cdots$ and $z^2 F(z) = f_0 z^2 + f_1 z^3 + \cdots$. Thus
$$\begin{aligned}zF(z) + z^2 F(z) &= 1z + (f_1 + f_0)z^2 + (f_2 + f_1)z^3 + \cdots \\ &= f_1 z + f_2 z^2 + f_3 z^3 + \cdots \\ &= (f_0 + f_1 z + f_2 z^2 + f_3 z^3 + \cdots) - f_0 = F(z) - 1.\end{aligned}$$
So for $|z| < \frac{1}{2}$, $1 = F(z) - zF(z) - z^2 F(z) = (1 - z - z^2)F(z)$, and $F(z) = \frac{1}{1-z-z^2}$.

(4) For $|z| < \frac{1}{2}$, $\frac{1}{z^{n+1}(1-z-z^2)} = \frac{F(z)}{z^{n+1}} = \frac{f_0 + \cdots + f_{n-1}z^{n-1} + f_n z^n + f_{n+1}z^{n+1} + \cdots}{z^{n+1}}$

$$= \frac{f_0}{z^{n+1}} + \frac{f_1}{z^n} + \cdots + \frac{f_n}{z} + f_{n+1} + f_{n+2}z + \cdots,$$

and so $\text{Res}(\frac{1}{z^{n+1}(1-z-z^2)}, 0) = f_n$.

(5) For $|z| = R > 2$: $|1 - z - z^2| \geqslant |z^2 + z| - 1 = |z||z+1| - 1 = R|z+1| - 1$

$$\geqslant R(|z|-1) - 1 = R(R-1) - 1 = R^2 - R - 1$$
$$> 2R - R - 1 = R - 1 > 2 - 1 > 0.$$

With $C_R(t) := Re^{it}$, $t \in [0, 2\pi]$, $|\int_{C_R} \frac{1}{z^{n+1}(1-z-z^2)} dz| \leqslant \frac{1}{R^{n+1}} \frac{1}{R^2-R-1} 2\pi R \xrightarrow{R \to \infty} 0$.

If $G(z) := \frac{1}{z^{n+1}(1-z-z^2)}$, then G has a pole at 0 of order $n+1$, a pole at $\frac{-1+\sqrt{5}}{2}$ of order 1, and a pole at $\frac{-1-\sqrt{5}}{2}$ of order 1. The Residue Theorem yields $\text{Res}(G, 0) + \text{Res}(G, \frac{-1+\sqrt{5}}{2}) + \text{Res}(G, \frac{-1-\sqrt{5}}{2}) = \frac{1}{2\pi i} \int_{C_R} G(z) dz$ for $R > 2$. Letting $R \to \infty$, $\text{Res}(G, 0) + \text{Res}(G, \frac{-1+\sqrt{5}}{2}) + \text{Res}(G, \frac{-1-\sqrt{5}}{2}) = \lim_{R \to 0} \frac{1}{2\pi i} \int_{C_R} G(z) dz = 0$, i.e., $f_n = \text{Res}(G, 0) = -\text{Res}(G, \frac{-1+\sqrt{5}}{2}) - \text{Res}(G, \frac{-1-\sqrt{5}}{2})$. If $\tau = \frac{\sqrt{5}+1}{2}$, then we get $\frac{1}{\tau} = \frac{\sqrt{5}-1}{2}$. So $\text{Res}(G, \frac{-1+\sqrt{5}}{2}) = \lim_{z \to \frac{1}{\tau}} (z - \frac{1}{\tau}) \frac{1}{z^{n+1}(1-z-z^2)} = \frac{1}{(\frac{1}{\tau})^{n+1}(-\sqrt{5})} = -\frac{\tau^{n+1}}{\sqrt{5}}$.

Also, $\text{Res}(G, \frac{-1-\sqrt{5}}{2}) = \frac{1}{(\frac{-1-\sqrt{5}}{2})^{n+1}\sqrt{5}} = (\frac{1-\sqrt{5}}{2})^{n+1}(\frac{1}{\sqrt{5}})$. Hence we obtain that

$$f_n = \frac{1}{\sqrt{5}}(\frac{1+\sqrt{5}}{2})^{n+1} - \frac{1}{\sqrt{5}}(\frac{1-\sqrt{5}}{2})^{n+1} = \frac{1}{\sqrt{5}}((\frac{1+\sqrt{5}}{2})^{n+1} - (\frac{1-\sqrt{5}}{2})^{n+1}).$$

Solution to Exercise 4.79.

(1) The poles of f are at the zeroes of $q(z) := z^{2022} - 1$, at $p_k := e^{i(0 + \frac{2\pi}{2022}k)}$, $k = 0, 1, \cdots, 2021$. These p_k are distinct, lying at the vertices of a regular 2022-gon inscribed in the unit circle, with a vertex at 1. We have

$$\lim_{z \to p_k} (z - p_k) f(z) = \lim_{z \to p_k} \frac{1}{\frac{z^{2022} - p_k^{2022}}{z - p_k}} = \frac{1}{\frac{dz^{2022}}{dz}|_{z = p_k}} = \frac{1}{2022 p_k^{2021}}$$

$$= \frac{p_k}{2022 p_k^{2022}} = \frac{p_k}{2022 \cdot 1} = \frac{p_k}{2022} \neq 0.$$

So $\lim_{z \to p_k} (z - p_k)^2 f(z) = 0$. This p_k is a pole of f or order 1, $k = 0, \cdots, 2021$.

(2) Near each pole p_k (in a punctured disc $D(p_k, r) \setminus \{p_k\}$, $r_k > 0$), f has a Laurent series $f(z) = \frac{c_{-1}^{(k)}}{z - p_k} + c_0^{(k)} + \cdots$. So $(z - p_k) f(z) = c_{-1}^{(k)} + c_0^{(k)}(z - p_k) + \cdots$. Hence $\text{Res}(f, p_k) = c_{-1}^{(k)} = \lim_{z \to p_k} (z - p_k) f(z) = \frac{p_k}{2022}$ by the calculation above.

(3) By the Residue Theorem, $\int_C f(z) dz = 2\pi i \sum_{k=0}^{2021} \text{Res}(f, p_k) = \frac{2\pi i}{2022} \sum_{k=0}^{2021} p_k$. But the even number of points p_k, $k = 0, \cdots, 2021$, are symmetrically arranged on the unit circle (with $p_0 = -p_{1011}$, $p_1 = -p_{1012}$, $\cdots$, $p_{1010} = -p_{2021}$), and so they all add up to 0. (Alternatively, $z^{2022} - 1 = \prod_{k=0}^{2021} (z - p_k)$, and as the coefficient of z^{2021} on the left-hand side is 0, the sum of the roots p_k is 0. See Exercise 1.29.)

Solution to Exercise 4.80.

(1) Let $f_1(z) = \frac{1}{1+z^2}$. Then f_1 has poles at i and $-i$, both of order 1. Hence

$$\int_0^\infty \frac{1}{1+x^2}dx = \tfrac{1}{2}2\pi i\,\mathrm{Res}(f_1,i) = \pi i \lim_{z\to i}\frac{z-i}{1+z^2} = \pi i \lim_{z\to i}\frac{1}{z+i} = \pi i\frac{1}{2i} = \frac{\pi}{2}.$$

(2) Let $f_2(z) := \frac{1}{(a^2+z^2)(b^2+z^2)}$. Then f_2 has poles at $ai, -ai, bi, -bi$, all of order 1.

As f_2 is even, $\int_0^\infty \frac{1}{(a^2+x^2)(b^2+x^2)}dx = \tfrac{1}{2}2\pi i(\mathrm{Res}(f_2,ai)+\mathrm{Res}(f_2,bi))$

$$= \pi i(\tfrac{1}{(b^2-a^2)2ai} + \tfrac{1}{(a^2-b^2)2bi}) = \frac{\pi}{2ab(a+b)}.$$

(3) Let $f_3(z) = \frac{1}{(1+z^2)^2}$. Then f_3 has poles at i and $-i$, both of order 2. We have

$$\int_0^\infty \frac{1}{(1+x^2)^2}dx = \tfrac{1}{2}2\pi i\,\mathrm{Res}(f_3,i) = \frac{\pi i}{1!}\lim_{z\to i}\frac{d}{dz}((z-i)^2\frac{1}{(z-i)^2(z+i)^2})$$

$$= \pi i \lim_{z\to i}\frac{-2}{(z+i)^3} = \pi i\frac{-2}{-8i} = \frac{\pi}{4}.$$

(4) Let $f_4(z) = \frac{1+z^2}{1+z^4}$. Then f_4 has poles at $p_1 = e^{\frac{\pi i}{4}}$, $p_2 = e^{\frac{3\pi i}{4}}$, $p_3 = e^{\frac{5\pi i}{4}}$, $p_4 = e^{\frac{7\pi i}{4}}$, all of order 1. We have $p_1+p_2 = \sqrt{2}i$ and $p_1p_2 = -1$. By the Residue Theorem,

$$\int_0^\infty \frac{1+x^2}{1+x^4}dx = \tfrac{1}{2}2\pi i(\mathrm{Res}(f_4,p_1)+\mathrm{Res}(f_4,p_2)) = \pi i(\frac{1+p_1^2}{4p_1^3}+\frac{1+p_2^2}{4p_2^3})$$

$$= \pi i(\frac{p_1}{4p_1^4}+\frac{1}{4p_1}+\frac{p_2}{4p_2^4}+\frac{1}{4p_2}) = \pi i(-\frac{p_1+p_2}{4}+\frac{1}{4p_1}+\frac{1}{4p_2})$$

$$= \pi i(-\frac{\sqrt{2}i}{4}+\frac{\sqrt{2}i}{4(-1)}) = \frac{\pi}{\sqrt{2}}.$$

Solution to Exercise 4.81. $\lim_{z\to i}(z-i)\frac{e^{-i\xi z}}{1+z^2} = \frac{e^\xi}{2i}\neq 0$, $\lim_{z\to -i}(z+i)\frac{e^{-i\xi z}}{1+z^2} = \frac{e^{-\xi}}{-2i}\neq 0$.

So by Corollary 4.17, $f(z) := \frac{e^{-i\xi z}}{1+z^2}$ has two poles, at i and at $-i$, both of order 1.

$1°$ $\xi < 0$. Let γ_+ be a path comprising a semicircle σ_+ centred at 0 of radius $r > 0$ in the upper half-plane, together with a line segment path from $-r$ to r, as on p. 110. Then $\int_{\sigma_+} f(z)dz+\int_{-r}^r f(x)dx = \int_{\gamma_+} f(z)dz = 2\pi i\,\mathrm{Res}(f,i) = 2\pi i\frac{e^\xi}{2i} = \pi e^\xi$. For $r > 1$ and $t \in [0,\pi]$, $|1+r^2e^{2it}| \geq r^2-1$, $|e^{-i\xi e^{it}}| = e^{\mathrm{Re}(-i\xi re^{it})} = e^{\xi r\sin t} \leq 1$. Thus $|\int_{\sigma_+}\frac{e^{-i\xi z}}{1+z^2}dz| \leq \pi r\max_{t\in[0,\pi]}\frac{|e^{-i\xi re^{it}}|}{|1+r^2e^{2it}|} \leq \pi r\frac{1}{r^2-1}\xrightarrow{r\to\infty} 0$. So if $\xi < 0$, then

$$\int_{-\infty}^\infty \frac{e^{-i\xi x}}{1+x^2}dx = \lim_{r\to\infty}\int_{-r}^r f(x)dx = \pi e^\xi.$$

$2°$ $\xi \geq 0$. Let γ_- be a path comprising a semicircle σ_- centred at 0 of radius $r > 0$ starting at $-r$ and ending at r in the lower half-plane, together with a line segment path from r to $-r$. Then

$$\int_{\sigma_-} f(z)dz - \int_{-r}^r f(x)dx = \int_{\gamma_-} f(z)dz = 2\pi i\,\mathrm{Res}(f,-i) = 2\pi i\frac{e^{-\xi}}{-2i} = -\pi e^{-\xi}.$$

For $r > 1$ and $t\in[-\pi,0]$, $|1+r^2e^{2it}| \geq r^2-1$, $|e^{-i\xi re^{it}}| = e^{\mathrm{Re}(-i\xi re^{it})} = e^{\xi r\sin t} \leq 1$. Thus $|\int_{\sigma_-}\frac{e^{-i\xi z}}{1+z^2}dz| \leq \pi r\max_{t\in[-\pi,0]}\frac{|e^{-i\xi re^{it}}|}{|1+r^2e^{2it}|} \leq \pi r\frac{1}{r^2-1}\xrightarrow{r\to\infty} 0$. So if $\xi \geq 0$, then

$$\int_{-\infty}^\infty \frac{e^{-i\xi x}}{1+x^2}dx = \lim_{r\to\infty}\int_{-r}^r f(x)dx = -(-\pi e^{-\xi}) = \pi e^{-\xi}.$$

Summarising, $\int_{-\infty}^\infty \frac{e^{-i\xi x}}{1+x^2}dx = \pi e^{-|\xi|}$.

Solution to Exercise 4.82.

(1) $\lim\limits_{z\to 0} z^3 f(z) = \lim\limits_{z\to 0} z^3 \dfrac{\cos(\pi z)}{z^2 \sin(\pi z)} = \lim\limits_{z\to 0} \cos(\pi z)\dfrac{1}{\frac{\sin(\pi z)-\sin 0}{\pi z - 0}}\dfrac{1}{\pi} = (\cos 0)\dfrac{1}{\sin' 0}\dfrac{1}{\pi} = \dfrac{1}{\pi} \neq 0.$

So 0 is a pole of f of order 3.

Let $\dfrac{\cos(\pi z)}{z^2 \sin(\pi z)} = \dfrac{c_{-3}}{z^3} + \dfrac{c_{-2}}{z^2} + \dfrac{c_{-1}}{z} + c_0 + c_1 z + \cdots$ in $D(0,1)\backslash\{0\}$. Then

$$c_{-3} = \lim\limits_{z\to 0} z^3 \dfrac{\cos(\pi z)}{z^2 \sin(\pi z)} = \dfrac{1}{\pi} \text{ (as found above).}$$

Thus $z^2\left(\dfrac{\cos(\pi z)}{z^2 \sin(\pi z)} - \dfrac{1}{\pi z^3}\right) = c_{-2} + c_{-1}z + \cdots$ in $D(0,1)\backslash\{0\}$, so that

$$c_{-2} = \lim\limits_{z\to 0} z^2\left(\dfrac{\cos(\pi z)}{z^2 \sin(\pi z)} - \dfrac{1}{\pi z^3}\right) = \lim\limits_{z\to 0} \dfrac{\pi z\cos(\pi z)-\sin(\pi z)}{\pi z \sin(\pi z)} = \lim\limits_{z\to 0} \dfrac{\frac{\pi z\cos(\pi z)-\sin(\pi z)}{(\pi z)^2}}{\frac{\sin(\pi z)}{\pi z}}.$$

For $z \in D(0,1)\backslash\{0\}$,

$$\dfrac{\pi z\cos(\pi z)-\sin(\pi z)}{(\pi z)^2} = \dfrac{\pi z(1-\frac{(\pi z)^2}{2!}+\frac{(\pi z)^4}{4!}-\cdots)-(\pi z-\frac{(\pi z)^3}{3!}+\frac{(\pi z)^5}{5!}-\cdots)}{(\pi z)^2} = -\dfrac{\pi z}{3} + \dfrac{(\pi z)^3}{30} + \cdots.$$

Thus $\lim\limits_{z\to 0} \dfrac{\pi z\cos(\pi z)-\sin(\pi z)}{(\pi z)^2} = 0$, and so

$$c_{-2} = \lim\limits_{z\to 0} \dfrac{\frac{\pi z\cos(\pi z)-\sin(\pi z)}{(\pi z)^2}}{\frac{\sin(\pi z)}{\pi z}} = \dfrac{\lim\limits_{z\to 0}\frac{\pi z\cos(\pi z)-\sin(\pi z)}{(\pi z)^2}}{\lim\limits_{z\to 0}\frac{\sin(\pi z)}{\pi z}} = \dfrac{0}{1} = 0.$$

Hence $z\left(\dfrac{\cos(\pi z)}{z^2 \sin(\pi z)} - \dfrac{1}{\pi z^3}\right) = c_{-1} + c_0 z + \cdots$ in $D(0,1)\backslash\{0\}$, so that

$$c_{-1} = \lim\limits_{z\to 0} z\left(\dfrac{\cos(\pi z)}{z^2 \sin(\pi z)} - \dfrac{1}{\pi z^3}\right) = \lim\limits_{z\to 0} \dfrac{\frac{\pi z\cos(\pi z)-\sin(\pi z)}{\pi^2 z^3}}{\frac{\sin(\pi z)}{\pi z}}.$$

For $z \in D(0,1)\backslash\{0\}$, as found above, $\dfrac{\pi z\cos(\pi z)-\sin(\pi z)}{\pi^2 z^3} = -\dfrac{\pi}{3} + \dfrac{\pi^3 z^2}{30} + \cdots.$

Thus $\lim\limits_{z\to 0} \dfrac{\pi z\cos(\pi z)-\sin(\pi z)}{\pi^2 z^3} = -\dfrac{\pi}{3}$. So

$$\mathrm{Res}(f,0) = c_{-1} = \lim\limits_{z\to 0} \dfrac{\frac{\pi z\cos(\pi z)-\sin(\pi z)}{\pi^2 z^3}}{\frac{\sin(\pi z)}{\pi z}} = \dfrac{-\frac{\pi}{3}}{1} = -\dfrac{\pi}{3}.$$

(2) For $n \in \mathbb{Z}\backslash\{0\}$,

$$\lim\limits_{z\to n} (z-n)f(z) = \lim\limits_{z\to n}(z-n)\dfrac{\cos(\pi z)}{z^2 \sin(\pi z)} = \lim\limits_{z\to n} \dfrac{\cos(\pi z)}{z^2}\dfrac{1}{\frac{\sin(\pi z)-\sin(n\pi)}{z-n}}$$

$$= \dfrac{\cos(\pi n)}{n^2}\dfrac{1}{\frac{d}{dz}\sin(\pi z)|_{z=n}} = \dfrac{(-1)^n}{n^2}\dfrac{1}{\pi(-1)^n} = \dfrac{1}{n^2\pi} \neq 0.$$

So n is a pole of f of order 1.

Let $\dfrac{\cos(\pi z)}{z^2 \sin(\pi z)} = \dfrac{c_{-1}}{z-n} + c_0 + c_1(z-n) + \cdots$ in $D(n,1)\backslash\{n\}$. Then

$$c_{-1} = \lim\limits_{z\to n}(z-n)\dfrac{\cos(\pi z)}{z^2 \sin(\pi z)} = \dfrac{1}{n^2\pi} \text{ (as found above).}$$

(3) We have $|\cos(\pi z)| = |\sin(i\pi y)|$ because

$$\cos(\pi z) = \cos(\pi(\pm(N+\tfrac{1}{2})+iy))$$
$$= \cos(\pm(\pi N + \tfrac{\pi}{2}))\cos(i\pi y) - \sin(\pm(\pi N + \tfrac{\pi}{2}))\sin(i\pi y)$$
$$= 0 - \pm(-1)^N \sin(i\pi y).$$

Similarly,

$$\begin{aligned}
\sin(\pi z) &= \sin(\pi(\pm(N + \tfrac{1}{2}) + iy)) \\
&= \sin(\pm(\pi N + \tfrac{\pi}{2}))\cos(i\pi y) + \cos(\pm(\pi N + \tfrac{\pi}{2}))\sin(i\pi y) \\
&= \pm(-1)^N \cos(i\pi y) + 0,
\end{aligned}$$

and so $|\sin(\pi z)| = |\cos(i\pi y)|$.

(4) The length of each side of the square is $(N + \tfrac{1}{2})|(1 + i) - (1 - i)| = 2N + 1$. So the length of S_N is $4(2N + 1)$. Let $M := \max\{A, 1\}$. Then $|f(z)| \leqslant \frac{M}{|z|^2}$ for all $z \in S_N$. Also, for each $z \in S_N$, $|z| \geqslant \max\{|\mathrm{Re}\,z|, |\mathrm{Im}\,z|\} \geqslant N + \tfrac{1}{2} > N$. So

$$\left| \int_{S_N} f(z)\,dz \right| \leqslant 4(2N + 1) \max_{z \in S_N} \frac{M}{|z|^2} \leqslant 4(2N + 1)\frac{M}{N^2} \xrightarrow{N \to \infty} 0.$$

(5) By the Residue Theorem,

$$\int_{S_N} f(z)\,dz = 2\pi i \sum_{|n| \leqslant N} \mathrm{Res}(f, n) = 2\pi i \left(-\tfrac{\pi}{3} + 2\sum_{n=1}^{N} \tfrac{1}{n^2 \pi}\right).$$

Passing to the limit as $N \to \infty$, and rearranging, we get $\sum_{n=1}^{\infty} \frac{1}{n^2} = \frac{\pi^2}{6}$.

Solutions to the exercises from Chapter 5

Solution to Exercise 5.1.

(1) We have for $(x, y) \in \mathbb{R}^2 \backslash \{(0,0)\}$ that $\frac{\partial u}{\partial x} = \frac{2x}{x^2+y^2}$ and $\frac{\partial^2 u}{\partial x^2} = \frac{2(y^2-x^2)}{(x^2+y^2)^2}$. Similarly, using the symmetry in x and y, $\frac{\partial u}{\partial y} = \frac{2y}{x^2+y^2}$ and $\frac{\partial^2 u}{\partial y^2} = \frac{2(x^2-y^2)}{(x^2+y^2)^2}$. Consequently $\frac{\partial^2 u}{\partial x^2} + \frac{\partial^2 u}{\partial y^2} = \frac{2(y^2-x^2)}{(x^2+y^2)^2} + \frac{2(x^2-y^2)}{(x^2+y^2)^2} = 0$. Since u is $C^2(\mathbb{R}^2 \backslash \{(0,0)\})$ and $\Delta u = 0$ in $\mathbb{R}^2 \backslash \{(0,0)\}$, u is harmonic there.

(2) We have $\frac{\partial u}{\partial x} = e^x \sin y$, $\frac{\partial^2 u}{\partial x^2} = e^x \sin y$, and $\frac{\partial u}{\partial y} = e^x \cos y$, $\frac{\partial^2 u}{\partial y^2} = e^x(-\sin y)$ in $\mathbb{R}^2$. So $\frac{\partial^2 u}{\partial x^2} + \frac{\partial^2 u}{\partial y^2} = e^x \sin y + e^x(-\sin y) = 0$. Since u is $C^2(\mathbb{R}^2)$ and $\Delta u = 0$ in $\mathbb{R}^2$, u is harmonic in $\mathbb{R}^2$.

Solution to Exercise 5.2. Let V be the real vector space of all real-valued functions defined on U, with pointwise operations. We will show that $\mathrm{H}(U)$ is a subspace of this vector space V, and hence a vector space with pointwise operations:

(1) The constant function $\mathbf{0}$ assuming value 0 everywhere on U belongs to $\mathrm{H}(U)$. Indeed $\mathbf{0} \in C^2(U)$ and $\frac{\partial^2 \mathbf{0}}{\partial x^2} + \frac{\partial^2 \mathbf{0}}{\partial y^2} = 0 + 0 = 0$.

(2) If $u, v \in \mathrm{H}(U)$, then
$$\frac{\partial^2(u+v)}{\partial x^2} + \frac{\partial^2(u+v)}{\partial y^2} = \frac{\partial^2 u}{\partial x^2} + \frac{\partial^2 v}{\partial x^2} + \frac{\partial^2 u}{\partial y^2} + \frac{\partial^2 v}{\partial y^2} = \left(\frac{\partial^2 u}{\partial x^2} + \frac{\partial^2 u}{\partial y^2}\right) + \left(\frac{\partial^2 v}{\partial x^2} + \frac{\partial^2 v}{\partial y^2}\right) = 0 + 0 = 0.$$

(3) If $\alpha \in \mathbb{R}$, $u \in \mathrm{H}(U)$, then $\frac{\partial^2(\alpha \cdot u)}{\partial x^2} + \frac{\partial^2(\alpha \cdot u)}{\partial y^2} = \alpha \frac{\partial^2 u}{\partial x^2} + \alpha \frac{\partial^2 u}{\partial y^2} = \alpha\left(\frac{\partial^2 u}{\partial x^2} + \frac{\partial^2 u}{\partial y^2}\right) = \alpha \cdot 0 = 0$.

The pointwise product of two harmonic functions need not be harmonic. We will show that every open set U admits a harmonic function u such that $u \cdot u = u^2$ is not harmonic in U. Any $u \in C^2(U)$ such that $\neg(\nabla u \equiv 0)$ will work, as we will see. Thus, we may take u to be a linear function, $u(x, y) = ax + by$, where $(a, b) \in \mathbb{R}^2 \backslash \{(0,0)\}$. Then $\frac{\partial u}{\partial x} = a$, $\frac{\partial u}{\partial y} = b$, and $\Delta u = 0 + 0 = 0$. So $u \in \mathrm{H}(U)$. But
$$\frac{\partial(u^2)}{\partial x} = 2u\frac{\partial u}{\partial x}, \quad \frac{\partial^2(u^2)}{\partial x^2} = 2\left(\frac{\partial u}{\partial x}\right)^2 + 2u\frac{\partial^2 u}{\partial x^2}, \text{ and similarly, } \frac{\partial^2(u^2)}{\partial y^2} = 2\left(\frac{\partial u}{\partial y}\right)^2 + 2u\frac{\partial^2 u}{\partial y^2}.$$
Thus $\Delta(u^2) = 2\|\nabla u\|_2^2 + 2u\Delta u = 2\|\nabla u\|_2^2 + 2u 0 = 2\|\nabla u\|_2^2$, which is not identically zero in U. (For the linear $u = ax + by$, $\|\nabla u\|_2^2 = a^2 + b^2 \neq 0$ for all $(x, y) \in U$.)

Solution to Exercise 5.3.

(1) Let $u = e^x \sin y$. We seek a v such that $u + iv$ is holomorphic. So the Cauchy-Riemann equations must be satisfied. Hence $\frac{\partial v}{\partial x} = -\frac{\partial u}{\partial y} = -e^x \cos y$. If we keep y fixed, we obtain by integrating that $v = -e^x \cos y + C(y)$, for some $C(y)$. As v and $-e^x \cos y$ are differentiable with respect to y, so is C. Thus $e^x \sin y + C'(y) = \frac{\partial v}{\partial y} = \frac{\partial u}{\partial x} = e^x \sin y$. So $C'(y) = 0$, giving $C(y) = K$. Taking $K = 0$, we try $v := -e^x \cos y$. Then
$$u + iv = e^x \sin y + i(-e^x \cos y) = -ie^x(\cos y + i \sin y) = -i\exp(x + iy) = -i\exp z,$$
where $z = x + iy$. Hence $u + iv = -i\exp z$, which is entire. Thus $v = -e^x \cos y$ is a harmonic conjugate for $u := e^x \sin y$.

(2) Let $u = x^3 - 3xy^2 - 2y$. We seek a v such that $u + iv$ is holomorphic. So the Cauchy-Riemann equations must be satisfied. In particular, we have that $\frac{\partial v}{\partial x} = -\frac{\partial u}{\partial y} = 6xy + 2$. Fixing y, and integrating with respect to x, we obtain $v = 6\frac{x^2}{2}y + 2x + C(y) = 3x^2y + 2x + C(y)$, for some $C(y)$. Differentiating with respect to y, $3x^2 + C'(y) = \frac{\partial v}{\partial y} = \frac{\partial u}{\partial x} = 3x^2 - 3y^2$ and so $C'(y) = -3y^2$, giving $C(y) = -3\frac{y^3}{3} + C = -y^3 + C$. Taking $C = 0$, we try $v = 3x^2y + 2x - y^3$. Thus

$$u + iv = x^3 - 3xy^2 - 2y + i(3x^2y + 2x - y^3)$$
$$= x^3 + 3x(iy)^2 + 3x^2(iy) + (iy)^3 - 2y + i2x$$
$$= (x + iy)^3 + 2i(x + iy) = z^2 + 2iz$$

for $z = x + iy$. So $u + iv = z^2 + 2iz$, which is entire. Hence $v := 3x^2y + 2x - y^3$ is a harmonic conjugate of $u := x^3 - 3xy^2 - 2y$.

(3) Let $u := x(1 + 2y)$. We seek a v such that $u + iv$ is holomorphic. So the Cauchy-Riemann equations must be satisfied. Hence $\frac{\partial v}{\partial x} = -\frac{\partial u}{\partial y} = -2x$. Fixing y, and integrating, we get $v = -2\frac{x^2}{2} + C(y) = -x^2 + C(y)$, for some $C(y)$. Thus $C'(y) = \frac{\partial v}{\partial y} = \frac{\partial u}{\partial x} = 1 + 2y$. So $C(y) = y + 2\frac{y^2}{2} + C = y + y^2 + C$. Taking $C = 0$, we try $v := -x^2 + y + y^2$. Then

$$u + iv = x(1 + 2y) + i(-x^2 + y + y^2) = x + iy + 2xy + i(y^2 - x^2)$$
$$= x + iy - i((x^2 - y^2) + i2xy) = x + iy - i(x + iy)^2 = z - iz^2$$

for $z = x + iy$. So $u + iv = z - iz^2$ is holomorphic, and $v := -x^2 + y + y^2$ is a harmonic conjugate of $u := x(1 + 2y)$.

Solution to Exercise 5.4. Let v be a harmonic conjugate of u. Then $f := u + iv$ is holomorphic in $\mathbb{C}\backslash\{0\}$. For $z = x + iy \in \mathbb{C}\backslash\{0\}$, $x, y \in \mathbb{R}$, we have

$$f'(z) = \frac{\partial u}{\partial x} + i\frac{\partial v}{\partial x} = \frac{\partial u}{\partial x} - i\frac{\partial u}{\partial y} = \frac{2x}{x^2+y^2} - i\frac{2y}{x^2+y^2} = 2\frac{x-iy}{x^2+y^2} = 2\frac{\bar{z}}{z\bar{z}} = \frac{2}{z}.$$

Thus $\frac{2}{z}$ has a primitive f in $\mathbb{C}\backslash\{0\}$, and so if $C(t) := e^{it}$, $t \in [0, 2\pi]$, then we obtain $4\pi i = \int_C \frac{2}{z}dz = f(1) - f(1) = 0$, which is absurd. Hence u has no harmonic conjugate in $\mathbb{C}\backslash\{0\}$.

Solution to Exercise 5.5. Set $u := x^3 + y^3$. If f were holomorphic, then u would be harmonic. But $\frac{\partial^2 u}{\partial x^2} + \frac{\partial^2 u}{\partial y^2} = \frac{\partial}{\partial x}(3x^2) + \frac{\partial}{\partial y}(3y^2) = 6x + 6y = 6(x + y) \neq 0$ for $x \neq -y$. Hence the answer is 'no'.

Solution to Exercise 5.6. As v is a harmonic conjugate of u, $f := u + iv \in \mathcal{O}(U)$. Then $f' \in \mathcal{O}(U)$, and $f' = \frac{\partial u}{\partial x} + i\frac{\partial v}{\partial x}$. As $0 \neq u^2 + v^2 = (u + iv)(u - iv) = f\bar{f} = |f|^2$, we have $f(z) \neq 0$ for all $z \in U$. Then $f'\frac{1}{f} \in \mathcal{O}(U)$, and its real part is harmonic in U. So $\mathrm{Re}(f'\frac{1}{f}) = \frac{\mathrm{Re}(f'\bar{f})}{|f|^2} = \frac{\mathrm{Re}((\frac{\partial u}{\partial x} + i\frac{\partial v}{\partial x})(u - iv))}{u^2 + v^2} = \frac{u\frac{\partial u}{\partial x} + v\frac{\partial v}{\partial x}}{u^2 + v^2}$ is harmonic in U.

Solution to Exercise 5.7. As v is a harmonic conjugate of u, $f := u + iv \in \mathcal{O}(U)$. Thus $\frac{f^2}{2} \in \mathcal{O}(U)$ too. As $\frac{f^2}{2} = \frac{u^2-v^2}{2} + iuv$, $uv = \mathrm{Im}\frac{f^2}{2}$, and being the imaginary part of a holomorphic function, uv is harmonic in U.

Solution to Exercise 5.8. Let $U \subset \mathbb{C}$ be open. It suffices to show that if u is harmonic in U, then so are $\frac{\partial u}{\partial x}$ and $\frac{\partial u}{\partial y}$. We know that $u \in C^\infty(U)$. We have

$$\frac{\partial^2}{\partial x^2}\left(\frac{\partial u}{\partial x}\right) + \frac{\partial^2}{\partial y^2}\left(\frac{\partial u}{\partial x}\right) = \frac{\partial}{\partial x}\left(\frac{\partial^2 u}{\partial x^2}\right) + \frac{\partial}{\partial y}\left(\frac{\partial^2 u}{\partial y \partial x}\right) = \frac{\partial}{\partial x}\left(-\frac{\partial^2 u}{\partial y^2}\right) + \frac{\partial}{\partial y}\left(\frac{\partial^2 u}{\partial x \partial y}\right)$$

$$= \frac{\partial}{\partial x}\left(-\frac{\partial^2 u}{\partial y^2}\right) + \frac{\partial}{\partial x}\left(\frac{\partial}{\partial y}\left(\frac{\partial}{\partial y}u\right)\right) = \frac{\partial}{\partial x}\left(-\frac{\partial^2 u}{\partial y^2} + \frac{\partial^2 u}{\partial y^2}\right) = \frac{\partial}{\partial x}(0) = 0.$$

Similarly, $\frac{\partial^2}{\partial x^2}\left(\frac{\partial u}{\partial y}\right) + \frac{\partial^2}{\partial y^2}\left(\frac{\partial u}{\partial y}\right) = \frac{\partial}{\partial y}\left(\frac{\partial^2 u}{\partial x^2}\right) + \frac{\partial}{\partial y}\left(\frac{\partial^2 u}{\partial y^2}\right) = \frac{\partial}{\partial y}\left(\frac{\partial^2 u}{\partial x^2} + \frac{\partial^2 u}{\partial y^2}\right) = \frac{\partial}{\partial y}(0) = 0.$

Solution to Exercise 5.9. By the mean-value property, for all $r \in (0, R)$, we have $2\pi u(z_0) = \int_0^{2\pi} u(z_0 + re^{it})dt$. Multiplying by r, and integrating from $r = 0$ to R,

$$\pi R^2 u(z_0) = 2\pi u(z_0) \int_0^R r\, dr = \int_0^R \int_0^{2\pi} u(z_0 + re^{it})dt\, dr = \iint_{D(z_0, R)} u\, dA.$$

Solution to Exercise 5.10. Let $z \in D(z_0, R)$. Then $D(z, R - |z - z_0|) \subset D(z_0, R)$. Using Exercise 5.9, and the nonnegativity of u, we have

$$u(z) = \frac{1}{\pi(R - |z - z_0|)^2} \iint_{D(z, R-|z-z_0|)} u\, dA \leqslant \frac{1}{\pi(R - |z - z_0|)^2} \iint_{D(z_0, R)} u\, dA$$

$$= \frac{\pi R^2}{\pi(R - |z - z_0|)^2} \frac{1}{\pi R^2} \iint_{D(z_0, R)} u\, dA$$

$$= \frac{\pi R^2}{\pi(R - |z - z_0|)^2} u(z_0) = \frac{1}{(1 - \frac{|z - z_0|}{R})^2} u(z_0).$$

Solution to Exercise 5.11. Using Exercise 5.10, we have for any two points $z, z_0 \in \mathbb{C}$ and any $R > |z - z_0|$ that

$$u(z) \leqslant \frac{1}{(1 - \frac{|z - z_0|}{R})^2} u(z_0).$$

Passing to the limit as $R \to \infty$, we obtain $u(z) \leqslant u(z_0)$. As $z, z_0 \in \mathbb{C}$ were arbitrary, we can swap their roles to obtain also $u(z_0) \leqslant u(z)$. Hence $u(z) = u(z_0)$ for all $z \in \mathbb{C}$, that is u is constant.

If u is simply bounded below, say by m, then $u - m \geqslant 0$ everywhere, and it is also clearly harmonic. So $u - m$ is constant, and consequently so is u.

On the other hand, if u is bounded above, then $-u$ is bounded below, and moreover harmonic, showing that $-u$ is constant. Thus u is constant as well.

Solution to Exercise 5.12. The function $\varphi := v - u$ is continuous on the compact set $\mathbb{D} \cup \mathbb{T}$, and so it has a maximiser, say z_0.

If $z_0 \in \mathbb{D}$, then z_0 is a maximiser of the harmonic function φ on the simply connected domain $\mathbb{D}$, and so by the Maximum Principle, φ is a constant in $\mathbb{D}$. By continuity, φ is constant, taking value say C, on $\mathbb{D} \cup \mathbb{T}$. Since the values of $\varphi|_\mathbb{T}$ are pointwise $\leqslant 0$, it follows that $C \leqslant 0$. So for all $z \in \mathbb{D}$, $v(z) - u(z) = \varphi(z) = C \leqslant 0$, i.e., $u(z) \geqslant v(z)$ for all $z \in \mathbb{D}$.

If $z_0 \in \mathbb{T}$, then for all $z \in \mathbb{D}$, $v(z) - u(z) = \varphi(z) \leqslant \varphi(z_0) = v(z_0) - u(z_0) \leqslant 0$, and so $u(z) \geqslant v(z)$ for all $z \in \mathbb{D}$.

Solution to Exercise 5.13.

(1) Let $\varphi(x) = c_0 + c_1 x + \cdots + c_d x^d$, where $d \in \mathbb{N}$ and $c_0, \cdots, c_d \in \mathbb{R}$. Then $\varphi(z) := \varphi(x + iy) = c_0 + c_1 z + \cdots + c_d z^d$ is entire, and so $h := \mathrm{Re}(\varphi(x + iy))$ is harmonic. Moreover for all $x \in \mathbb{R}$, $h(x, 0) = \mathrm{Re}(\varphi(x + i0)) = \mathrm{Re}(\varphi(x)) = \varphi(x)$.

(2) We have $\frac{i}{z+i}$ is holomorphic in the upper half-plane, and so its real part is harmonic there. Moreover, $h(x, 0) = \frac{\mathrm{Re}(i(x - i(0+1)))}{x^2 + (0+1)^2} = \frac{1}{x^2+1} = \varphi(x)$ for all $x \in \mathbb{R}$.

Solution to Exercise 5.14. As $\mathbb{C}$ is simply connected, there exists an $f \in \mathcal{O}(\mathbb{C})$ such that $u = \mathrm{Re}\, f$. Then $\exp(-f)$ is entire too. For all $z \in \mathbb{C}$, $u(z) \geqslant 0$, and so $|\exp(-f(z))| = e^{-\mathrm{Re}\, f(z)} = e^{-u(z)} \leqslant e^0 = 1$. By Liouville's Theorem, $\exp(-f)$ is a constant. Hence $|\exp(-f)|$ is constant too, that is, e^{-u} is constant. Consequently, the real logarithm $\log(e^{-u}) = -u$ is constant, and so u is constant as well.

Solution to Exercise 5.15.

(1) For $z = re^{i\theta}$ ($r > 0$, $\theta \in \mathbb{R}$), $\exp(-\frac{1}{z^4}) = \exp(-\frac{e^{-i4\theta}}{r^4})$. Taking $r = \frac{1}{n}$, $n \in \mathbb{N}$, and $4\theta = -\pi$, i.e., with $z_n := \frac{1}{n}\exp(-i\frac{\pi}{4}) =: x_n + iy_n$, $x_n, y_n \in \mathbb{R}$, we have $u(x_n, y_n) = \exp(-n^4 e^{i\pi}) = e^{n^4}$. Now as $n \to \infty$, we have $(x_n, y_n) \to (0, 0)$, but $\neg\big(\lim_{n \to \infty} u(x_n, y_n) = 0\big)$, showing that u is not continuous at $(0, 0)$.

(2) $u(x, 0) = \exp(-\frac{1}{(x+0i)^4}) = e^{-\frac{1}{x^4}}$, $u(0, y) = \exp(-\frac{1}{(0+yi)^4}) = \exp(-\frac{1}{i^4 y^4}) = e^{-\frac{1}{y^4}}$.

(3) We have $\frac{\partial u}{\partial x}(0, 0) = \lim_{x \to 0} \frac{u(x, 0) - u(0, 0)}{x - 0} = \lim_{x \to 0} \frac{e^{-\frac{1}{x^4}} - 0}{x} = \lim_{x \to 0} \frac{e^{-\frac{1}{x^4}}}{x} = 0$.

(For the last equality, note that for $x \neq 0$, $e^{\frac{1}{x^4}} = 1 + \frac{1}{x^4} + \frac{1}{2!}(\frac{1}{x^4})^2 + \cdots > \frac{1}{x^4}$ and so $0 \leqslant |\frac{e^{-\frac{1}{x^4}}}{x}| \leqslant |x|^3$.) Similarly, $\frac{\partial u}{\partial y}(0, 0) = 0$. Thus

$$\frac{\partial^2 u}{\partial x^2}(0, 0) = \lim_{x \to 0} \frac{\frac{\partial u}{\partial x}(x, 0) - \frac{\partial u}{\partial x}(0, 0)}{x - 0} = \lim_{x \to 0} \frac{\frac{d}{dx}e^{-\frac{1}{x^4}} - 0}{x} = \lim_{x \to 0} \frac{e^{-\frac{1}{x^4}}\frac{4}{x^5}}{x} = \lim_{x \to 0} \frac{4e^{-\frac{1}{x^4}}}{x^6} = 0$$

(as $e^{\frac{1}{x^4}} = 1 + \frac{1}{x^4} + \frac{1}{2!}(\frac{1}{x^4})^2 + \cdots > \frac{1}{2x^8}$ for $x \neq 0$, which gives $0 \leqslant |\frac{e^{-\frac{1}{x^4}}}{x^6}| \leqslant 2|x|^2$). Similarly, $\frac{\partial^2 u}{\partial y^2}(0, 0) = 0$. Hence $\frac{\partial^2 u}{\partial x^2}(0, 0) + \frac{\partial^2 u}{\partial y^2}(0, 0) = 0 + 0 = 0$.

Solution to Exercise 5.16.

(1) Let $z_0 \in D_1$. Then $\varphi(z_0) \in D_2$. Let Ω be a disc with centre $\varphi(z_0)$ and radius $\epsilon > 0$ small enough so that $\Omega \subset D_2$. Since Ω is simply connected, there exists an $H \in \mathcal{O}(\Omega)$ such that $h|_\Omega = \mathrm{Re}\, H$. Then $\varphi^{-1}\Omega \subset D_1$ is an open set in $\mathbb{R}^2$ containing z_0. The composition of $\varphi|_{\varphi^{-1}(\Omega)} : \varphi^{-1}(\Omega) \to \Omega$ and $H : \Omega \to \mathbb{C}$ is holomorphic, and so $\mathrm{Re}(H \circ \varphi|_{\varphi^{-1}(\Omega)})$ is harmonic in $\varphi^{-1}(\Omega)$. For $z \in \varphi^{-1}(\Omega)$, $(\mathrm{Re}(H \circ \varphi|_{\varphi^{-1}(\Omega)}))(z) = (\mathrm{Re}\, H)(\varphi(z)) = h(\varphi(z)) = (h \circ \varphi)(z)$. Hence $(h \circ \varphi)|_{\varphi^{-1}(\Omega)}$ is harmonic in $\varphi^{-1}(\Omega)$. As $z_0 \in D_1$ was arbitrary, $h \circ \varphi$ is harmonic in D_1.

(2) If $h : D_2 \to \mathbb{R}$ is harmonic, then by the first part, $h \circ \varphi : D_1 \to \mathbb{R}$ is harmonic. If $h \circ \varphi : D_1 \to \mathbb{R}$ is harmonic, then as $\varphi^{-1} : D_2 \to D_1$ is holomorphic, by the first part, $(h \circ \varphi) \circ \varphi^{-1} : D_2 \to \mathbb{R}$ is harmonic. But

$$(h \circ \varphi) \circ \varphi^{-1} = h \circ (\varphi \circ \varphi^{-1}) = h \circ \mathrm{id}_{D_2} = h,$$

where id_{D_2} is the identity map $D_2 \ni z \mapsto z \in D_2$. So $h : D_2 \to \mathbb{R}$ is harmonic.

(3) By the triangle inequality in $\Delta PO'B$ (see the picture below), for s ($\equiv P$) in $\mathbb{H}$,
$$|s+1| = PA = PO' + O'A = PO' + O'B > PB = |s-1|,$$
where we have used the fact that O' is on the perpendicular bisector of AB to get the third equality. It is evident that $PA > PB$ holds when P lies on the right of 0 along the line joining A to B. So $\varphi(s) \in \mathbb{D}$ for all $s \in \mathbb{H}$.

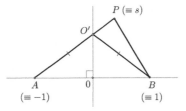

φ is holomorphic: $\frac{\cdot-1}{\cdot+1} \in \mathcal{O}(\mathbb{C}\backslash\{-1\})$, and as $\mathbb{H} \subset \mathbb{C}\backslash\{-1\}$, we have $\frac{\cdot-1}{\cdot+1} \in \mathcal{O}(\mathbb{H})$. Define $\psi : \mathbb{D} \to \mathbb{H}$ by $\psi(s) = \frac{1+z}{1-z}$, $z \in \mathbb{D}$. (This expression for ψ, which is a candidate for φ^{-1}, is obtained by solving for s in the equation $z = \varphi(s) = \frac{s-1}{s+1}$.)

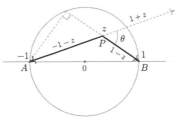

The angle subtended by the diameter AB at any point of the circle is $90°$. So for any P ($\equiv z$) in $\mathbb{D}$, $\angle APB > 90°$. So
$$\mathrm{Re}(\psi(z)) = \mathrm{Re}\frac{1+z}{1-z} = |\psi(z)| \cos\theta = |\psi(z)| \cos(\pi - \angle APB) > 0.$$
Thus $\psi(z) \in \mathbb{H}$ for all $z \in \mathbb{D}$. This can also be seen analytically, since if $z \in \mathbb{D}$, then $z = re^{i\theta}$ for an $r \in [0,1)$ and $\theta \in \mathbb{R}$, and so
$$\mathrm{Re}\frac{1+z}{1-z} = \mathrm{Re}\frac{1+re^{i\theta}}{1-re^{i\theta}} = \frac{\mathrm{Re}((1+r\cos\theta+ir\sin\theta)(1-r\cos\theta+ir\sin\theta))}{(1-r\cos\theta)^2+r^2(\sin\theta)^2} = \frac{1-r^2}{(1-r\cos\theta)^2+r^2(\sin\theta)^2} > 0.$$
Also, $\psi \in \mathcal{O}(\mathbb{D})$, since $\frac{1+\cdot}{1-\cdot} \in \mathcal{O}(\mathbb{C}\backslash\{1\})$, and $\mathbb{D} \subset \mathbb{C}\backslash\{1\}$.

For $s \in \mathbb{H}$, $(\psi \circ \varphi)(s) = \frac{1+\frac{s-1}{s+1}}{1-\frac{s-1}{s+1}} = \frac{s+1+s-1}{s+1-s+1} = \frac{2s}{2} = s = \mathrm{id}_\mathbb{H}(s)$, where $\mathrm{id}_\mathbb{H}$ is the identity map on $\mathbb{H}$. This shows that φ is injective. Also, for $z \in \mathbb{D}$, we have
$(\varphi \circ \psi)(z) = \frac{\frac{1+z}{1-z}-1}{\frac{1+z}{1-z}+1} = \frac{1+z-1+z}{1+z+1-z} = \frac{2z}{2} = z = \mathrm{id}_\mathbb{D}(z)$, where $\mathrm{id}_\mathbb{D}$ is the identity map on $\mathbb{D}$. This shows that φ is surjective. So φ is a bijection and $\varphi^{-1} = \psi$.

Some real analysis background

We recall here a few of the real analysis results we use in the book. Proofs of results listed here, or other preliminaries from real analysis needed in the book can be found in most undergraduate real analysis textbooks, for example [1], [14], [16].

One variable real analysis. Let S be a subset of $\mathbb{R}$. A real number u is said to be an *upper bound of S* if for all $x \in S$, $x \leqslant u$. A number $u_* \in \mathbb{R}$ is called a *least upper bound of S* (or a *supremum of S*) if u_* is an upper bound of S, and for every upper bound u of S, $u_* \leqslant u$. We denote the supremum of S by $\sup S$. If $\sup S \in S$, then we call u_* the *maximum of S*.

A real number ℓ is said to be a *lower bound of S* if for all $x \in S$, $\ell \leqslant x$. A number $\ell_* \in \mathbb{R}$ is called a *greatest lower bound of S* (or an *infimum of S*) if ℓ_* is a lower bound of S, and for every lower bound ℓ of S, $\ell \leqslant \ell_*$. We denote the infimum of S by $\inf S$. If $\inf S \in S$, then we call ℓ_* the *minimum of S*.

Least Upper Bound Property of $\mathbb{R}$: If $S \subset \mathbb{R}$ is such that $S \neq \varnothing$ and S has an upper bound, then the supremum/least upper bound $\sup S$ of S exists.

Archimedean Property of $\mathbb{R}$: If $x, y \in \mathbb{R}$ and $x > 0$, then there exists an $n \in \mathbb{N}$ such that $nx > y$.

An *interval* is a set consisting of all the real numbers between two given real numbers, or of all the real numbers on one side or the other of a given number. So an interval is a set of any of the following forms, where $a, b \in \mathbb{R}$ and $a < b$:

$(a, b) = \{x \in \mathbb{R} : a < x < b\}$

$[a, b] = \{x \in \mathbb{R} : a \leqslant x \leqslant b\}$

$(a, b] = \{x \in \mathbb{R} : a < x \leqslant b\}$

$[a, b) = \{x \in \mathbb{R} : a \leqslant x < b\}$

$(a, \infty) = \{x \in \mathbb{R} : a < x\}$

$[a, \infty) = \{x \in \mathbb{R} : a \leqslant x\}$

$(-\infty, b) = \{x \in \mathbb{R} : x < b\}$

$(-\infty, b] = \{x \in \mathbb{R} : x \leqslant b\}$

$(-\infty, \infty) = \mathbb{R}$

Cauchy-Schwarz inequality: If $n \in \mathbb{N}$, and $x_1, \cdots, x_n, y_1, \cdots, y_n$ are real numbers, then $(x_1 y_1 + \cdots + x_n y_n)^2 \leqslant (x_1^2 + \cdots + x_n^2)(y_1^2 + \cdots + y_n^2)$.

A *real sequence* is a function $f : \mathbb{N} \to \mathbb{R}$. We use the notation $(a_n)_{n \in \mathbb{N}}$ for the sequence $\mathbb{N} \ni n \mapsto a_n \in \mathbb{R}$. A sequence $(a_n)_{n \in \mathbb{N}}$ is said to be *convergent with limit* L ($\in \mathbb{R}$) if for every $\epsilon > 0$, there exists an $N \in \mathbb{N}$ such that for all $n \in \mathbb{N}$ with $n > N$, we have $|a_n - L| < \epsilon$. Then we write $\lim_{n \to \infty} a_n = L$. If there is no $L \in \mathbb{R}$ such that $\lim_{n \to \infty} a_n = L$, then $(a_n)_{n \in \mathbb{N}}$ is called *divergent*. A real sequence is *monotone* if it is *increasing* (for all $n > N \in \mathbb{N}$, $a_{n+1} \geqslant a_n$) or if it is *decreasing* (for all $n > N \in \mathbb{N}$, $a_{n+1} \leqslant a_n$). A real sequence is *bounded* if there exists an $M > 0$ such that for all $n \in \mathbb{N}$, $|a_n| \leqslant M$. Every monotone and bounded real sequence is convergent.

Sandwich Theorem: Let $(a_n)_{n \in \mathbb{N}}$, $(b_n)_{n \in \mathbb{N}}$ be real sequences that converge with the same limit, that is, $\lim_{n \to \infty} a_n = \lim_{n \to \infty} b_n$. Let $(c_n)_{n \in \mathbb{N}}$ be a real sequence, and let $N \in \mathbb{N}$ be such that for all $n > N \in \mathbb{N}$, $a_n \leqslant c_n \leqslant b_n$. Then $(c_n)_{n \in \mathbb{N}}$ is also convergent with the same limit, that is, $\lim_{n \to \infty} a_n = \lim_{n \to \infty} c_n = \lim_{n \to \infty} b_n$.

Let $(a_n)_{n \in \mathbb{N}}$ be a sequence and let $n_1 < n_2 < n_3 < \cdots$ be a strictly increasing sequence of natural numbers. Then $(a_{n_k})_{k \in \mathbb{N}}$ is called a *subsequence of* $(a_n)_{n \in \mathbb{N}}$. Any subsequence of a convergent real sequence is convergent with the same limit.

Bolzano-Weierstrass Theorem: Every bounded real sequence has a convergent subsequence.

A real sequence $(a_n)_{n \in \mathbb{N}}$ is *Cauchy* if for every $\epsilon > 0$, there exists an $N \in \mathbb{N}$ such that for all $m, n \in \mathbb{N}$ such that $m, n > N$, we have $|a_n - a_m| < \epsilon$. Every convergent sequence is Cauchy, and a consequence of the least upper bound property of $\mathbb{R}$ is that every Cauchy sequence is convergent.

For a real sequence $(a_n)_{n \in \mathbb{N}}$, if $s_n := a_1 + \cdots + a_n$, $n \in \mathbb{N}$, then $(s_n)_{n \in \mathbb{N}}$ is called the *sequence of partial sums of* $(a_n)_{n \in \mathbb{N}}$. If $(s_n)_{n \in \mathbb{N}}$ converges, we say that 'the *series* $\sum_{n=1}^{\infty} a_n$ *converges*', and we write *the sum of the series* $\sum_{n=1}^{\infty} a_n = \lim_{n \to \infty} s_n$. If $(s_n)_{n \in \mathbb{N}}$ does not converge we say that 'the series $\sum_{n=1}^{\infty} a_n$ *diverges*'.

The real geometric series $\sum_{n=0}^{\infty} r^n$ converges if and only if $|r| < 1$, and its sum is $\sum_{n=0}^{\infty} r^n = \frac{1}{1-r}$. The harmonic series $\sum_{n=1}^{\infty} \frac{1}{n}$ diverges. In fact for $s \in \mathbb{R}$, $\sum_{n=1}^{\infty} \frac{1}{n^s}$ converges if and only if $s > 1$.

The real series $\sum_{n=1}^{\infty} a_n$ is said to *converge absolutely* if $\sum_{n=1}^{\infty} |a_n|$ converges. If a real series converges absolutely, then it converges.

Leibniz Alternating Series Theorem: Let $(a_n)_{n \in \mathbb{N}}$ be a real, decreasing sequence such that $a_n \geqslant 0$ for all $n \in \mathbb{N}$, that converges to 0. Then the series $\sum_{n=1}^{\infty} (-1)^n a_n$ converges.

There are three important tests for the convergence of a real series:

- The comparison test (where we compare with a series whose convergence status is known).
- The ratio test (where we look at the behaviour of the ratio $\frac{a_{n+1}}{a_n}$ of terms).
- The root test (where we look at the behaviour of $\sqrt[n]{|a_n|}$).

We have summarised these in the following table.

		Comparison	Ratio	Root
Absolute convergence	$\Leftarrow$	$\|a_n\| \leqslant c_n$ for all large n; $\sum\limits_{n=1}^{\infty} c_n$ converges.	$\left\|\frac{a_{n+1}}{a_n}\right\| \leqslant r < 1$ for all large n.	$\sqrt[n]{\|a_n\|} \leqslant r < 1$ for all large n.
Divergence	$\Leftarrow$	$a_n \geqslant d_n \geqslant 0$ for all large n; $\sum\limits_{n=1}^{\infty} d_n$ diverges.	$\left\|\frac{a_{n+1}}{a_n}\right\| \geqslant 1$ for all large n.	$\sqrt[n]{\|a_n\|} \geqslant 1$ infinitely often.

Real power series: Let $(c_n)_{n \geqslant 0}$ be a real sequence. Then either $\sum\limits_{n=0}^{\infty} c_n x^n$ is absolutely convergent for all $x \in \mathbb{R}$ (the power series is then said to have an infinite radius of convergence), or there exists a unique $r \geqslant 0$ (called the radius of convergence of the power series) such that $\sum\limits_{n=0}^{\infty} c_n x^n$ is absolutely convergent for $x \in (-r, r)$ and $\sum\limits_{n=0}^{\infty} c_n x^n$ diverges for $x \in \mathbb{R} \backslash [-r, r]$.

Let I be an interval in $\mathbb{R}$, $c \in I$ and $f : I \to \mathbb{R}$. The function f is *continuous at c* if for every $\epsilon > 0$, there exists a $\delta > 0$ such that for all $x \in I$ satisfying $|x - c| < \delta$, $|f(x) - f(c)| < \epsilon$. The function f is *continuous (on I)* if for every $x \in I$, f is continuous at x.

Intermediate Value Theorem: If $f : [a, b] \to \mathbb{R}$ is continuous and $y \in \mathbb{R}$ lies between $f(a)$ and $f(b)$, (that is, $f(a) \leqslant y \leqslant f(b)$ or $f(b) \leqslant y \leqslant f(a)$), then there exists a $c \in [a, b]$ such that $f(c) = y$.

Extreme Value Theorem: If $f : [a, b] \to \mathbb{R}$ is continuous, then

(1) The set S, where $S := \{f(x) : x \in [a, b]\} =: f([a, b]) = $ range of f, is bounded.

(2) $\sup S$ and $\inf S$ exist.

(3) $\sup S$ and $\inf S$ are attained, i.e., there exist $c, d \in [a, b]$ such that $f(c) = \sup S = \max S$ and $f(d) = \inf S = \min S$.

If $I \subset \mathbb{R}$ is an interval, $f : I \to \mathbb{R}$, and $c \in I$, then f is *differentiable at c* if there exists an $L \in \mathbb{R}$ such that $\lim\limits_{x \to c} \frac{f(x) - f(c)}{x - c} = L$, that is, for every $\epsilon > 0$, there exists a $\delta > 0$ such that whenever $x \in I$ satisfies $0 < |x - c| < \delta$, we have $\left|\frac{f(x) - f(c)}{x - c} - L\right| < \epsilon$. The number L is unique, and we denote this unique number by $f'(c)$ or by $\frac{df}{dx}(c)$, and call it the *derivative of f at c*. If f is differentiable at every $x \in I$, then f is called *differentiable on I*.

If f is differentiable at $c \in I$, then f is continuous at c. One may wonder how badly behaved continuous functions might be with respect to the notion of differentiability. It turns out that there are functions that are continuous *everywhere*, but differentiable *nowhere*. For example, the *blancmange function* b is continuous on $\mathbb{R}$, but not differentiable at any $x \in \mathbb{R}$, and is constructed as follows. We start from the *sawtooth* function f_1, and construct $f_2, f_3, \cdots$ by setting $f_n(x) = \frac{1}{2^{n-1}} f_1(2^{n-1} x)$, $x \in \mathbb{R}$, $n \geqslant 2$. The blancmange function b is obtained by adding these: $b(x) = \sum\limits_{n=1}^{\infty} f_n(x)$, $x \in \mathbb{R}$.

Mean Value Theorem: Let $f : [a, b] \to \mathbb{R}$ be continuous on $[a, b]$ and differentiable on (a, b). Then there is a point $c \in (a, b)$ such that $\frac{f(b) - f(a)}{b - a} = f'(c)$.

The Riemann integral. A partition P of $[a, b]$ is a finite set $P = \{x_0, x_1, \ldots, x_{n-1}, x_n\}$ such that $a = x_0 < x_1 < \cdots < x_{n-1} < x_n = b$. The set of all partitions P of $[a, b]$ will be denoted by $\mathcal{P}$. Let $f : [a, b] \to \mathbb{R}$ be a bounded function and let P be a partition of $[a, b]$. An *upper sum* $U(f, P)$ *of f associated with P* is $U(f, P) := \sum_{k=0}^{n-1} (\sup_{x \in [x_k, x_{k+1}]} f(x))(x_{k+1} - x_k)$.

A *lower sum* $L(f, P)$ *of f associated with P* is $L(f, P) := \sum_{k=0}^{n-1} (\inf_{x \in [x_k, x_{k+1}]} f(x))(x_{k+1} - x_k)$.

A bounded function $f : [a, b] \to \mathbb{R}$ is called *Riemann integrable* if $\sup_{P \in \mathcal{P}} L_P(f) = \inf_{P \in \mathcal{P}} U_P(f)$.

We denote this common value by $\int_a^b f(x)dx$, and call it the *Riemann integral of f*. All functions continuous on $[a, b]$ are Riemann integrable. Let f, g be Riemann integrable on $[a, b]$ and $\alpha \in \mathbb{R}$. Then the following hold:

- $\int_a^b f(x) + g(x)dx = \int_a^b f(x)dx + \int_a^b g(x)dx$.

- $\int_a^b \alpha \cdot f(x)dx = \alpha \int_a^b f(x)dx$.

- If for all $x \in [a, b]$, $f(x) \geqslant 0$, $\int_a^b f(x)dx \geqslant 0$.

- If for all $x \in [a, b]$, $f(x) \leqslant g(x)$, then $\int_a^b f(x)dx \leqslant \int_a^b g(x)dx$.

- $|f|$ is also Riemann integrable and $|\int_a^b f(x)dx| \leqslant \int_a^b |f(x)|dx$.

- If f is continuous, $f \geqslant 0$ on $[a, b]$, and $\int_a^b f(x)dx = 0$, then $f \equiv 0$ on $[a, b]$.

Fundamental Theorem of Calculus: Let $f : [a, b] \to \mathbb{R}$ be continuously differentiable on $[a, b]$. Then $\int_a^b f'(x)dx = f(b) - f(a)$.

Multivariable real analysis. In $\mathbb{R}^m$, we use the Euclidean metric, $d_2(\mathbf{x}, \mathbf{y}) = \|\mathbf{x} - \mathbf{y}\|_2$ for $\mathbf{x}, \mathbf{y} \in \mathbb{R}^m$. Here $\|\mathbf{v}\|_2 := \sqrt{v_1^2 + \cdots + v_m^2}$ for $\mathbf{v} = (v_1, \cdots, v_m) \in \mathbb{R}^m$.

A sequence $(\mathbf{x}_n)_{n \in \mathbb{N}}$ in $\mathbb{R}^m$ is said to *converge to* $\mathbf{x} \in \mathbb{R}^m$ if for every $\epsilon > 0$, there exists an $N \in \mathbb{N}$ such that for all $n > N$, $\|\mathbf{x}_n - \mathbf{x}\|_2 < \epsilon$. A sequence $(\mathbf{x}_n)_{n \in \mathbb{N}}$ is convergent to $\mathbf{x} \in \mathbb{R}^m$ if and only if converges componentwise to $\mathbf{x}$, that is, for all $k \in \{1, \cdots, m\}$, $(\mathbf{x}_n^{(k)})_{n \in \mathbb{N}}$ converges to $(\mathbf{x}^{(k)})_{n \in \mathbb{N}}$. Here $\mathbf{x} =: (\mathbf{x}^{(1)}, \cdots, \mathbf{x}^{(m)})$ and $\mathbf{x}_n =: (\mathbf{x}_n^{(1)}, \cdots, \mathbf{x}_n^{(m)})$.

A set $U \subset \mathbb{R}^m$ is *open* if for every $\mathbf{x} \in U$, there exists an $r > 0$ such that the ball $B(\mathbf{x}, r) := \{\mathbf{y} \in \mathbb{R}^m : \|\mathbf{y} - \mathbf{x}\|_2 < r\} \subset U$. A set $F \subset \mathbb{R}^m$ is *closed* if $\mathbb{R}^m \backslash F$ is open. Any finite intersection of open sets is open, and arbitrary unions of open sets are open. Taking complements, it follows that a finite union of closed sets is closed, and arbitrary intersections of closed sets are closed. A set $F \subset \mathbb{R}^m$ is closed if and only if for every convergent sequence $(\mathbf{x}_n)_{n \in \mathbb{N}}$ such that $\mathbf{x}_n \in F$ $(n \in \mathbb{N})$, we have that $\lim_{n \to \infty} \mathbf{x}_n \in F$. A set $K \subset \mathbb{R}^m$ is *compact* if every sequence in K has a convergent subsequence with its limit belonging to K. A set $K \subset \mathbb{R}^m$ is compact if and only if K is closed and bounded.

Let $\mathbb{R}^n, \mathbb{R}^m$ be equipped with the Euclidean distance. Let $S \subset \mathbb{R}^n$, and $\mathbf{c} \in S$. Then f is *continuous at* $\mathbf{c}$ if for every $\epsilon > 0$, there exists a $\delta > 0$ such that whenever $\mathbf{x} \in S$ satisfies $\|\mathbf{x} - \mathbf{c}\|_2 < \delta$, we have $\|f(\mathbf{x}) - f(\mathbf{c})\|_2 < \epsilon$. Also, f is said to be *continuous on* S if for each $\mathbf{x} \in S$, f is continuous at $\mathbf{x}$. The function $f : S \to \mathbb{R}^m$ is continuous on S if and only if for every V open in $\mathbb{R}^n$, $f^{-1}(V) := \{\mathbf{x} \in S : f(\mathbf{x}) \in V\}$ is open in S, that is, for each $\mathbf{x} \in f^{-1}(V)$, there exists an $r > 0$ such that $\{\mathbf{y} \in S : \|\mathbf{y} - \mathbf{x}\|_2 < r\} \subset f^{-1}(V)$. The function $f : S \to \mathbb{R}^m$ is continuous at $\mathbf{c} \in S$ if and only if for every sequence $(\mathbf{x}_k)_{k \in \mathbb{N}}$ in S such that $(\mathbf{x}_k)_{k \in \mathbb{N}}$ converges to $\mathbf{c}$, $(f(\mathbf{x}_k))_{k \in \mathbb{N}}$ converges to $f(\mathbf{c})$. If $f : S \to \mathbb{R}^m$ is continuous at $\mathbf{c} \in S$, $T \subset \mathbb{R}^m$ is such that $f(S) \subset T$, and if $g : T \to \mathbb{R}^k$ is continuous at $f(\mathbf{c}) \in T$, then the composition map $f \circ g : S \to \mathbb{R}^k$, defined by $(f \circ g)(\mathbf{x}) := f(g(\mathbf{x}))$ $(\mathbf{x} \in S)$, is continuous at $\mathbf{c} \in S$. If $S \subset \mathbb{R}^n$ and $T \subset \mathbb{R}^m$, then S, T are said to be *homeomorphic* if there exist continuous maps $f : S \to \mathbb{R}^m$ and $g : T \to \mathbb{R}^n$ such that $f(S) \subset T$, $g(T) \subset S$, $f \circ g = \mathrm{id}_T$ and $g \circ f = \mathrm{id}_S$. Here, for a set X, id_X denotes the *identity map*, given by $X \ni x \mapsto x =: \mathrm{id}_X(x)$. A function $f : S \to \mathbb{R}^m$ is continuous if and only if each of its components $f_1, \cdots, f_m : S \to \mathbb{R}$ are continuous. Here $f(\mathbf{x}) =: (f_1(\mathbf{x}), \cdots, f_m(\mathbf{x}))$ for all $\mathbf{x} \in S$. A function $f : S \to \mathbb{R}^n$ is *uniformly continuous* if for every $\epsilon > 0$, there exists a $\delta > 0$ such that for all $\mathbf{x}, \mathbf{y} \in S$ satisfying $\|\mathbf{x} - \mathbf{y}\|_2 < \delta$, we have $\|f(\mathbf{x}) - f(\mathbf{y})\|_2 < \epsilon$. If $K \subset \mathbb{R}^n$ is compact and $f : K \to \mathbb{R}^m$ is continuous, then f is uniformly continuous.

Weierstrass's Theorem: If $K \subset \mathbb{R}^n$ is a compact set and $f : K \to \mathbb{R}^m$ is a continuous function on K, then $f(K)$ is a compact subset of $\mathbb{R}^m$, and there exist $\mathbf{x}_{\min}, \mathbf{x}_{\max} \in K$ such that $f(\mathbf{x}_{\min}) = \inf\{f(\mathbf{x}) : \mathbf{x} \in K\}$ and $f(\mathbf{x}_{\max}) = \sup\{f(\mathbf{x}) : \mathbf{x} \in K\}$.

Let $U \subset \mathbb{R}^n$ be open, and $\mathbf{c} \in U$. A function $f : U \to \mathbb{R}^m$ is *(real) differentiable at* $\mathbf{c}$ if there exists a linear transformation $L : \mathbb{R}^n \to \mathbb{R}^m$ such that $\lim_{\mathbf{x} \to \mathbf{c}} \frac{\|f(\mathbf{x}) - f(\mathbf{c}) - L(\mathbf{x} - \mathbf{c})\|_2}{\|\mathbf{x} - \mathbf{c}\|_2} = 0$, that is, for every $\epsilon > 0$, there exists a $\delta > 0$ such that whenever $\mathbf{x} \in U$ satisfies $0 < \|\mathbf{x} - \mathbf{c}\|_2 < \delta$, we have $\frac{\|f(\mathbf{x}) - f(\mathbf{c}) - L(\mathbf{x} - \mathbf{c})\|_2}{\|\mathbf{x} - \mathbf{c}\|_2} < \epsilon$. Then (the unique such linear transformation) L is called the *derivative of f at* $\mathbf{c}$, and we write $f'(\mathbf{c}) = L$. Let $f_1, \ldots, f_m$ denote the components of f and let $\mathbf{c} = (c_1, \cdots, c_n)$. If $\frac{\partial f_i}{\partial x_j}(\mathbf{c}) := \lim_{x_j \to c_j} \frac{f_i(c_1, \cdots, c_{j-1}, x_j, c_{j+1}, \cdots, c_n) - f_i(c_1, \cdots, c_{j-1}, c_j, c_{j+1}, \cdots, c_n)}{x_j - c_j}$ exists, then we call $\frac{\partial f_i}{\partial x_j}(\mathbf{c})$ the (i, j)th partial derivative f at c. If $f : U \to \mathbb{R}^m$ is differentiable at $\mathbf{c}$, then all the partial derivatives of f at $\mathbf{c}$, namely, $\frac{\partial f_i}{\partial x_j}(\mathbf{c})$ $(i = 1, \cdots, m; j = 1, \cdots, n)$ exist, and the matrix $[f'(\mathbf{c})]$ of the linear transformation $f'(\mathbf{c})$ with respect to the standard bases for $\mathbb{R}^n$ and $\mathbb{R}^m$ is given by

$$[f'(\mathbf{c})] = \begin{bmatrix} \frac{\partial f_1}{\partial x_1}(\mathbf{c}) & \cdots & \frac{\partial f_1}{\partial x_n}(\mathbf{c}) \\ \vdots & & \vdots \\ \frac{\partial f_m}{\partial x_j}(\mathbf{c}) & \cdots & \frac{\partial f_m}{\partial x_n}(\mathbf{c}) \end{bmatrix}.$$

By the standard basis for $\mathbb{R}^n$, we mean the set $\{\mathbf{e}_1, \cdots, \mathbf{e}_n\}$, where for $k \in \{1, \cdots, n\}$, $\mathbf{e}_k \in \mathbb{R}^n$ has all components zero, except for the kth component, which is equal to 1.

If all the partial derivatives exist at each $\mathbf{x} \in U$ and the maps $\mathbf{x} \mapsto \frac{\partial f_i}{\partial x_j}(\mathbf{x}) : U \to \mathbb{R}$ $(i = 1, \ldots, m; j = 1, \ldots, n)$ are continuous on U, then f is differentiable at each $\mathbf{x} \in U$.

Schwarz Theorem: If $U \subset \mathbb{R}^2$, $f : U \to \mathbb{R}$, and at each $(x, y) \in U$, the partial derivatives $\frac{\partial f}{\partial x}(x, y)$, $\frac{\partial f}{\partial y}(x, y)$, $\frac{\partial^2 f}{\partial x \partial y}(x, y) := \frac{\partial}{\partial x} \frac{\partial f}{\partial y}(x, y)$, $\frac{\partial^2 f}{\partial y \partial x}(x, y) := \frac{\partial}{\partial y}(x, y) \frac{\partial f}{\partial x}(x, y)$ exist, and define continuous functions on U, then $\frac{\partial^2 f}{\partial x \partial y}(x, y) = \frac{\partial^2 f}{\partial y \partial x}(x, y)$ for all $(x, y) \in U$.

Some relevant set theory. A *relation* R on a set S is a subset of the Cartesian product $S \times S := \{(a, b) : a, b \in S\}$. If $(a, b) \in R$, then we write aRb. A relation R on a set S is called an *equivalence relation* if it is *reflexive* (that is, for all $a \in S$, aRa), *symmetric* (that is, if aRb, then bRa), and *transitive* (that is, if aRb and bRc, then aRc). If R is an equivalence relation of a set S, then the *equivalence class of a*, denoted by $[a]$, is defined to be the set $[a] = \{b \in S : aRb\}$.

A set S is *finite* if it is empty or there exists an $n \in \mathbb{N}$ and a bijection $f : \{1, \cdots, n\} \to S$. If no such bijection exists, then S is called *infinite*. If S is an infinite set, then S is *countable* if there is a bijective map from $\mathbb{N}$ onto S. If S is infinite and not countable, then it is called *uncountable*. Every infinite subset of a countable set is countable. If A, B are countable, then $A \times B$ is also countable. A countable union of countable sets is countable. A countable union of finite sets is finite or countable. The sets $\mathbb{N}, \mathbb{Z}, \mathbb{Q}$ are examples of countable sets, while $\mathbb{R}$ and any interval in $\mathbb{R}$ are examples of uncountable sets.

Notes

Chapter 1: The remark on the historical development of complex numbers is based on [11]. Exercise 1.2 is taken from [17]. Exercises 1.8, 1.8, 1.14, 1.24 are taken from [11]. Exercises 1.34 and 1.52 are taken from [2].

Chapter 2: The section on the geometric meaning of the complex derivative follows [11]. Exercises 2.21 and 2.23 are taken from [11].

Chapter 3: The proof of Theorem 3.9 follows [2]. Exercises 3.4, 3.5, 3.15, 3.27, 3.36 are taken from [2]. Exercises 3.12, 3.19, 3.26, 3.25, 3.30, 3.32 are taken from [11]. Exercise 3.29 is taken from [15]. Exercises 3.33, 3.43, 3.47 are taken from [5]. Exercise 3.37 is taken from [9]. The proof of the Cauchy Integral Theorem given in the Appendix is based on [3].

Chapter 4: Section 4.8 follows [2]. Exercise 4.15 is based on problem 6 of the International Mathematics Olympiad, 1979. Exercises 4.17, 4.40, 4.61, 4.64, 4.65, 4.66, 4.67, 4.68 are taken from [5]. Exercise 4.23 is taken from [22]. Exercises 4.27, 4.35 are taken from [24]. Exercises 4.28, 4.43 are taken from [15]. Exercise 4.49 is taken from [2]. Exercise 4.70 is taken from [12].

Chapter 5: Exercises 5.4, 5.5 and the proof of Theorem 5.2 is taken from [2]. Exercise 5.13 is based on [6].

Bibliography

[1] T. Apostol. *Calculus II*. 2nd edition. John Wiley, 1969.

[2] M. Beck, G. Marchesi, D. Pixton, and L. Sabalka. *A first course in complex analysis*. http://math.sfsu.edu/beck/papers/complex.pdf, 2008.

[3] J. Conway. *Functions of one complex variable I*. 2nd edition. Springer, 1978.

[4] S. Fisher. *Complex variables*. 2nd edition. Dover, 1999.

[5] F. Flanigan. *Complex variables*. Dover, 1972.

[6] F. Flanigan. Classroom notes: Some half-plane Dirichlet problems: A bare hands approach. *American Mathematical Monthly*, 80:59-61, no. 1, 1973.

[7] B. Gelbaum and J. Olmsted. *Counterexamples in analysis*. Dover, 1964.

[8] J. Gilman, I. Kra, and R. Rodriguez. *Complex analysis. In the spirit of Lipman Bers*. Springer, 2007.

[9] J. Howie. *Complex analysis*. Springer, 2003.

[10] D. Minda. The Dirichlet problem for a disk. *American Mathematical Monthly*, 97:220-223, no. 3, 1990.

[11] T. Needham. *Visual complex analysis*. Oxford University Press, 1997.

[12] R. Ash, and W. Novinger. *Complex analysis*. 2nd edition. Dover, 2007.

[13] R. Remmert. *Theory of complex functions*, Springer, 1991.

[14] W. Rudin. *Principles of mathematical analysis*. 3rd edition. McGraw-Hill, 1976.

[15] W. Rudin. *Real and complex analysis*. 3rd edition. McGraw-Hill, 1987.

[16] A. Sasane. *The how and why of one variable calculus*. Wiley, 2015.

[17] A. Shastri. *Basic complex analysis of one variable*. Macmillan Publishers India, 2011.

[18] W. Shaw. *Complex analysis with* MATHEMATICA, Cambridge University Press, 2006.

[19] J. Shurman. *Course materials for Mathematics 311: Complex analysis*. http://people.reed.edu/~jerry/311/mats.html , 2009.

[20] I. Stewart and D. Tall. *Complex Analysis*. 2nd edition. Cambridge University Press, 2018.

[21] D. Tall. *Functions of a complex variable*. Dover, 1970.

[22] L. Volkovyskiĭ, G. Lunts, and I. Aramanovich. *A collection of problems on complex analysis*. Dover, 1991.

[23] Wikipedia page on Liouville's theorem in Differential algebra.

[24] F. Wikström. *Funktionsteori. Övningsbok*. Studentlitteratur, 2016.

Index

Printed in the USA
CPSIA information can be obtained
at www.ICGtesting.com
LVHW012006160324
774517LV00004B/564